Telemetric Studies of Vertebrates

SYMPOSIA OF THE ZOOLOGICAL SOCIETY OF LONDON
NUMBER 49

Telemetric Studies

of Vertebrates

*(The Proceedings of a Symposium held at
The Zoological Society of London
on 21 and 22 November 1980)*

Edited by

C. L. CHEESEMAN

*Ministry of Agriculture, Fisheries and Food,
Agricultural Science Service, Worplesdon,
Surrey, England*

and

R. B. MITSON

*Ministry of Agriculture, Fisheries and Food,
Fisheries Laboratory, Lowestoft, Suffolk,
England*

Published for

THE ZOOLOGICAL SOCIETY OF LONDON

BY

ACADEMIC PRESS

1982

ACADEMIC PRESS INC. (LONDON) LTD
24/28 Oval Road, London NW1 7DX

United States Edition published by
ACADEMIC PRESS INC.
111 Fifth Avenue, New York, New York, 10003

British Library Cataloguing in Publication Data

Telemetric studies of vertebrates.—(Symposia of
    the Zoological Society of London, ISSN 0084-5612;
    49)
    1. Telemeter(Physiological apparatus)—Congresses
    I. Cheeseman, C. L. II. Mitson, R. B. III. Series
    591.1'028   QP55

    ISBN 0-12-613349-2

Printed in Great Britain at the Alden Press
Oxford London and Northampton

Contributors

ARNOLD, G. P., *Ministry of Agriculture, Fisheries and Food, Fisheries Laboratory, Lowestoft, Suffolk NR33 0HT, England* (p. 75)

BEACH, M. H., *Ministry of Agriculture, Fisheries and Food, Fisheries Laboratory, Lowestoft, Suffolk NR33 0HT, England* (p. 31)

BERTRAM, B. C. R., *Curator of Mammals, The Zoological Society of London, Regent's Park, London NW1 4RY, England* (p. 34)

BIRKS, J. D. S., *Department of Zoology, University of Durham, Science Laboratories, South Road, Durham DH1 3LE, England* (p. 231)

BUTLER, P. J., *Department of Zoology and Comparative Physiology, University of Birmingham, P.O. Box 363, Birmingham B15 2TT, England* (p. 107)

HARDEN JONES, F. R., *Ministry of Agriculture, Fisheries and Food, Fisheries Laboratory, Lowestoft, Suffolk NR33 0HT, England* (p. 75)

HARRIS, S. S., *Department of Zoology, The University of Bristol, Woodland Road, Bristol BS8 1UG, England* (p. 301)

HERSTEINSSON, P., *Animal Behaviour Research Group, Department of Zoology, University of Oxford, South Parks Road, Oxford OX1 3PS, England* (p. 259)

HIRONS, G. J. M., *The Game Conservancy, Fordingbridge, Hampshire SP6 1EF, England* (pp. 129 and 139)

KENWARD, R. E., *Institute of Terrestrial Ecology, Monks Wood Experimental Station, Abbots Ripton, Huntingdon PE17 2LS, England* (pp. 129 and 175)

KRUUK, H., *Institute of Terrestrial Ecology, Banchory Research Station, Hill of Brathens, Glassel, Banchory, Kincardineshire AB3 4BY, Scotland* (p. 291)

KUECHLE, V. B., *Director, Bioelectronics Laboratory, J. F. Bell Museum of Natural History, University of Minnesota, Minneapolis, Minnesota 55415, USA* (p. 1)

LINN, I. J., *Department of Biological Sciences, University of Exeter, Hatherly Laboratories, Prince of Wales Road, Exeter EX4 4PS, England* (pp. 197 and 231)

MACDONALD, D. W., *Animal Behaviour Research Group, Department of Zoology, University of Oxford, South Parks Road, Oxford OX1 3PS, England* (p. 259)

MORAN, P. L., *Microelectronics Research Centre, University College, Cork, Eire* (p. 47)

OWEN, R. B., Jr, *School of Forest Resources, University of Maine, Orono, Maine, USA* (p. 139)

PARISH, T., *Institute of Terrestrial Ecology, Banchory Research Station, Hill of Brathens, Glassel, Banchory, Kincardineshire AB3 4BY, Scotland* (p. 291)

REEVE, N. J., *Department of Zoology, Royal Holloway College, Alderhurst, Bakeham Lane, Englefield Green, Surrey TW20 9TY, England* (p. 207)

SKIFFINS, R. M., *Home Office Directorate of Radio Technology, Waterloo Bridge House, Waterloo Road, London SE1 8UA, England* (p. 19)

SLADE, M. G., *E & EE Department, Royal Military College of Science, Shrivenham, Wilts, England* (p. 61)

SOLOMON, D. J., *Ministry of Agriculture, Fisheries and Food, Fisheries Laboratory, Lowestoft, Suffolk NR33 0HT, England* (p. 95)

STEBBINGS, R. E., *Institute of Terrestrial Ecology, Monks Wood Experimental Station, Abbots Ripton, Huntingdom PE17 2LS, England* (p. 161)

STORETON–WEST, T. J., *Ministry of Agriculture, Fisheries and Food, Fisheries Laboratory, Lowestoft, Suffolk NR33 0HT, England* (p. 31)

VAN ORSDOL, K. G., *Department of Applied Biology, University of Cambridge, Pembroke Street, Cambridge CB2 3DX, England* (p. 325)

WIDÉN, P., *The National Swedish Environment Protection Board, Grimsö Wildlife Research Station, S-770 31 Riddarhyttan, Sweden* p. 153)

WILCOX, P., *Department of Biological Sciences, University of Exeter, Hatherly Laboratories, Prince of Wales Road, Exeter EX4 4PS, England* (p. 197)

WOAKES, A. J., *Department of Zoology and Comparative Physiology, University of Birmingham, P.O. Box 363, Birmingham B15 2TT, England* (p. 107)

ZIESEMER, F., *Staatliche Vogelschutzwarte Schleswig-Holstein, Olshausenstr. 40-60, D-2300 Kiel, West Germany* (p. 129)

Organizers and Chairmen of Sessions

ORGANIZERS

C. L. CHEESEMAN and R. B. MITSON, on behalf of the Zoological Society of London

CHAIRMEN OF SESSIONS

M. J. DELANY, *Undergraduate School of Environmental Sciences, University of Bradford, Bradford, West Yorkshire BD7 1DP, England*

J. W. R. GRIFFITHS, *Department of Electrical and Electronic Engineering, Loughborough University of Technology, Loughborough, Leics, England*

I. J. LINN, *Department of Biological Sciences, University of Exeter, Hatherly Laboratories, Prince of Wales Road, Exeter EX4 4PS, England*

H. G. LLOYD, *Ministry of Agriculture, Fisheries and Food, Government Buildings, Spa Road East, Llandrindod Wells, Powys, Wales*

Preface

This Symposium was jointly organized by the Zoological Society of London, which is gratefully acknowledged for providing the financial and administrative support, the Mammal Society and the Institution of Electronic and Radio Engineers. The amalgamation of biologists and electronic engineers gave members of both disciplines an idea not only of what has been achieved technically and in the field, but also of the potential for future development of telemetry.

In facilitating the study of shy, secretive and otherwise inaccessible animals, telemetry has become an invaluable technique for the field biologist. Made possible by the development of microelectronic technology, telemetry provides a source of detailed information on individual animals and gives a better comprehension of animal movements and behaviour than previous methods such as capture—mark—recapture.

In wildlife studies, telemetry is still in an embryonic phase of development. A variety of commercial equipment has become available in recent years, originating mainly from North America. However, even in that continent the manufacture of equipment remains largely a cottage industry and it is still necessary for the user either to adapt off-the-shelf equipment to suit his own requirements, or to build a system (from scratch). Unless a biologist has considerable skill in electronic engineering and possesses the necessary constructional and test equipment, or has access to technical staff with such skills, the latter option is out of the question. Commercial equipment, however, is relatively expensive because of the small demand and those on a limited budget can often achieve a compromise, perhaps by purchasing a receiver and constructing their own transmitters. The latter approach was used by many of the contributors to this Symposium.

As telemetry has become more widely used, some new terminology has emerged and inevitably there has been confusion over the meaning of some terms. In this volume the word telemetry is used in its literal sense (i.e. tele — at a distance, meter — measure) and this embraces the most elementary tracking or position fixing as well as the measurement of physiological or environmental variables. The chapter entitled *Techniques for Monitoring the Behaviour of Grey Squirrels by Radio* by R. E. Kenward contains some useful suggestions on terminology.

The first five chapters of this volume are concerned mainly with engineering aspects such as microelectronic technology and radio propagation tests. These give some idea of the scope for further miniaturization and improved performance of equipment. In the United Kingdom there is legislation controlling the use and performance of radio equipment for telemetry. To assist the newcomer in applying for a licence to operate a system there is a chapter from the Home Office which details this aspect. The remaining 16 chapters cover the application of both radio and underwater acoustic telemetry in studies of a variety of vertebrate species. These deal comprehensively with the methodology, covering the construction and use of equipment in a way that will assist beginners. Most of all these papers give a fascinating glimpse of the exciting new biological information obtained by using telemetry.

Worplesdon and Lowestoft   C. L. CHEESEMAN
November 1981   R. B. MITSON

Contents

Contributors   v
Organizers and Chairmen of Sessions vii
Preface ix

State of the Art of Biotelemetry in North America

V. B. KUECHLE

Synopsis.   1
Introduction   1
Transmitters   2
   Design   2
   Tuning and testing   3
   Reliability   4
   Power sources   5
   Power output   7
   Sensors   7
   Transmitters for monitoring by satellite 11
   Attachment 11
Receivers and receiving systems 12
   Introduction 12
   Automatic data recording systems 13
   Receiving antennas 15
Fish tags. 15
Future directions of progress 17
   Transmitters 17
   Receivers and receiver systems 17
Acknowledgements 17
References 17

Regulatory Control of Telemetric Devices Used in Animal Studies

R. M. SKIFFINS

Synopsis. 19
Introduction 19
Radio frequency bands 20
   Exclusive bands for biomedical telemetry 20
   Shared radio frequencies 20
   Future frequency allocation 21
   Frequency assignment 21
Equipment standards 22
Licensing 23
   Operational licences 23
   Test and development licences 23

Acknowledgement   23
Appendix 1.   24
Appendix 2.   29

Design Considerations and Performance Checks on a Telemetry Tag System

M. H. BEACH and T. J. STORETON-WEST

Synopsis.   31
Introduction   31
Transmitter tag   32
Receiver.   34
Aerial   37
    Whip aerial (simple dipole).   37
    H-Adcock array   38
    Yagi-array   38
Equipment performance specification for the United Kingdom. . . .   38
Compliance with the specification   39
    Transmitter tag   39
    Receiver.   41
Conclusion   42
Acknowledgements   42
References   42
Appendix   43

Microelectronic Technology and its Application to Telemetry

P. L. MORAN

Synopsis.   47
Introduction   47
Silicon technology   49
    Processing   49
    Testing and yield   51
    Applications   53
Hybrid microcircuit technology   53
    Processing   54
    Thick vs thin film.   55
Microwiring   55
Summary   57
Acknowledgements   59
References   59
Further reading   59

Signal-to-Noise Ratio Enhancement in Receivers for Use in Radio Location

M. G. SLADE

Synopsis.   61
Introduction   61

The matched filter 62
   System realization 63
Single-channel tracking system 66
   Phase comparison. 67
Multi-channel tracking system 67
Summary 69
References 69
Appendix 70
Discussion 71

General Discussion 73

Acoustic Telemetry and the Marine Fisheries

F. R. HARDEN JONES and G. P. ARNOLD

Synopsis. 75
Introduction 75
Our need for telemetry 76
The efficiency of the otter trawl 80
Tracking plaice in the open sea 80
The acoustic transponding compass tag 85
A new requirement 88
Other telemetry systems 89
References 91
Discussion 93

Tracking Fish with Radio Tags

D. J. SOLOMON

Synopsis. 95
Introduction 95
   Approach 95
   Advantages of radio tracking 96
   The River Fowey experiment 96
Performance 97
Tags and tagging 99
   Tags 99
   Insertion of tags 100
Tracking. 101
   Approach and methods 101
   Equipment 102
Interpretation of results 103
Acknowledgements 104
References 104
Discussion 105

Telemetry of Physiological Variables from Diving and Flying Birds

P. J. BUTLER and A. J. WOAKES

Synopsis. 107
Introduction 107

Diving 108
   Historical background 108
   Telemetry systems 110
   Naturally diving birds 111
Flight 119
   Historical background 119
   Free-flying birds 121
Acknowledgements 124
References 124
Discussion 128

Devices for Telemetering the Behaviour of Free-living Birds

R. E. KENWARD, G. J. M. HIRONS and F. ZIESEMER

Synopsis. 129
Introduction 129
Methods. 130
Results and discussion 133
   Thermistor package 133
   Posture-sensing package. 134
   Application. 135
Acknowledgements 136
References 136

Radio Tagging as an Aid to the Study of Woodcock

G. J. M. HIRONS and R. B. OWEN, JR

Synopsis. 139
Introduction 139
   The problem 139
   Background 140
   Study area 140
Methods. 141
   Radio equipment 141
   Activity recording 142
   Techniques of radio location 142
   The effect of radio packages on woodcock 143
Some results from radio tagging woodcock 144
   Breeding biology 144
   Mating system 145
   Feeding behaviour 145
Conclusions 148
References 149
Appendix 149
Discussion 151

Radio Monitoring the Activity of Goshawks

P. WIDÉN

Synopsis. 153

Contents

Introduction 153
Study area 154
Material and methods 154
Results and discussion 156
Acknowledgements 158
References 159
Discussion 160

Radio Tracking Greater Horseshoe Bats with Preliminary Observations on Flight Patterns

R. E. STEBBINGS

Synopsis 161
Introduction 161
Conservation and research problems 161
Radio equipment 163
Radio tagging 164
Radio tracking 167
Bat movements 168
Conclusions 170
Acknowledgements 172
References 172
Discussion 173

Techniques for Monitoring the Behaviour of Grey Squirrels by Radio

R. E. KENWARD

Synopsis 175
Introduction 175
Radio equipment 175
  General considerations for receivers and transmitters 175
  The squirrel collar 177
  Factors affecting length of tag life 179
The effect of radio packages on study animals 181
  Effects recorded in mammal studies 181
  Effects of squirrel collars 182
Data collection 186
  Terminology 186
  Radio biotelemetry techniques 187
  Radio location and surveillance techniques 190
Concluding remarks 192
Acknowledgements 193
References 193
Appendix 194
Discussion 196

A Semi-automated System for Collecting Data on the Movements of Radio Tagged Voles

I. J. LINN and P. WILCOX

Synopsis.   197
Introduction   197
Description and operation of the system   198
    Receiving equipment   198
    Data recording.   199
    Calibration of aerials and calculation of position   199
    Transmitters   201
Analysis of data   202
Summary of important features of the system   203
Limitations.   203
Acknowledgements   204
References   204
Discussion   205

The Home Range of the Hedgehog as Revealed by a Radio Tracking Study

N. J. REEVE

Synopsis.   207
Introduction   207
The study site   208
Methods   208
    Capture   208
    Marking   210
    The radio tracking equipment.   212
    Attachment of the transmitter   212
    The radio location of hedgehogs in the field   213
Results and discussion   215
    Population size   215
    Home range area   215
    Territoriality   217
    Distance travelled as a representation of home range   219
    Average speed   222
    Differential use of home range area   223
    Seasonal changes in home range   225
Discussion of some practical aspects of measuring distance travelled . .   225
Acknowledgements   227
References   228
Discussion   230

Studies of Home Range of the Feral Mink, *Mustela vison*

J. D. S. BIRKS and I. J. LINN

Synopsis.   231
Introduction   232
Study areas.   232

The River Teign 233
Slapton Ley 233
The River Exe in Exeter 233
Methods. 234
The effects of radio tracking on mink 235
Results 236
Optimum tracking duration 236
Home range size and shape. 238
Denning behaviour 240
Home range use intensity 244
Movements within the home range 244
Activity 247
Discussion of results 248
Home range size and shape. 249
Denning behaviour 250
Intensity of use 251
Movements 252
Activity 253
Seasonality 254
Acknowledgements 255
References 255
Discussion 257

Some Comparisons Between Red and Arctic Foxes, *Vulpes vulpes* and *Alopex lagopus*, as Revealed by Radio Tracking

P. HERSTEINSSON and D. W. MACDONALD

Synopsis. 259
Introduction 259
Similarities and differences 260
Morphology 260
Habitat and diet 262
Movements and social organization 266
General behaviour 267
Two case studies 268
Methodology 268
The red fox on Boars Hill 272
The Arctic fox at Ófeigsfjördur 275
An ecological basis of red and Arctic fox societies 282
Conclusion 284
References 284
Discussion 288

The Uses of Radio Tracking Combined with Other Techniques in Studies of Badger Ecology in Scotland

T. PARISH and H. KRUUK

Synopsis. 291
Introduction 291
Methods. 292

Results   294
Discussion of results   297
Acknowledgements   297
References   298
Discussion   298

Activity Patterns and Habitat Utilization of Badgers (*Meles meles*) in Suburban Bristol: A Radio Tracking Study

S. HARRIS

Synopsis.   301
Introduction   302
Methods.   302
Results   304
    Distribution of setts   304
    Activity patterns   306
    Food   310
    Habitat utilization   313
Summary   318
Acknowledgements   320
References   321
Discussion   322

Ranges and Food Habits of Lions in Rwenzori National Park, Uganda

K. G. VAN ORSDOL

Synopsis.   325
Introduction   325
Study areas   327
Methods.   329
Results   330
    Ranges   330
    Pride size and biomass   332
    Pride association and ranging patterns   333
    Nocturnal movements   335
    Food intake   336
Discussion   337
Conclusions   339
Acknowledgements   340
References   340

Leopard Ecology as Studied by Radio Tracking

B. C. R. BERTRAM

Synopsis.   341
Introduction   341
Methods.   342
Results   343
    Range   343

Contents

Comparison with lion ranges 345
Food taken. 346
Discussion of results 348
Acknowledgements 350
References 350
Discussion 352

Symp. zool. Soc. Lond. (1982) No. 49, 1—18

State of the Art of Biotelemetry in North America

V. B. KUECHLE

J. F. Bell Museum of Natural History, University of Minnesota, Minneapolis, USA

SYNPOSIS

In telemetry from animals, the state of the art is in applying existing technology from a variety of sources to the problem. Currently, primary emphasis is on factors that increase the quantity and quality of the data collected. Factors influencing the reliability and quality of transmitters and receivers, such as batteries, attachment design, component selection and encapsulation design, are discussed. Also covered are new design techniques that enhance the quality of data on a second level, such as tags that telemeter temperature or pressure and the automatic or remote data collection systems. A short summary of the state of the ar in satellite tracking of wildlife is also given. Finally, future directions in telemetry are discussed.

INTRODUCTION

The use of telemetry started in earnest in the early 1960s after the transistor became readily available. Since that time the use of telemetry has extended to projects all over the world and has undergone an evolution in equipment design and in technique of use. This chapter will outline the current state of the art in North America and indicate some of the probable future directions of progress.

During the last ten years, much of the work in radio telemetry has moved from laboratories funded to do both the development and the field work, to groups that are using radio telemetry in fieldwork but are buying their equipment from commercial sources. As a consequence they are buying known equipment or equipment from catalogs, so much of the impetus for new development is gone. In addition, most of the commercial providers of the equipment are small business operations with little funding available for engineering development, and they usually have very small engineering staffs. Thus although there continues to be improvement in equipment design, it is occurring at a reduced rate.

TRANSMITTERS

Design

A wide range of frequency bands are being used for radio tracking. The greatest use is in the 150–151 MHz and 164–165 MHz bands. According to U.S. regulations only two bands are authorized for general use in wildlife tracking: they are the 40.66–40.70 MHz and 216–220 MHz bands (Federal Communications Commission, 1976). Many groups, however, operate at other frequency bands with special use permits. Biomedical devices may also be operated under wireless microphone provisions.

Most transmitters follow the general design shown in Fig. 1. Depending on the frequency used, the crystal is operated at its third, fifth or seventh overtone mode. If the output frequency is below about 85 MHz the crystal is operated at the output frequency and the second stage is used as an amplifier. If the frequency is above 85 MHz the second stage is used as a frequency doubler or tripler. In either case the first stage is tuned to the crystal overtone frequency and the output stage tuned to the desired output frequency.

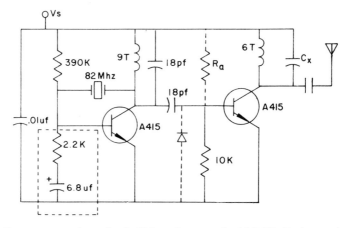

FIG. 1. Common transmitter circuit. Values shown are for 164 MHz. Resistor and capacitor enclosed in dashed lines form pulsing circuit. Resistor R_a and diode are used in some low voltage circuits.

At lower voltages (1.4, 2.8 V), the second stage is sometimes operated closer to class A mode than to class C by the addition of biasing resistors. A diode is also sometimes used in the base of the second stage to reduce base storage time. In all cases it is desirable to operate the first stage collector peak voltage as close to twice the supply voltage as possible. The second stage then acts as an impedance transformer with little or no voltage gain. Matching to the antenna is done by tapping the output coil down from the collector. For a quarter wavelength antenna the coil is tapped near the center. Most transmitter design is done by "cut and try"; very little theoretical evaluation of the potential design is done. This is partly because the technical capability to do so is not available, but probably a more important reason is that the nonlinearity of the devices and the conditions imposed on them by the user render the theoretical design almost useless on an animal. The transistors used are small signal radio frequency types such as the 2N918. Transistor selection is not critical if the device has adequate gain at low current levels. Most telemetry groups have a favorite transistor type that is used in all their designs.

Tuning and Testing

In most cases the transmitter is tuned for maximum output after assembly, by using a receiver, or some other device that can measure the relative output of the transmitter in conditions as close as possible to those in which it will be used in the field. The goal is to take as many factors into consideration as possible. The present author uses a short antenna connected to a preamplifier, the output displayed on an oscilloscope from which a measure of transmitter power is made. It should be emphasized that building transmitters of this type is as much an art as a science, and that generally the greater the experience the better the transmitter. For example, with experience one can recognize how the transmitter should be tuned to compensate for the effects of encapsulation (potting) and packaging.

Each group probably has its own system of component selection and testing; in our case, all crystals are checked for reliable operation in a test circuit. This indicates that the crystal has a low series resistance and that it does not have spurious emissions near the desired response. We use this test rather than the actual measurement of spurious emission because it is faster and easier. Most of our selection criteria are the result of experience. Transmitters are tested at each stage of construction; if they do not behave normally they are not used. Those selected for use are subject to further checks by running them for at least 24 h, after which they are temperature-cycled and checked again. If a large number of transmitters are used they are further checked for frequency drift *vs* temperature to ensure that frequencies will not overlap. This consists of a check of frequency at three different temperatures within the likely temperature range. Almost all frequency drift occurs because of changes in the crystal, and it is also found that crystals of the same manufacturing batch have a similar temperature drift characteristic.

Since the design of all transmitters is about the same, their performance from the various sources should also be similar. In fact comparative tests made by several laboratories indicate that this is the case.

Reliability

Thus, the major emphasis in design is on making the transmitters more reliable, and this improved reliability is really the major change that has occurred over the last ten years. The improvements in reliability can be grouped into component selection, potting, and testing. Improvements in batteries or power sources will be discussed later.

The most significant factor in transmitter reliability is the encapsulation, or potting, of the transmitters. It is thought that more transmitters fail because of poor potting material or technique than for any other reason. Unfortunately the materials which are the best in terms of reliability are also often the most difficult to use. The goal in any potting is to protect the transmitter from conditions in its environment and it is also important that the potting material be compatible with the attachment materials and the components. As an example, in our early work we used dental acrylic because it was convenient and readily available. Our first problem occurred when transmitters failed after about three weeks in the field. After much testing we found that the solder flux being used reacted with the potting material, causing a conductive bridge across the transistor leads. This was corrected by carefully cleaning the transmitter after assembly. A second problem occurred when transmitters again failed prematurely, this time while on ducks, owing to breakage of the P.V.C. (polyvinyl chloride) tubing used over the harness wires. The cause was found to be leaching of the plastizer from the tubing by the dental acrylic, causing the tubing to become brittle and break with a subsequent breaking of the harness wires. Since that time we have exclusively used resins designed for electrical encapsulation. These resins have several advantages. They are usually less costly and are specified in greater detail with important parameters such as conductivity and moisture absorption indicated. To ensure the adequate penetration of parts a low vicosity material is necessary and this requirement is what makes most people avoid electrical resins. It would be much more convenient to use a paste-like material such as dental acrylic that would not run off. These materials are desirable because molds are not required and because a minimum amount of material can be used, reducing weight. Unfortunately, however, it is extremely difficult to get adequate waterproofing using paste-like materials. Several techniques are used in potting transmitters to ease the mold problem. For the larger transmitters, the entire package is made in the shape of a cylinder; potting can then be done inside a tube or syringe barrel. Another technique is to pot in two stages. In the first stage the electronic parts are covered with an electrical resin, and then harness or attachment parts are added using a paste adhesive, thus making the mold simpler and more universal. Use of translucent materials aids in the inspection of the quality of encapsulation.

Some manufacturers enclose all parts in metal cans which can be solder-sealed and for reliable operation these should be evacuated and backfilled with dry nitrogen. The problem of what to do with the leads as they emerge from the package is still present, although less critical. This technique has the additional advantage of not changing the transmitter tuning by potting. Either technique is reliable provided adequate quality control is used. The importance of quality control in encapsulation techniques cannot be over-emphasized. It is the single most important cause of transmitter failure.

Power Sources

Batteries

Most battery powered transmitters in North America now use lithium cells as power sources. They have a greater power to weight ratio and perform much better in cold temperatures. A wide range of sizes is currently available (Table I), although small sizes with sufficient current capability are not readily available. Additionally at least one U.S. company now makes custom lithium batteries at about twice the cost of off-the-shelf types (Battery Engineering Inc., 1980). For good reliability, only hermetically-sealed cells should be used. Some provision also must be made for cell expansion in cold weather or the expansion will crack the potting. In small transmitters (under about 8 g) mercury batteries are still used. Silver oxide batteries have fallen into disuse because of the high and unstable price of silver.

Solar

Solar power transmitters are being used in a number of studies. Although we have no accurate figures on the number being used, we estimate that less than 10% of the transmitters are solar powered.

TABLE I

Summary of the most commonly used batteries

Battery number	Type	Voltage	Weight (g)	Life days (mA)	Height (mm)	Diameter (mm)
Lithium						
BR 1/2 A	National[a]	3.0	9.5	31	22.5	16.8
BR 2/3 A	National	3.0	13.5	50	33.4	16.8
BR C	National	3.0	47.1	208	50.0	26.0
440	Power Conv[b]	2.8	11.3	42	33.2	16.2
400-(AA)	Power Conv	2.8	12.5	50	49.5	14.0
660-2(3/4c)	Power Conv	2.8	32.6	100	41.4	24.1
660 (C)	Power Conv	2.8	40.0	125	50.8	24.1
660-3(1 1/4c)	Power Conv	2.8	48.0	158	59.7	24.1
550 D	Power Conv	2.8	83.0	333	61.0	33.2
660-4	Power Conv	2.8	100.0	416	49.5	41.6
660-5	Power Conv	2.8	221.0	1042	113.0	41.6
660-5AS	Power Conv	2.8	221.0	1250	113.0	41.6
LO-37S	Mallory[c]	2.9	6.5	19	23.3	13.9
LO-32S	Mallory	2.9	12.0	40	35.0	16.5
LO-28	Mallory	2.9	43.0	146	50.0	24.4
LO-29S	Mallory	2.9	48.0	167	50.0	26.0
LO-27S	Mallory	2.9	58.0	208	60.0	26.0
LO-30S	Mallory	2.9	62.0	279	60.0	30.0
LO-26S	Mallory	2.9	85.0	375	60.0	34.0
LO-25S	Mallory	2.9	95.0	400	50.9	38.6
LO-50S	Mallory	2.9	207.0	917	114.3	38.6
Mercury						
RM-212	Mallory	1.4	0.3	0.7	3.3	5.6
RM-312	Mallory	1.4	0.6	1.8	3.6	7.8
RM-675	Mallory	1.4	2.6	7.5	5.3	11.6
RM-625	Mallory	1.4	4.2	14.6	6.0	15.6
RM-630	Mallory	1.4	4.8	14.6	6.0	15.6
RM-640	Mallory	1.4	8.0	20.8	11.2	15.9
RM-660	Mallory	1.4	7.6	25.0	7.6	17.4

[a] Matsushita Electric, Japan.
[b] Power Conversion Inc., Mount Vernon, New York.
[c] P. P. Mallory, Inc., Tarrytown, New York.

Two areas of application seem to predominate: the first is with mid-sized birds such as grouse that can carry transmitters in the 25-g range. With solar cells life times of two years or more could be expected, whereas using lithium batteries at normal drains four to six months are typical. In fact the life of the animal or harness is most often the limiting factor. Solar applications typically use as the storage device a 20 mAh capacity nickel cadmium battery (Church, 1980), which is a rechargeable cell. Such cells are often unreliable, partly because it is difficult to control the rate of charge, or regulate the discharge cycle. Fortunately, with failures of this type the transmitter will operate at least intermittently for a period, allowing an opportunity to recapture the animal. The second application is the use of small solar-powered transmitters as ear tags on large animals, where because of growth rate, or other characteristics, neck collars cannot be used. A number of these transmitters have been tried on moose (*Alces alces*) and caribou (*Rangifer caribou*). However, keeping the orientation correct so that sufficient light hits the transmitter solar cell has been a problem. Transmitters powered by primary batteries such as lithium cells are generally considered more reliable than solar cell systems.

Power Output

Typically the power output of the transmitters just discussed varies from about 1 mW to 10 mW. In many applications a higher output power is desirable. To achieve this a different design must be used and an example of the design used by this laboratory is shown in Fig. 2, where the component values depend on frequency; nominal values for 164 MHz are given. The power output of this circuit is 250—300 mW using a 5.6 V source and this can be easily increased by increasing the battery voltage. The design is conventional; the first stage operates as an overtone oscillator, the second stage as a doubler or tripler and the final stage as a power amplifier. Tuning of this circuit is rather critical if maximum power output is to be achieved. Additionally, tuning is effected by the potting. Compensation for this and for different antennas is achieved by using the two variable capacitors which are adjusted after potting. These transmitters have been used on seals, manatees and polar bears. In the case of polar bears two transmitters were used on each animal, one high power, the other low. All high-powered transmitters operated at the same frequency; each of the second, low power, transmitters operated at a unique frequency to identify the individual animal. Thus search time can be optimized since only one frequency needs to be monitored. Once the transmission is found, all possible frequencies are scanned to identify the individual animal.

Sensors

Although various types of transducers have been added to transmitters, the number of applications remains low. The addition of temperature measuring circuits has been the most common; a typical circuit is shown in Fig. 2 where a thermistor is used. In this and most similar circuits, the pulse rate is varied as a function of the measured

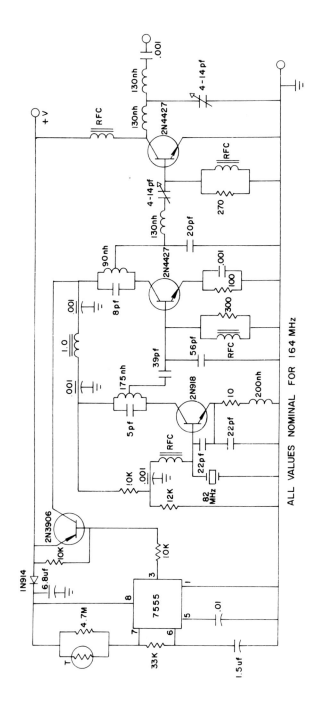

FIG. 2. Transmitter circuit used where higher power is needed.

parameter. Most do use thermistors and thus are non-linear which is a disadvantage for automatic data recording, since a separate curve must be made for each transmitter. Base/emitter junctions and other linear devices have been tried; most, however, operate at current levels which are too high for application in lower power devices.

Pressure, heart rate and a number of other functions have been measured, almost all by means of the same basic transmitter circuit. In the example shown (Fig. 3) the amplifier is a voltage to current converter; the current being converted to a pulse rate or width. To save power the bridge can be turned off between pulses as shown in this circuit. A variation of these circuits was used on elk, where an implanted transmitter sent the heart-rate signal to a collar transmitter for retransmission (Weeks, Long & Cupal, 1977).

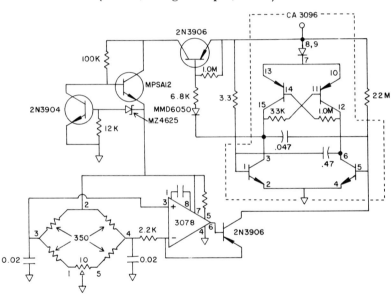

FIG. 3. Circuit diagram of transmitter used to measure pressure (depth).

The most common sensing transmitter is one used to detect the cessation of movement and it is most often used as an indicator of mortality. The circuit used by this laboratory is shown in Fig. 4, but most others are similar. A mercury switch is used as the motion sensor and is used to reset a counter. If activity stops the counter is no longer reset, and will count up to its full value; on reaching its full value the decode "out" is set, which in turn changes the pulse rate of the transmitter. The clock inhibit line is used to stop the counter until another reset is received. The delay time, which is the time

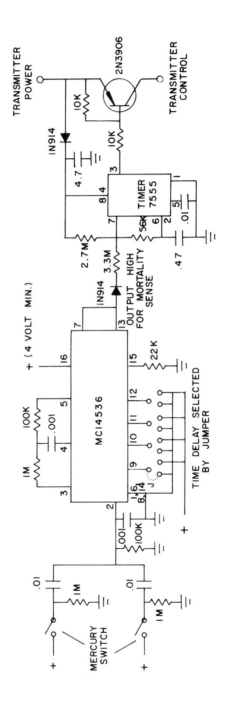

FIG. 4. Circuit diagram of mortality sensing transmitters.

from the last reset to mode change, is usually from two to four hours. Variations of this circuit are used; some have several switches mounted in different positions to reduce the sensitivity to different collar orientations. Some earlier circuits used a mercury switch or other device to momentarily change the pulse rate (Swanson, Kuechle & Sargeant, 1976; Knowlton, Martin & Haug, 1968). These are satisfactory if there is sufficient time to monitor each individual.

Transmitters for Monitoring by Satellite

Tags for monitoring by satellite have been proposed for a great number of applications. In North America, using Nimbus VI, tags have been used on polar bears, sea turtles and dolphins (Kolz, Lentfer & Fallek, 1980; Kuechle, DeMaster & Siniff, 1979). The polar bear and sea turtle experiments were moderately successful but those with dolphins were not because the dolphins did not surface long enough to give adequate transmission times. While the Argos system is currently operational, and animal transmitters have been approved as "operational platforms", no North American groups appear to have programs to track animals using Argos. This is probably because most animals can be tracked by conventional means at lower cost, while others are either too small to carry the tag needed for satellite monitoring, or have other characteristics that make the use of these tags difficult. Whales, dolphins, and long range migrating birds are examples of cases where satellite tracking would be helpful, but where application has proved difficult.

Attachment

Attachment of transmitters to animals remains one of the most nebulous and difficult areas. Attachment becomes a greater problem as more of the equipment is ordered from catalogs rather than being custom designed. In many cases an investment is made in material before the attachment is tested. Thus as problems are encountered it often becomes a case of make-do with available material.

Most neck collars are made from fabric impregnated rubber belting material with the transmitter and batteries fitted at the bottom. The antenna is usually run up the side between two layers of belting material which are riveted or sewn together. Molded urethane collars are also used. They have the advantage of ease of battery replacement, but require molds which can be expensive.

A few projects use implanted transmitters; however, few data are available on their success (Melquist & Hornocker, 1979). Our experience with white tail deer and badger indicates that implantation of transmitters is no small problem. The major difficulty is in keeping the transmitter in place, and experience has shown that the range of an internal transmitter is much less than that of an external transmitter. This is to be expected because of the need to transmit through the body tissue and also because usually the antenna must be kept short, thus reducing efficiency. A notable exception to these comments is the use of implanted transmitters on fishes where the range of internal and external transmitters is about the same for low frequencies (Winter *et al.*, 1978).

RECEIVERS AND RECEIVING SYSTEMS

Introduction

The basic design of receivers has remained the same over the last ten years or more. Most of these receivers are of the double conversion superheterodyne type. Enhancement in terms of output options and controls is varied. An example is the use of a frequency synthesizer for tuning rather than variable coil/capacitor (LC) or crystal tuning. These receivers, when used for transmitter searching, allow faster progress if a large number of transmitters are to be tuned. Precision of tuning frequency is also better than in LC types. Their primary advantage is, however, that scanning and memory functions can be easily added; for example, if transmitters are to be located by aircraft, all the transmitter frequencies can be programmed into the receiver memory. The time that the receiver is to remain tuned to each frequency can also be preset, depending on transmitter pulse rate, transmitter range and flying speed, thus increasing the probability of finding a signal. If the operator hears a signal he notes its presence or, if he wishes, he can stop the scan to localize more precisely. This results in more accurate sampling and reduces the work of the operator.

Most of the receivers have similar specifications. Bandwidths vary from about 500 Hz to 5 kHz. Experience of aural signal detection has shown no advantage in reducing the bandwidth below 3.5 kHz, probably because the ears are able to act as a bandpass filter (Hamilton, 1957). Noise figures are also reduced to the level where further reduction will yield no advantage because atmospheric and antenna noise will be higher than receiver noise. A key factor in receiver design now being recognized is the need for a wide dynamic range. This is the range of signal level that can be handled by the receiver. If it is too low, the receiver will tend to overload at high signal levels, making measurement or localization of the signal difficult. The typical dynamic range is 30 dB, although ideally it should be about 60 dB for optimum performance.

Many receivers have a meter to give an indication of signal level which can usually be output to a meter jack, so that a paper chart recorder can be used to monitor activity. A high dynamic range receiver is needed for successful operation of these activity recorders, but the interpretation of data is often difficult because the activity categories are not discrete enough. Most studies use only three classes, inactive, active, or absent. Strip chart recorders can be used, with scanning receivers to scan a number of transmitters; however, deciding which animal channel is where on the strip chart is difficult. A better approach is to use multiple pen recorders so that each animal channel will always be on the same pen. If the activity can be modulated into an on/off mode the system can be made very reliable by the use of phase-lock-loop devices as detectors because, being correlation detectors, they offer high noise rejection and a wide dynamic range. We have used this scheme in a number of applications with a 20-pen Esterline Angus event recorder to note the presence or absence of animals.

Automatic Data Recording Systems

A few automatic data recording systems are in use; they are illustrated by the following examples. The first is the Radio Tracking System at Cedar Creek (Cochran et al., 1965). Its operation has been described in the literature; only a few of the pertinent points will be discussed here. This system uses two towers with antennas that rotate once every 45 s. The antennas are two Yagis spaced by two wavelengths and fed in phase. Received signals are fed to a central station where pairs of receivers are tuned to individual transmitters to determine the bearing of the animal in relation to the station. An ideal antenna pattern, stored in the computer, is matched to the signal from the animal in a correlation process; the point of maximum correlation is the bearing to the animal. With the bearing from the second tower the x, y, co-ordinates are determined, the entire process being controlled by a minicomputer. The x, y co-ordinates along with a index of signal quality are output to magnetic tape for later data reduction. The system can sample animals in groups of ten every 1.5 min and although it is quite simple in design, it has proven reliable over its 15 years of operation. Various other techniques have been tried in attempts to develop a more portable and less costly automatic location system. Most of these have employed electron-ically rotated antennas (Marten, Evens & Bowers, 1971). A few have been used in the field, but none has seen wide use. The principal problem is the high signal level needed to determine an accurate location. A typical signal requirement is -100 to -110 dBm — con-trast this with the usual signal detection by the ear of -143 dBm. The problem with signal level is caused in large part by the short on-time of the transmitters which is typically 20–30 ms. With such short times it becomes a matter of instantaneously determining the location, thus precluding the use of any signal enhancement pro-cesses. One application where these systems have found use is in tracking ocean mammals which surface for only short periods of time. In such cases the human operator suffers many of the same handicaps, i.e. short signal duration, as the automatic system. Additionally it takes the human operator some time to determine the direction of signal arrival. When the transmitter is used in the ocean a saltwater switch is used to turn it off whilst it is submerged. When the animal surfaces the transmitter can be kept on for a longer period, because the duty cycle is already low, since the animal is sub-merged most of the time.

Another example of an automatic system was one built to locate fishes as they swam along experimental channels, in order to deter-mine their response to temperature and various other pollutants (Kuechle, Reichle *et al.*, 1979). In this case the transmitters were confined to a linear dimension. To determine the location of the fishes, antennas were placed along the channels at 33-m intervals, and could be fed into a common signal cable by command from a central station. The signal level at each antenna was measured and stored, and after all the antennas were scanned, the antenna with the maximum signal level was determined. Average noise level was also measured to give an indication of signal quality. The entire process was controlled by a RCA 1802 microprocessor. This also controlled the sampling interval, receiver tuning and time keeping. Data in this experiment were output to a printer which recorded time of day, animal number, antenna with the maximum signal and an index of signal quality. In this sytem the fishes are located to ± half the distance between the antennas.

It can be seen that the automatic location systems are special applications; to our knowledge no general system exists that can be taken to a field situation to determine transmitter location. The systems currently in use can determine proximity by measuring signal strength, or they require high signal levels. The only exception is the one in use at Cedar Creek; however, it requires too much support to be considered portable.

Receiving Antennas

Antennas will not be discussed in general, for their characteristics are well known. The most widely used in North America is the multi-element Yagi. Loops are used for lower frequencies and sometimes for close-in location.

Aircraft are often used in North America to track wider ranging animals. A complete description of antenna patterns, together with mounting and operating recommendations for aircraft use, are in preparation (Gilmer *et al.*, in press). In general, high-winged aircraft are used with the antennas mounted below and to the front of the wings, pointed away from the aircraft and slightly downward. Location is determined by switching between the right and left antennas to determine whether the transmitter is to the left or right. Further resolution can be achieved by flying smaller and smaller search patterns. Experience is a big factor in locating transmitters from aircraft.

FISH TAGS

Radio frequency (RF) tags have been used to track fishes for some time. The success of RF tags depends on the conductivity of the water; as this goes up, range goes down. Figure 5 gives theoretical and measured results for several conductivities; a typical surface range for the transmitter circuit of Fig. 1 is about 3 km. The importance of the relationship of depth and conductivity can readily be seen. Another important factor in fish tag transmitters is that the tuning and antenna matching be adjusted for operation in water. Antenna loading in water is very different from the loading in air and must be compensated.

Acoustic tracking is used where the underwater attenuation is high, precluding the use of RF tags. This includes all salt water and fresh water with conductivity of about 0.50 S/m (mho/m) and above. Between 0.01 and 0.05 S/m, which tags (acoustic or RF) should be used, depends on the depth of the transmitter (see Fig. 5).

Ranges for the acoustic transmitters in fresh water are typically 0.5 to 1.5 km but depend on the output power and the losses due to absorption. Higher power transmitters have been used: for example we have used transmitters which develop sound pressure levels of 165

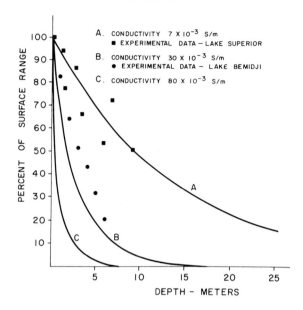

FIG. 5. Percentage of surface range versus depth for several water conductivities. Solid lines are from theoretical calculations.

dB re 1 μPa at 1 m which yielded ranges of about 3.5 km. The pressure level of tags in the USA is normally 145 to 155 dB re 1 μPa at 1 m.

Location is usually by means of directional hydrophones. A few applications have used a three-hydrophone array and time of arrival to determine location (Kuechle, Thomas *et al.*, in press). Location can be graphically determined or a hand-held programmable calculator can be used. The latter system is restricted to fixed locations because accuracy depends on the three hydrophones remaining in a fixed position.

Potting of transmitters is done by inserting the electronics in a plastic case, filling with oil or other inert substance and sealing the cap, or by complete encapsulation using epoxy resins or urethanes. There is a preference for urethanes because of their good acoustic transmission characteristics. A good description of acoustic tracking is given by Stasko (1975, 1977).

FUTURE DIRECTIONS OF PROGRESS

Transmitters

Little change in transmitter design is expected. Most designs are reliable and operating near optimum output levels. More forms of modulation are likely to be added to transmitters, especially if they give better indication of what the animal is doing. There is also some interest in having all or many of the transmitters on the same frequency with some form of coding to identify individual animals. The primary limitation upon developments in animal transmitters is size and weight restrictions, although cost is also a factor.

Receivers and Receiver Systems

Receivers will probably undergo some improvement in reliability and ease of operation. Improvements that will yield greater range are unlikely. Most improvements will probably occur in the recording of data, perhaps on digital cassettes, so that data can be easily transferred to a computer for analysis. Some advances in unattended location systems are likely but no significant move toward the recording of physiological data is expected, although most of the technology is available. One great hindrance to the improvement of equipment and techniques is that very few laboratories are actively engaged in engineering research and, whilst commercial organizations do some design and development, most have very small engineering staffs with little time available for development.

ACKNOWLEDGEMENTS

The original work reported here was supported by the National Science Foundation, Division of Polar Programs, U.S. Fish and Wildlife Service, Jamestown, North Dakota and U.S. Department of Energy.

REFERENCES

Battery Engineering Inc. (1980). *Data sheets*. 80 Oak Street, Newton, Mass., U.S.A.

Church, K. E. (1980). Expanded radio tracking potential in wildlife investigations with the use of solar transmitters. In *Handbook on biotelemetry and radio tracking*: 247–250. Amlaner, C. J. & Macdonald, D. W. (Eds). Oxford: Pergamon Press.

Cochran, W. W., Warner, D. W., Tester, J. R. & Kuechle, V. B. (1965). Automatic radio tracking system for monitoring animal movements. *BioScience* 2: 98–100.

Federal Communications Commission Rules and Regulations Part II (August, 1976). Paragraph 15.161–15.194 and Paragraph 5.108. U.S. Gov. Printing Off., Washington, D. C.

Gilmer, D. S., Cowardin, L. M., Duval, R. L., Mechlin, L. M., Shaiffer, C. W. & Kuechle, V. B. (In press). Procedures for the use of aircraft in wildlife biotelemetry studies. *Techn. Rep. U.S. Fish Wildl. Serv.*

Hamilton, P. M. (1957). Noise masked thresholds as a function of tonal duration and masking noise band width. *J. acoust. Soc. Am.* 4: 506–511.

Knowlton, F. F., Martin, D. E. & Haug, J. C. (1968). A telemetric monitor for determining animal activity. *J. Wildl. Mgmt* 4: 943–948.

Kolz, A. C., Lentfer, J. W. & Fallek, H. G. (1980). Satellite radio tracking of polar bears instrumented in Alaska. In *Handbook on biotelemetry and radio tracking*: 743–752. Amlaner, C. J. & Macdonald, D. W. (Eds). Oxford: Pergamon Press.

Kuechle, V. B., DeMaster, D. P. & Siniff, D. B. (1979). State of the art and needs of the earth platform. *Proc. Int. Symp. Remote Sensing Environmt* 13: 505–518.

Kuechle, V. B., Reichle, R. A., Zinnel, K. E. & Ross, M. J. (1979). Use of a microprocessor to locate fish in experimental channels. *Int. Conf. Wildl. Biotelem.* 2: 33–39.

Kuechle, V. B., Thomas, J., Ross, M. J. & Eagle, T. (In press). Preliminary results on the development of satellite tracking and automatic position/activity monitoring techniques for the Antarctic cod and the Weddell seal. *Antarct. J. U.S.*

Marten, H., Evens, W. E. & Bowers, C. A. (1971). Methods for radio tracking marine mammals in the open sea. *Proc. IEEE Conf. Eng. Ocean Environmt:* 44–49. New York: IEEE Press.

Melquist, W. E. & Hornocker, M. G. (1979). Development and use of telemetry technique for studying river otter. *Proc. Int. Conf. Wildl. Telemet.* 2: 104–114.

Stasko, A. B. (1975). Underwater biotelemetry, an annotated bibliography. *Tech. Rep. Fish. Mar. Serv., Can.* No. 534: 1–31.

Stasko, A. B. (1977). Review of underwater biotelemetry, with emphasis on ultrasonic techniques. *J. Fish. Res. Bd Can.* 34: 1261–1285.

Swanson, George A., Kuechle, V. B & Sargeant, Alan B. (1976). A telemetry technique for monitoring diel waterfowl activity. *J. Wildl. Mgmt* 1: 187–190.

Weeks, R. W., Long, F. M. & Cupal, S. S. (1977). An improved repeater heart rate telemetry system for use on wildlife. *Proc. Int. Conf. Wildl. Biotelem.* 1: 2–8.

Winter, J. D., Kuechle, V. B., Siniff, D. B. & Tester, J. R. (1978). Equipment and methods for radio tracking fish agricultural experiment station. *Misc. Rep. Agr. exp. Stn Univ. Minnesota* No. 152, 18 pp. (mimeo).

Symp. zool. Soc. Lond. (1982) No. 49, 19—30

Regulatory Control of Telemetric Devices Used in Animal Studies

R. M. SKIFFINS

Home Office Directorate of Radio Technology, London SE1, England

SYNOPSIS

Under the Wireless Telegraphy Act 1949, control of the civil use of radio is vested in the Secretary of State for the Home Office through the exercise of the licensing requirement laid down in the Act. By means of this statutory responsibility the Home Office Radio Regulatory Department maintains regulatory control over all uses of civil land mobile radio. For a given application, licensing control will cover the choice of frequencies, the standard of radio equipment to be used and other operational conditions.

This chapter outlines how these controls affect the use of radio-telemetric devices used particularly for the study of animal behaviour patterns.

INTRODUCTION

In the United Kingdom, control of the civil use of radio is vested in the Secretary of State for the Home Office who is advised on radio regulatory matters by the Radio Regulatory Department of the Home Office. This department comprises R1 and R2 divisions which are responsible for frequency policy, regulatory and licensing matters, and the Directorate of Radio Technology which provides the necessary technical advice.

The principal objects of statutory control of the uses of radio are to permit efficient and equitable use of the radio frequency spectrum and to provide freedom from harmful interference.

Throughout the world the allocation of frequency bands to different radio services is governed by the International Radio Regulations which have treaty force. These regulations are the outcome of World Administrative Conferences of the International Telecommunications Union (ITU), a specialized agency of the United Nations.

Within the constraints imposed by these Regulations the United Kingdom, like other countries, is free to allocate and sub-divide frequency bands for different uses as it thinks fit. In this regard, matters affecting the use of private land mobile radio are the responsibility of the Home Office, which is advised by a committee known as the Mobile Radio Committee. It is the executive function of the Radio Regulatory Department of the Home Office to process applications from private mobile radio (PMR) users, assign suitable radio frequencies, define the requirements for the equipments to be used and issue the appropriate licenses.

Each of these functions will be examined in relation to the use of radio telemetry devices for ecological and biological research.

RADIO FREQUENCY BANDS

Exclusive Bands for Biomedical Telemetry

In April 1965, the first exclusive frequency band for medical and biological telemetry applications was allocated between 102.2 MHz and 102.4 MHz. Narrowband and wideband equipments were authorized, the former, for which a more stringent specification was applied, being permitted a higher power and hence greater range.

At around the same time, use of the band 300 kHz to 30 MHz was authorized for very low power devices intended to be wholly contained within the body (e.g. radio pills). No protection could be given, however, against interference from other radio users.

Changes consequent upon an extension of the frequency modulation (FM) broadcasting band II up to 100 MHz, required that a new solution be found for the medical and biological services and in 1974, a further band between 104.6 MHz and 105 MHz was made available.

Continued use of the 102.2 MHz to 102.4 MHz band was permitted for existing services or where existing equipment was to be used, but from 1 January 1976, all new equipments were required to operate in the new band. Between 104.6 MHz and 105 MHz, the maximum effective radiated power (ERP) permitted is 1 mW for wideband equipment and 10 mW for narrowband equipment.

Shared Radio Frequencies

In addition to the frequencies given above, biomedical telemetry equipment may operate in those bands allocated to the general purpose low power telemetry and telecontrol services. Two bands are available, at very high frequency (VHF) between 173.2 MHz and 173.35 MHz and at ultra-high frequency (UHF) between 458.5 MHz and 458.8 MHz.

In the VHF band, both wideband and narrowband equipments are permitted with a maximum ERP of 1 mW. In the UHF band, narrowband equipment only may operate with a maximum ERP of 0.5 W. Different technical requirements apply to the equipments used in these bands as described on pp. 22–23.

These frequencies are used extensively in the United Kingdom for industrial telemetry and telecontrol purposes. Biological telemetry services operating in these bands cannot therefore be afforded the degree of protection possible on the normal medical and biological telemetry frequencies.

Future Frequency Allocation

As a result of a decision at the World Administrative Radio Conference (WARC) of the ITU in 1979, the FM sound broadcasting band is to be extended up to 108 MHz. Owing to this change any continued usage of the band 102.2 to 102.4 MHz may be subject to interference from broadcasting after 1 January 1982 when revisions to the Radio Regulations made at the Conference generally come into effect. Future use of the 104.6–105 MHz band for animal telemetry is not immediately threatened by this change because mobile radio services in the United Kingdom remain protected internationally until 31 December 1995. However it may be necessary to reallocate mobile services in the 104–108 MHz band before 1995 owing to national broadcasting requirements.

Final proposals for the re-location of the services in the 104.6 MHz to 105 MHz band have yet to be completed, but it is hoped that provision can be made for an equal or greater amount of frequency spectrum at VHF below 220 MHz. Where possible, at least ten years' notice will be given of these changes of frequency.

Frequency Assignment

A suitable frequency within the appropriate frequency band is assigned by the Directorate of Radio Technology on receipt of a completed licence application. Care is taken to ensure that, as far as is practicable, the proposed system will not cause interference to, or suffer interference from existing biomedical systems, or in the case of systems assigned in a shared radio frequency band, that the best available frequency is chosen.

Information is recorded of the location or area in which each system will be used, and the purpose for which the system is required.

EQUIPMENT STANDARDS

Under the terms of the Wireless Telegraphy Act 1949, no radio equipment may be installed and used in the United Kingdom except under the authority of a licence granted by the Secretary of State. It is a condition of such a licence that the equipment used must meet certain minimum standards. These standards are necessary to ensure that a satisfactory grade of service is achieved and that undue interference is not caused to other radio users.

The minimum standards of performance are contained in specifications prepared by the Home Office, Radio Regulatory Department, in consultation with the relevant manufacturers. Representative production samples of equipment are tested by the Directorate of Radio Technology to ensure compliance with the appropriate specification, equipments so complying being certified as type-approved for use in the United Kingdom.

Three performance specifications are currently used for type-approval testing of biological telemetry equipment:

(i) MPT 1312. Medical and Biological Telemetry Devices. (Published January 1978.)

This specification covers the requirements for equipment operating in the band 104.6 to 105 MHz and both wideband and narrowband (25 kHz channelling) equipments are covered by the specification. Under the specification, limits are set for transmitter frequency error, transmitter output power, adjacent channel power and out-of-band power, and transmitter and receiver spurious emissions. This specification has superseded the existing General Post Office specifications W6803 and W6804 for Class II and Class III devices.

(ii) W6802, Medical and Biological Telemetry Devices, Class I (1968).

Following the issue of MPT 1312, specification W6802 was retained for Class I devices operating in the range 300 kHz to 30 MHz. Measurements are required of transmitter power, and transmitter and receiver spurious emissions only.

(iii) MPT 1309, Transmitters and Receivers for Use in the Bands Allocated to Low Power Telemetry in the PMR Service.

This specification covers equipments to be used in the 173 MHz and 458 MHz low power telemetry bands. The parameters specified are essentially the same as those contained in MPT 1312.

Applications for type approval testing should be addressed to:

    Mobile Services Laboratory
    Directorate of Radio Technology
    Home Office
    Waterloo Bridge House
    Waterloo Road
    London SE1 8UA   Tel: 01-275 3150

No direct charge is made for this service.

The application should state to which specification the equipment is to be tested and should be accompanied by a description of the equipment including drawings and test results.

LICENSING

Operational Licences

The use of biological telemetry equipment requires a licence under the Wireless Telegraphy Act 1949. The licence is renewable annually and a fee of £7.50 p.a. is charged. The licence includes a schedule containing technical details of the system covered by the licence. A sample of the licence is given in Appendix 1.

Applications for a licence must be made on form BR15, a sample of which is given in Appendix 2. Copies of this form are obtainable from R2 Division (Licensing Branch), Radio Regulatory Department.

Test and Development Licences

The development of radio apparatus must be authorized by a testing and development licence. This requirement applies even where work is to be carried out under suppressed radiation conditions (i.e. no electromagnetic energy capable of detection outside the premises where the work is carried out shall avoidably be emitted). Application for a testing and development licence should be made to the Radio Regulatory Department, R2 Division.

ACKNOWLEDGEMENT

This paper is published with the kind permission of the Director of Radio Technology, Home Office.

R. M. Skiffins

APPENDIX 1.

HOME OFFICE
WIRELESS TELEGRAPHY ACT 1949
TELEMETERING AND TELECONTROL LICENCE

Date of Issue:
Date of Renewal: in each year
Fee on Issue: £
Fee on Renewal: £

1.   (1)   *Licence*

(hereinafter called "the Licensee"), having paid to the Secretary of State an issue fee of £   , is hereby licensed, subject to the terms, provisions and limitations herein contained:

*( ) to establish (a) sending and receiving station(s) for wireless telegraphy (hereinafter called "the Control Station(s)") at the location(s) referred to in Part 1 of the Schedule hereto;

*( ) to establish (a) sending and receiving station(s) for wireless telegraphy (hereinafter called "the Remote Station(s)") at the location(s) referred to in Part 2 of the Schedule hereto;

*( ) to establish (a) sending and receiving station(s) for wireless telegraphy (hereinafter called "the Station(s)") at the location(s) referred to in the Schedule hereto;

*( ) to establish (a) sending and receiving station(s) for wireless telegraphy and (a) receiving station(s) for wireless telegraphy which sending station(s) and receiving station(s) are hereinafter collectively called "the Station(s)" at the location(s) referred to in Parts 1 and 2 respectively of the Schedule hereto;

( )* to use the Station(s):—

* to use the Control Station(s) and the Remote Station(s) hereinafter collectively called "the Station(s)") for the purpose(s) only:—

(i)   of sending and receiving (between the Station(s)) by wireless tele-graphy by automatic or other means signals (not being messages having a verbal significance) which serve only for the purpose of gaining, recording or indicating information consisting only of measurements;

*(ii) of actuating or controlling the operation of the apparatus comprised in the Remote Station(s) or any associated machinery or apparatus not being machinery or apparatus for wireless telegraphy by means of the emission of electro-magnetic energy from the Control Station(s) and the reception of such energy by the Remote Station(s);

*(iii) of sending and receiving between the Stations spoken messages in connection only with the maintenance of the apparatus comprised in the Stations.

RLn HO 97/1 200 7/74 P

---

* Delete the inapplicable alternatives

Appendix 1 (*cont.*)

(2)   *Limitations*. The foregoing Licence to establish and use the Station(s) is
subject to the following limitations:

(a) The Station(s) shall be used only with emissions at the frequencies and
of the classes and characteristics respectively specified in the Schedule
hereto; and with such power and aerial characteristics as are specified in
the Schedule hereto in relation to the class and characteristics of the
emission in use.

(b) When the Station(s) are being operated by other than automatic means
the Station(s) shall be operated only by persons authorised by the Licensee
in that behalf.

2.   *Non-Interference*. The apparatus comprised in the Station(s) shall be so
designed, constructed, maintained and used that the use of the Station(s) does
not cause any avoidable interference with any wireless telegraphy, and in particular and without prejudice to the generality of this condition, with Government, commercial and broadcast wireless telegraphy services.

3.   *Operators and Access to Apparatus*. The Licensee shall not permit or suffer
any unauthorised person to operate the Station(s) or to have access to the
apparatus comprised therein. The Licensee shall ensure that persons operating
the Station(s) observe the terms, provision and limitations of this Licence at all
times.

4.   *Apparatus*. The apparatus comprised in the Station(s) shall at all times
comply with such of the performance specifications which at the date of installation of the apparatus shall have been most recently published by Her Majesty's
Stationery Office on behalf of the Secretary of State as are applicable to such
Stations, subject however to such modifications of the said specifications in
favour of the Licensee as the Secretary of State may from time to time permit.

5.   *Inspection*. The Station(s) and this Licence shall be available for inspection
at all reasonable times by a person acting under the authority of the Secretary of
State.

6.   *Station(s) to Close Down*. The Station(s) shall be closed down at any time
on the demand of a person acting under the authority of the Secretary of State.

7.   *International Requirements*. The Licensee and all persons operating the
Station(s) shall observe and comply with the relevant provisions of the Telecommunication Convention.

8.   *Period of Licence, Renewal, Revocation and Variation*. This Licence shall
continue in force for one year from the date of issue and thereafter so long as
the licensee pays to the Secretary of State in advance in each year on or before
the anniversary of the date of issue a renewal fee of £   ; provided that the
Secretary of State may at any time after the date of issue revoke this Licence or
vary the terms, provisions or limitations thereof by a notice in writing served on
the Licensee. Any notice given under this clause may take effect either forthwith
or on such subsequent date as may be specified in the notice.

RLn HO 97/1 200 7/74 P

Appendix 1 (*cont.*)

9. This Licence is not transferable.

10. *Return of Licence*. This Licence shall be returned to the Secretary of State when it has expired or been revoked.

11. *Previous Licences revoked*. Any Licence however described which the Secretary of State has previously granted to the Licensee in respect of the Station(s) is hereby revoked.

12. *Interpretation*. (1) In this Licence:— (a) the expression "Secretary of State" shall mean the Secretary of State for the Home Department; (b) the expression "the Telecommunication Convention" means the International Tele-communication Convention signed at Montreux on the 12th day of November 1965 and the Radio Regulations and Additional Radio Regulations in force thereunder and includes any Convention and Regulations which may from time to time be in force in substitution for or in amendment of the said Convention or the said Regulations; (c) except where the context otherwise requires, other words and expressions have the same meaning as they have in the Wireless Tele-graphy Act 1949, or in the Regulations made under Part 1 thereof.

for the Secretary of State for the Home Department

Appendix 1 (cont.)

THE SCHEDULE

NAME
ADDRESS
REF. NO

Date of Issue:

Location(s) of Station(s)	Frequency and Maximum Frequency Tolerance (para. D applies)	Bandwidth of Emission (para. D applies)	Class of Emission (para. C applies)	Pulse Characteristics (pulse repetition frequency, pulse duration, pulse rise time — para. E applies)	Effective Radiated Power in the Direction of Maximum Radiation (Watts) (paras. A & B apply)	Aerial Characteristics (paras A & B apply)

Appendix 1 (*cont.*)

For the purposes of the Schedule:

A. Effective Radiated Power (ERP) is the power supplied to the antenna multiplied by the relative gain of the antenna in a given direction.

    The ERP shall be expressed in terms of the peak envelope power (P_p), the mean power (P_m), or the carrier power (P_c), whichever is appropriate, taking into account the class of emission used. These powers shall have the meanings assigned to them in the Telecommunication Convention.

B. ERP and the aerial characteristics will be assessed either by measurements or by calculation from the characteristics of the types of apparatus used, at the discretion of the Minister. For this purpose the appropriate recommendation of the International Radio Consultative Committee (CCIR), should be used as a guide.

C. The symbols used to designate the classes of emission have the meanings assigned to them in the Telecommunication Convention.

D. "Bandwidth" and "frequency tolerance" have the meanings assigned to them in the Telecommunication Convention.

E. Pulse Repetition Frequency (PRF) is the reciprocal of the minimum interval separating corresponding points (eg 50% of the peak amplitude) of successive pulses.

    Pulse duration (length) of any specific pulse is the interval between the first and the last instance at which the instantaneous amplitude reaches 50% of the peak amplitude.

    Pulse rise time is the time taken during any specific pulse for the amplitude to increase from 10% to 90% of the peak amplitude.

APPENDIX 2.

RD/DZ/ BR15

APPLICATION FOR A LICENCE FOR A LOW POWER TELEMETRY OR TELECONTROL SYSTEM

This form should be completed by the applicant and sent to:

The Cashier
Home Office
Tolworth Tower
Ewell Road
SURBITON, Surrey, KT6 7DS

NOTES

1.   The use of radio transmitting systems is licensable under the Wireless Telegraphy Act of 1949. The Licence fee in respect of telemetry and telecontrol equipment is £7.50 p.a.

The appropriate fee should be forwarded with this application form. Cheques, Giro transfer forms, postal orders and money orders should be made payable to "The Accounting Officer, Home Office, and crossed "A/C Payee". Bank notes should be registered.

2.   Equipment used must be of a type approved by the Home Office and must operate on a frequency assigned by the Home Office. Further information is contained in the leaflet BR14, obtainable from the Home Office Radio Regulatory Division.

1. FULL NAME of individual or Company in whose name the licence is required.	. . . .
2. Address for correspondence.	. . . .
3. Address at which system will be used, if different from 2 above.	. . . .
4. Address of Registered Office, if a Limited Company (if different from 2 above).	. . . .
5. Purpose of use.	. . . .

Appendix 2 (*cont.*)

6.   Maximum range over which the   . . . .
apparatus will operate.

7.   (a) Manufacturer of equipment   . . . .
      (b) Type number of equipment   . . . .

8.   Frequency band in which system   . . . .
will operate.

9.   Number of Transmitting Stations to
be used.   . . . .

10.   Number of Receiving Stations to be
used.   . . . .

I enclose cheque/money order/postal order/bank notes/Giro transfer form (our account NO 5120314) in accordance with Note 1.

Signed. . . . .

MPT 383   Date. . . . .

Symp. zool. Soc. Lond. (1982) No. 49, 31–45

Design Considerations and Performance Checks on a Telemetry Tag System

M. H. BEACH and T. J. STORETON-WEST

Ministry of Agriculture, Fisheries and Food, Fisheries Laboratory, Lowestoft, Suffolk, England

SYNOPSIS

The design and performance of a radio telemetry system for tracking small mammals and fish is described. Details are given of the design considerations for the transmitter tags and receiver units, the advantages and disadvantages of the various aerial arrays, the electronic test equipment, and the measurement of the system parameters necessary to ensure compliance with the Home Office Performance Specification (MPT 1312) issued in 1978.

INTRODUCTION

The purpose of this paper is to describe performance checks on a radio tracking system designed and developed within the laboratories of the Ministry of Agriculture, Fisheries and Food (MAFF), formerly by the late G. E. Ashwell of the Worplesdon Laboratory, and latterly the Fisheries Laboratory, Lowestoft. The earlier work was published by Taylor & Lloyd (1978) and gave details of a variety of designs intended for use on rats, foxes and badgers.

At about the time the work was transferred to Lowestoft the Performance Specification for medical and biological devices issued by the Home Office (MPT 1312, 1978) allocated a frequency band of 104.6–105 MHz (bandwidth of 0.4 MHz) for devices used in the United Kingdom. This required that all new equipment be changed from the previous 102.3 MHz to comply with this new band and the opportunity was taken to investigate the whole system in relation to its technical performance, construction and testing techniques.

TRANSMITTER TAG

The original tag presented problems when attempts were made to produce it to a consistent standard and the alterations in the circuitry, due to the change in frequency, required type approval by the Home Office. Variations in tag encapsulation and the proximity of the animal gave capacitance problems at the tuning stage which could cause the output from the tag to cease completely. Improvements were achieved by changes in component values and a new component layout on a printed circuit board to aid construction and reduce circuit failures. The circuit was miniaturized and the case tailored to fit both the electronics and the neck of the animal to be tracked. Consideration was given to weight, restriction of animal movement, ruggedness, waterproof qualities, the ease of labelling for subsequent identification and recovery, and the possibility of opening the case to fit new batteries.

The basic circuit used in current production of the transmitter tag is shown in Fig. 1; the crystal-controlled oscillator being modulated by the multi-vibrator which switches the oscillator supply via TR3 to produce a pulse of radio frequency approximately 100 ms long every second (50 ms for the fish tag). The radiating element is either a small iron dust core (25 mm long × 6 mm diameter, animal tag) or ferrite core (16 mm long × 6 mm diameter, fish tag) wound with either five or three turns, respectively, of self-fluxing insulated copper wire (22 SWG) to form the inductor in the tuned collector

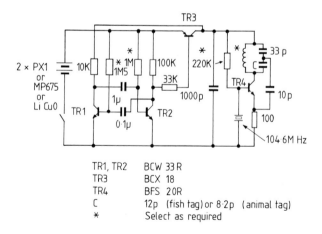

FIG. 1. The circuit used for the transmitter tag.

load of the oscillator and the loaded loop aerial. The circuit construction uses microminiature transistors, resistors and chip capacitors on a single-sided fibreglass printed circuit board.

Power is provided by a battery pack of either two mercury or two lithium cells in series, and the circuit arranged so that oscillation ceases at a battery pack end-life of about 1.5 V (0.75 V per battery); this is ensured by the rapid fall-off in supply voltage that occurs at the end of the battery discharge characteristic. The size of the animal determines the size of the battery pack that can be carried, and since the pack is the largest component of the system it dominates the design and layout of the tag. To achieve the ampere—hour rating specified for a battery pack it is important that rigorous procedures are followed. First, batteries should be stored at a low temperature (about 1°C) and then visually checked for electrolyte leakage. If there is any observable deposit the battery should be rejected. Secondly, an absolute minimum of heat should be applied to a battery, particularly a pressurized lithium cell, since it is easily damaged and potentially explosive. To minimize risks associated with heat, MAFF battery packs are assembled using gold-plated Kovar ribbon which is welded to individual cells using the parallel-gap technique. Thirdly, before assembly into a radio tag an instantaneous high current test is performed and the battery pack accepted only if a minimum current level is achieved (e.g. 400 mA minimum instantaneous short circuit current for SP 675 cells with model 8 Avometer on the 1A DC range).

The operation of the oscillator section of the transmitter tag was studied theoretically using an equivalent circuit. This enabled the crystal parameters to be specified and crystals are now specially manufactured so that the performance of the oscillator is both predictable and reliable. The radiating section was also studied to see if other aerial arrangements would improve the efficiency. A short monopole was the most efficient of the aerials tested (up to 15% radiation efficiency) but can only be used where the size of animal can accommodate its length. A loaded loop aerial is currently used because, although having a low efficiency (c. 1%), it does have the advantage of being within the tag case and the predictable separation between aerial and animal produces less variation in the capacitive coupling. An integral aerial and the appropriate case allows a standard circuit to be used that is suitable for a variety of mammals or fish, whereas a separate aerial requires a waterproof connection to the circuit and possible support by attachment to the back of the animal.

The tag case was studied to enable it to be produced easily and to a consistent standard. Direct dipping of the tag components in an epoxy resin produced difficulties since final tuning was impossible and circuit faults could not be rectified. Circuit encapsulation in Elvax wax and subsequent moulding in epoxy resin to the required shape gave problems of capacitance and component failure due to the high temperature required. The method currently used consists of casting the case in two halves using a mould and Stycast 1263 resin. Recesses are provided for the circuit board and battery pack and the two halves finally glued together by means of Stycast adhesive when all adjustments have been made. This method allows the case to be cut open if a new battery pack or circuit repair is necessary. A lightweight, polystyrene, cylindrical case has been developed to allow the transmitter (length, 35 mm; diameter, 16 mm; weight in air, 3 g) to be used on an adult salmon (*Salmo salar* L.). This tag has also been used on rabbits (*Oryctolagus cuniculus* L.). The salmon tag must be as small as possible with well-rounded surfaces to allow insertion through the mouth into the stomach.

To date the Lowestoft laboratory has produced some 145 transmitter tags: 61 badger (*Meles meles* L.), 23 fox (*Vulpes vulpes* L.), six coypu (*Myocastor coypus* (Molina)), six squirrel (*Sciurus carolinensis* Gmelin), three rabbit and 46 salmon.

RECEIVER

The change from the 102 to the 104 MHz frequency band, brought about by the introduction of MPT 1312, necessitated a change in receiver design. Again the opportunity was taken not only to improve the circuit but also to attempt to make the entire unit satisfy the operational requirements in the field as closely as possible. Circuit improvements took account of frequency coverage and resolution, sensitivity, and a filtered audio response to maximize the ease of detection over a limited audio bandwidth. The design of the receiver case had to take account of weight and balance when carried during long tracking periods, the ease of operation of controls, waterproof qualities, and battery replacement.

The original Ashwell receiver was of the double conversion, superheterodyne type. It comprised two radio frequency (RF) stages followed by a mixer stage which converted from the 102.3 MHz band to the first intermediate frequency (IF) of 10.7 MHz using a crystal-controlled oscillator. The second conversion was achieved using a variable crystal oscillator (VXO) and provided a second IF of 455 kHz;

the signals at this frequency were amplified, detected and fed to an audio amplifier. An aerial trimmer was necessary to match the aerial to the receiver input. Figure 2A shows the interior and front panel of the Ashwell receiver.

The design and construction of the new (Mariner Radar) receivers was achieved in collaboration with the Lowestoft laboratory. Figure 2B shows details of the Mariner 54 receiver and Fig. 2C the later Mariner 57 receiver. The requirement for the model 54 receiver was to obtain a sensitivity similar to, or better than, the original Ashwell unit. Hence the same basic circuit was employed with careful attention given to component layout, RF stages and method of detection. The receiver input was designed to match the aerial impedance of 50 Ω which made the aerial trimming capacitor unnecessary.

Radical circuit changes were employed in the Mariner 57 compared with the Ashwell and model 54 receivers. The RF and first mixer stages were improved with a VXO being employed at the first conversion stage, rather than at the second as in previous receivers, to allow a narrower bandwidth at the first IF frequency of 10.7 MHz; this resulted in noise reduction and improved sensitivity. The second conversion stage was changed to a single crystal-controlled oscillator, and ring modulation was used with a filtered audio response to improve noise rejection and signal detection.

The Mariner 57 receiver is tuneable over a 50 kHz bandwidth with a calibrated dial giving a resolution of at least 1 kHz which enables a particular tag frequency to be readily located. The Ashwell and model 54 receivers were tuneable using switched crystal oscillators to give six 5 kHz bands with coarse tuning throughout each band. Electrical tests performed on all three receivers, using an HP 8640B calibrated signal generator within a shielded system, showed there to be little difference between the Ashwell receiver and the Mariner 54 receiver. The most sensitive receiver was the Mariner 57 and a "working sensitivity" of 0.015 μV (-143 dBm) was obtained, which, assuming a transmitter tag output power of 100 nW, an aerial gain of 0 dB over a half-wave dipole and no path losses (only spreading losses), gives a calculated absolute maximum range of 43 km. However, a transmitter on a mammal might only be detectable at a range of 2 km, and on a fish at a depth of 1 m in fresh water in a wooded valley, at a range of less than 1 km (Solomon, this volume, p. 95).

Figure 2 (D–P), shows a variety of transmitter tags. D, E and F were early animal tags, D being a fox tag and E and F being rat tags. The remaining transmitter tags comprise various animal tags at G, H, I and J, a fox tag K and a coypu tag L. M is the current fox tag and N

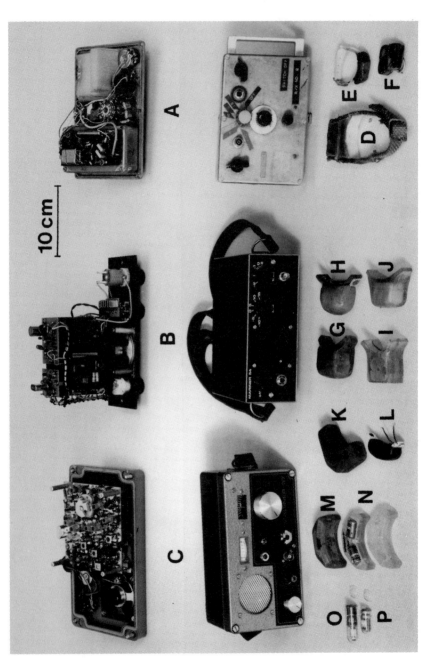

FIG. 2. Types of radio receivers and transmitter tags used. (A) The interior and exterior of the Ashwell receiver; (B) and (C) the Mariner 54 and

the component halves of the same tag. O and P are two salmon tags with O having the larger battery capacity (operating life of 5—6 months) and P the smaller (2—2½ months).

AERIAL

Three types of aerial are employed with the tracking system (Fig. 3) and the fourth aerial, a standard dipole (Fig. 3C), is used for output power measurements.

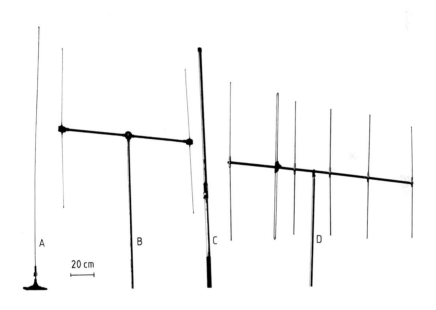

FIG. 3. Types of aerial used. (A) A whip aerial with magnetic mount; (B) an H-Adcock array; (C) a standard dipole for calibration measurements; and (D) a 6-element Yagi-array.

Whip Aerial (Simple Dipole)

This aerial (Fig. 3A) is omnidirectional, easily portable and can be attached to a car with a magnetic mount. It will give no directional indication, other than by signal increase when moving towards the transmitter, but can be useful for initial detection if a vehicle has access to the vicinity of the transmitter. A gain of 3 dB is realizable with a $5\lambda/8$ aerial ($\simeq 1.8$ m) and 0 dB with a $\lambda/4$ ($\simeq 0.72$ m) aerial (referred to a half-wave dipole).

H-Adcock Array

This aerial (Fig. 3B) is currently used in MAFF projects. It is compact, reasonably robust, collapsible and easily erected being an array of two driven dipoles connected to be anti-phase. Thus, while offering little gain, a high directionality is achieved using its null facility and an angular resolution of about ±5° is possible if a tripod and compass rose are employed. A 180° directional ambiguity exists but can usually be resolved by either triangulation or the constraints of the site on the transmitter location.

Yagi-array

This is a multi-element aerial using directing and reflecting elements to increase the gain and directivity. The array shown in Fig. 3D is a 6-element version with folded dipole. Direction indication is by signal strength and inaccuracies occur due to its wide beamwidth, about ±22° for the array shown. This aerial is cumbersome in wooded areas or undergrowth but where it can be used has the advantage of about 8.7 dB gain (6-element type) over a half-wave dipole.

EQUIPMENT PERFORMANCE SPECIFICATION FOR THE UNITED KINGDOM

Performance Specification MPT 1312 (Home Office, 1978) covers the minimum requirements for low-power medical and biological telemetry devices for use in the band 104.6 to 105 MHz. For MAFF purposes it was desirable to distribute the frequencies throughout an allocated band, rather than to adopt precisely defined frequencies with adjacent guard bands, and the three lowermost 50 kHz frequency bands in the 104.6 to 104.8 MHz range were allotted. The Specification defines the allowable output power, transmitter frequency error, out-of-band power and spurious emissions, and receiver spurious emissions. The test conditions are carefully defined and include a specified range of power supply voltage, temperature and relative humidity throughout which the transmitter or receiver must comply with the emission requirements. Construction, labelling and maladjustment of too readily available controls is also considered.

COMPLIANCE WITH THE SPECIFICATION

Transmitter Tag

Every tag produced by the Lowestoft laboratory is tested to ensure that it complies with each of the categories defined in Specification MPT 1312 (Home Office, 1978), and is issued with a specification sheet and identification code. The tests carried out are described below.

Radiated power output

This is determined by using one of two methods, the first involving direct measurement and the second substitution of the unknown tag for a calibrated tag. Direct measurement is used to establish a standard tag and to check its output, and the substitution method, which is quicker, for most of the tags produced subsequently.

Both methods are undertaken at a range site which must be reasonably flat and free from multipath conditions and RF noise sources. A standard dipole and spectrum analyser are required for the direct method (Fig. 4A) and the addition of a signal generator for the substitution method (Figs 4B and 4C). Further details of each of these two methods and worked examples are given in Appendix 1; a selected tag was found to have a radiated power output of 38.02 nW by direct measurement and of 38.37 nW by the substitution method.

Spectral output

A spectrum analyser (Tektronix 7L 13) is used to examine the fundamental frequency and harmonic content (Fig. 5). Particular attention is given to the harmonic content since spurious emissions in excess of 250 nW in the frequency band between 30 and 1000 MHz are not permitted. All spectral irregularities are checked, since it is only by ensuring that most of the RF power is radiated at the fundamental frequency that maximum output power and battery life is achieved.

Frequency

The transmitter frequency is measured in either of two ways. In the first the ouput mode is changed to continuous wave (CW) by a circuit alteration and the frequency measured using an inductive loop connected to a frequency counter (Marconi 2438). This method is inconvenient since access is required to the tag circuit which is encapsulated at this stage and the use of an inductive loop can alter the output frequency; furthermore, a measurement in the CW mode might not be applicable to the designed pulse mode. In the second

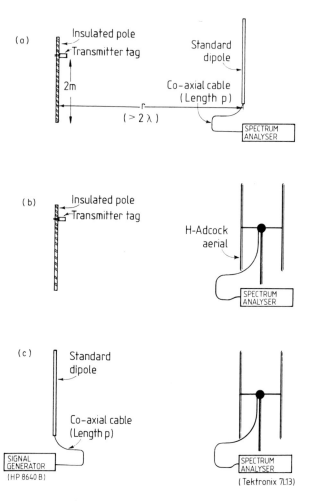

FIG. 4. Measurement of RF power output. (A) The arrangement for determination of RF power by direct measurement of field strength; (B) and (C) the arrangement for the measurement of field strength by the substitution method, with (B) using the transmitter tag and (C) a "driven" standard dipole substituted for the tag.

method a signal generator (HP 8640B) is used and the receiver (Mariner 57) employed as a null detector. The output from the signal generator is mixed with the output from the tag and detected by the receiver. When both frequencies are made identical by tuning the signal generator a zero beat results, and the frequency can be read from the signal generator or separate frequency counter.

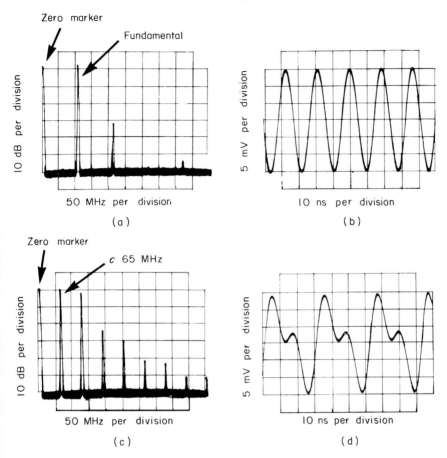

FIG. 5. Acceptable and unacceptable RF outputs. (a) and (b) An acceptable output spectrum and waveform, the fundamental frequency is at 104.7 MHz (and at about 130 nW radiated power) with a second harmonic 33 dB down, a negligible third harmonic and the fourth 57 dB down; and (c) and (d) an unacceptable output spectrum and waveform, the crystal is oscillating at a lower overtone (c. 65 MHz) than the fifth, and harmonic radiations up to the eighth are evident. With the receiver tuned within the 104.6–105 MHz band no transmissions would be detected.

Pulse duration and repetition rate

The pulse duration and repetition rate are measured directly using an inductive loop and high frequency oscillosope.

Receiver

The receiver is checked for spurious emissions and for frequency stability within the specified temperature range. MPT 1312 (Home Office, 1978) requires that spurious emissions shall not exceed 2 nW.

Radiation from the local oscillator and other frequencies generated within the receiver are measured using a spectrum analyser connected directly to the aerial input socket. Spurious emissions are measured using an inductive loop both around and in the vicinity of the receiver unit.

CONCLUSION

The tests described above have been used not only to ensure compliance with the Home Office (1978) Specification but also to produce a high-performance tracking system. Many aspects have been studied in an attempt to make the system rugged, reliable, consistent in its performance and easy to maintain. In many instances, particularly when tracking fish, the constraints of the small size of some of the animals used and the nature of the environment itself are more severe than the Home Office requirements, and so the search for smaller tags of lighter weight with greater RF output and longer life continues. Electronic circuit size, with the exception of the crystal, is no longer a problem but difficulties remain in respect of transmitter aerial size, in connection with radiation efficiency, and battery size/ weight for a given output power and operating life. However, there is little doubt that telemetry tags will get smaller whilst offering even greater monitoring facilities.

ACKNOWLEDGEMENTS

The work described here would not have been possible without the considerable efforts of P. C. W. Beatty and W. G. E. Hone formerly of MAFF, and the extensive contribution by J. G. French of Mariner Radar. Gratitude is also due to the biologists on the staff of the MAFF laboratories at Worplesdon, Tolworth and Lowestoft for their advice and encouragement.

REFERENCES

Hewlett Packard (1976). Spectrum analysis: field strength measurement. *Spectrum Analyzer Series, Application Note* 150—10 (September issue): 1—2.
Home Office (1978). *MPT 1312. Performance specification: medical and biological telemetry devices.* London: Her Majesty's Stationery Office.
Taylor, K. D. & Lloyd, H. G. (1978). The design, construction and use of a radio-tracking system for some British mammals. *Mammal Rev.* 8: 4: 117—141.

APPENDIX: THE MEASUREMENT OF THE RADIATED POWER OUTPUT FROM A TAG

Measurement of Tag Output Power by Direct Measurement

The procedure is as follows:

(i) The transmitter tag is attached to an insulated pole at least 2 m above the ground and at a distance r (greater than 2λ) from a standard receiving aerial (Fig. 4A).

(ii) A standard dipole (Jaybeam 7034) is connected by a known length p of screened cable to a spectrum analyser (Tektronix 7L 13).

(iii) A reading is obtained from the spectrum analyser in dBm (1 mW into 50 Ω), corrected for cable loss and aerial reflection loss, and referred to dB re $1\,\mu V\ Q$.

(iv) An aerial factor R (Hewlett Packard, 1976) is derived for the standard dipole whose characteristics are supplied by Jaybeam, i.e. omnidirectionality and voltage standing wave ratio (VSWR).

(v) From (iii) and (iv) the transmitter output power is calculated.

Example based on the MAFF "standard" tag against which all production tags are compared:
The field strength E_{dB}, is the sum of the adjusted spectrum analyser reading and the aerial factor:

$$E_{dB} = Q + R \qquad (1)$$

where
E_{dB} = rms of field strength (dB re $1\,\mu V/m$)
Q = corrected analyser reading (dB re $1\,\mu V$)
R = aerial factor (dB/m)
Q is corrected for cable loss and aerial reflection loss (due to VSWR).

Cable Loss
For the cable used (URM 43) the attenuation loss at 100 MHz was 0.13 dB/m and the length p was 30.03 m. Hence,

$$\text{cable loss} = 3.90 \text{ dB}$$

Aerial Loss
For the standard dipole the VSWR was 1.56 (from Jaybeam curve, aerial designed for 100 MHz but used at 102.3 MHz). Now

$$A = S + T \qquad (2)$$

and

$$\Phi = \frac{S}{T} \qquad (3)$$

where

A = power incident on aerial
S = power reflected due to aerial mismatch
T = power transmitted to aerial feeder cable

and since

$$\Phi = \left(\frac{\text{VSWR} - 1}{\text{VSWR} + 1}\right)^2$$

from equations (2) and (3)

$$T = 0.954 \times A$$

$$\text{aerial loss} = 0.20 \text{ dB}$$

The spectrum analyser reading was -75 dBm. Hence correcting for losses

$$= -75 + 3.90 + 0.20 = -70.90 \text{ dBm}$$

and referring to $1 \mu V$ by adding 107 dB (50 Ω impedance),

$$Q = 36.1 \text{ dB re } 1 \mu V$$

Aerial Factor, R (Hewlett Packard, 1976)

$$R = 20 \log f - G - 29.78 \tag{4}$$

where

R = aerial factor in dB/metre
f = frequency in MHz
G = aerial gain in dB relative to isotropic radiator (i.e. 2.14 dB for the Jaybeam 7034).

Hence

$$R = 20 \log 102.3 - 2.14 - 29.78$$

Thus

$$R = 8.28 \text{ dB/m}$$

and from equation (1)

$$E_{dB} = 44.38 \text{ dB re } 1 \mu V/m \tag{5}$$

Thus field strength, E, from equation (5)

$$= \text{antilog}\left(\frac{44.38}{20}\right)$$

$$E = 165.58 \mu V/m$$

But

$$P_R = \frac{E^2 r^2}{30} \qquad (6)$$

where

P_R = radiated power in watts
E = rms value of field strength in volts/metre
r = distance in metres (Fig. 4A)

and from equation (6), with $r = 6.45$ m

$$P_R = (165.58)^2 \times (6.45)^2 / 30$$

Hence

$$\text{radiated power} = 38.02 \text{ nW}$$

Measurement of Tag Output Power by Substitution

The procedure is as follows:

(i) The transmitter is orientated on the insulated pole to give maximum signal on the spectrum analyser (Fig. 4B).

(ii) The transmitter tag and insulated pole are replaced by the standard dipole (Jaybeam 7034) which is driven from a high quality signal generator (HP 8640B) via a known length p of screened cable (Fig. 4C).

(iii) The signal generator output is adjusted until the reading on the spectrum analyser is exactly the same as that produced by the transmitter tag.

(iv) The tag output power is calculated from the signal generator output, the attenuation losses in the screened cable, reflection losses at the aerial and the aerial gain:

Signal generator output	$= -42.2$ dBm
Cable attenuation for 30.02 m of URM 43	$= -3.9$ dB
Aerial reflection loss (as for previous example)	$= -0.2$ dB
Aerial gain (over isotropic radiator)	$= 2.14$ dB

Hence equivalent tag transmitter output

$$= -42.2 - 3.9 - 0.2 + 2.14 \text{ dBm} = -44.16 \text{ dBm}$$

or

$$\text{radiated power } (P_R) = 38.37 \text{ nW}$$

Symp. zool. Soc. Lond. (1982) No. 49, 47–59

Microelectronic Technology and its Application to Telemetry

P. L. MORAN

Microelectronics Research Centre, University College, Cork, Eire

SYNOPSIS

In this chapter microelectronic technologies are reviewed and analysed with regard to their possible application in telemetry systems. In particular the trade-off between development cost, production quantities and technological performance is discussed and, whilst no one technology is used universally, general guidelines are presented as to the most suitable technology for various situations.

INTRODUCTION

The science of telemetry is by definition the measurement of some quantity at a distance from an operator and the relaying of this information to the operator. A typical system is shown in Fig. 1. The input transducer could be measuring such quantities as temperature, strain or acceleration, or perhaps a combination of several such quantities. Often the system is merely responding to a received signal and this response gives an indication of position. In favourable circumstances, it is possible to use thin wires to relay the information back to the operator and also to supply the power to the system. More generally the information must be transmitted using either

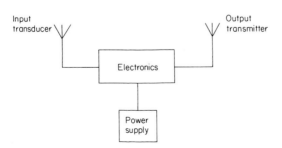

FIG. 1. Typical telemetry system diagram.

radio or acoustic techniques and so it is necessary to carry the power supply "on board" in the form of a battery. In some cases it is possible that solar cells may be employed or power may be sent in the form of a radio transmission but the situations in which such techniques may be used are very limited.

It can be seen, therefore, that even before the electronics system is taken into account the overall size and weight of the telemetry package could be quite large, the limit on the size reduction being governed by quantities that cannot be altered without compromising performance. The electronics part of the system plays a very important role, although at its simplest, the signals are merely amplified and changed so that they are suitable for transmission. More complex systems may convert the signals into digital form to improve the tolerance to noise and possibly carry out signal processing to reduce the bandwidth of the information and hence reduce the power requirements for relaying the signal to the operator. Recent developments in microelectronic technology make it possible to gather the required information at predetermined times, process and store it *in situ*, then relay it to the operator when transmission conditions are most favourable — again reducing the overall power requirements of the system.

Unfortunately for the telemetry system designer and manufacturer, his market position is not very strong for taking advantage of the advances in microelectronics technology. Manufacturers are reluctant to produce non-standard components when relatively few are required, because the production lines of the semiconductor and other devices are highly automated. The telemetry designer must, therefore, either use standard components and packages (some examples of which are shown in Fig. 2) or involve himself in microelectronic technology. For size comparison some of the "chips" contained within the packages are also shown in Fig. 2, where it is evident that a great reduction in size and weight can be obtained through employing the unpackaged devices. Most of the production problems occur in the packaging and testing of the semiconductor devices.

In this chapter a brief outline of the various microelectronic technologies is presented, together with a summary of their relative advantages and the problems likely to be encountered.

FIG. 2. A selection of representative electronic packages and the devices contained within them.

SILICON TECHNOLOGY

Processing

It is impossible to review silicon technology comprehensively in the space available here. Only the most relevant points will be considered. Further information can be obtained from standard texts (Meyer, Lynn & Hamilton, 1968; Fitchen, 1970).

The starting point for all silicon devices is the basic slice which is from 2 in (50 mm) diameter up to 4 in (100 mm) diameter. Its purity is such that the concentration of unwanted material is generally less than one part in 10^9 and the concentration of the so-called "dopants" will vary from about one part in 10^3 to one part in 10^7. Therefore one of the first requirements is extreme purity and all operating areas and processing chemicals must be of a similar standard. The slice, which is a single crystal, is as flat as possible and is polished to an optical finish.

The production of a finished slice can be divided into three different types of process: pn-junction formation, oxide growth and metallization. A pn-junction is the fundamental semiconductor

structure used in so-called bipolar devices, produced by first doping
selected regions with one material, and then repeating the process
with different parameters and dopant. The term doping refers to the
process of altering the electrical properties of the silicon by the controlled introduction of other elements. If, for instance, phosphorus
is added to the silicon, then considerably more electrons will be
present in the silicon than holes, and the material is called n-type.
Similarly, using a different element will produce a p-type region in
the silicon with an excess of holes. The presence of two types of con-
ducting species is an important property of the semiconductor
material. A typical cross-section of a transistor in an integrated
circuit is shown in Fig. 3 in which the p and n regions are marked.
Doping is carried out at temperatures of around $1000°C$ by exposing
the polished silicon surface to a gaseous dopant which enters the
surface and diffuses in a well controlled manner to give the required
vertical profile. Since the rate of diffusion is a rapidly increasing
function of temperature, when the device is cooled to room temperature the dopant is firmly held in the host crystal lattice, thus giving
longevity to the basic device.

The growth of silicon dioxide on the surface of the silicon is an

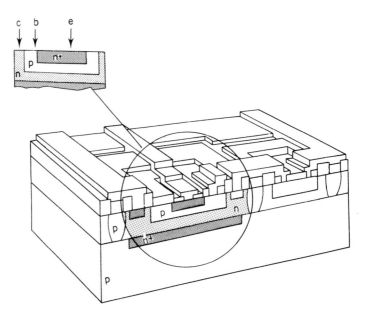

FIG. 3. Cross-section of a bipolar integrated circuit. c, collector; e, emitter; b, base; p, p-
type doping; n, n-type doping.

extremely important part of the process; it is used in two ways. First, it provides "windows" at the surface of the silicon through which the selective dopings are made. Then, after the various doping stages have been completed, a layer of oxide is grown on the surface to act as a passivation layer, and to provide windows, through which contact can be made to the devices underneath. The passivation layer is important as it covers and seals the surface of the slice hence preventing surface effects and contamination masking the performance of the basic pn structures.

Next, a metallization layer is produced by coating the entire slice with aluminium, by boiling the latter in a vacuum chamber, then defining the required pattern using photolithographic techniques. The metallization is used to make a contact area for an external connection and to join this contact area to the active device in the case of a transistor, and, in addition, to interconnect the individual devices in the case of an integrated circuit.

Although the basic slice is at this stage complete, a further layer of glass is often added to protect the aluminium interconnections. A photograph of a typical slice is shown in Fig. 4. The dimensions of a representative commercial device would be approximately 5 μm for the transistor geometry and aluminium interconnections, the oxide thickness would be to the order of 100 nm (hence the need for a filtered atmosphere) and the bonding pads about 100 μm. A single low power transistor "chip" would occupy between 0.25 and 0.5 mm^2 of the slice area, whilst a complex integrated circuit would be between 2 and 4 mm^2 or more and contain perhaps 10^4 transistors. All the devices on a slice will be of the same design and ought to be of the same performance. It can be seen therefore that single transistors are an inefficient use of the silicon slice. Each slice could contain from a few hundred complex integrated circuits up to perhaps 5×10^4 single transistors.

Testing and Yield

The slice will now be subjected to a simple direct current (DC) performance check and obvious rejects marked. It will then be broken into the individual chips, and each unmarked device will be assembled into a package, the interconnections being made by welding 25 μm diameter gold or aluminium wires to the bonding areas. At this stage the devices will be parametrically checked and graded. It is important to realize that it is not possible to carry out full checks on the unencapsulated device and that in some instances the package contributes to the overall performance.

FIG. 4. A slice of integrated circuits.

It is relatively easy to demonstrate that the yield, y, of an integrated circuit process will be approximately

$$y = \exp -kA_{\mathrm{D}} \tag{1}$$

where k is a process constant and A_{D} is the device area.

This is due to the fact that the faults on the slice tend to occur at random and the inclusion of just one fault on a circuit will render it faulty. Thus if the device area is reduced, the number of devices per slice will increase, so the manufacturer will produce more devices, with a higher yield for a given amount of labour and capital equipment (this argument applies until the size is reduced so much that it becomes impossible to make the devices). The economics of the process are such that whilst simple devices require yields approaching 100%, the more complex and custom-designed devices often only have a yield of 10 to 15%. These low yield circuits may contain in the region of 10^4 individual transistors and perhaps 24 to 40 connections to the package. It is therefore an extremely difficult task to carry out a full parametric test on the device.

It is of course impossible to "correct" a faulty design by modifying the completed slice so a great deal of effort must be put into the design of the circuit to ensure that the electrical function is correct; considerably more effort than would be required if the circuit were being made by other techniques. For this reason the cost of a custom-designed integrated circuit is very high if a relatively low number are required. There are, however, a number of processes based on standard arrays of transistors where only the aluminium interconnections are unique. Such techniques reduce the development costs considerably but do not yield an optimum device.

Application

The silicon integration process gives rise to excellent active devices but the quality and range of passive components (capacitors, resistors, inductors) are strictly limited. Hence the most spectacular success of silicon integration has been in digital circuitry where all transistor designs are feasible. Because of the nature of the process it is most applicable when large quantities of any one design are required, thus the major growth areas have been in elementary logic gates, counters, memory circuits etc., and simple amplifiers. The recent development of the so called microprocessor (although it dates back to 1970) is an attempt to provide custom devices through the ability to program the device and also to use the production lines efficiently by manufacturing just the one component. These devices are, in computing terms, of low performance; however, they are sufficiently inexpensive to be used in consumer equipment and complex enough to provide a wide range of useful functions. Their low cost also makes the inclusion of several devices, each dedicated to a specific task, quite acceptable. They are unfortunately not readily available unencapsulated, mainly because of testing problems.

HYBRID MICROCIRCUIT TECHNOLOGY

In comparison to the rather spectacular developments in silicon technology there has been a steady progress in other areas to overcome some of the limitations of the passive components of silicon technology. In most telemetry applications good quality passive components are required. Although the thin film process superficially appears superior, the thick film has been commercially more successful and is now dominant.

Processing

In the thick film process, a paste or ink is deposited onto an inert, rigid substrate, usually alumina, by a screen printing process. This paste may be either a conductor (gold, silver, or one of these metals alloyed with palladium or platinum), a resistor (usually ruthenium dioxide and glass), or an insulator (glass plus a "filler"). The paste is then sintered to the substrate by passing it through a tunnel furnace with a peak temperature in the region of 900°C for a few minutes. Further prints may be added to give several layers of conductor and insulator or a number of different resistor formulations. The process seems deceptively simple but requires a considerable degree of process control and skill to achieve consistent results.

Generally, components other than resistors such as miniature capacitors and inductors, will be added at a later stage, by soldering or glueing. An important part of the technology is to attach unencapsulated integrated circuits to the substrate and to wire them in to the hybrid circuit. The deposited resistors can be adjusted to within 1% of the desired value by laser machining, either prior to component attachment or after assembly with the circuit operating. Resistance values may be in the range of $1\,\Omega$ to $10\,\text{M}\Omega$, and have stabilities in the middle of this range of better than 0.005% per °C and 0.05% drift. Typical film circuits constructed and assembled by this technique are shown in Fig. 5.

FIG. 5. Thick film circuits.

Thin film circuits are made by evaporating either a resistor (nickel chromium mixture) or a conductor (gold) in a vacuum over the entire area of the substrate (usually polished glass), patterns being defined by photo-engraving. Other components are added at a later stage. Resistor material is determined by the process, different values being attained by altering the geometry of the resistor, and these values may be adjusted by laser trimming to lie in the approximate range $10 \, \Omega$ to $100 \, k\Omega$. Although the stability is very high, usually better than 0.0015% per °C, the process is extremely critical and is usually left to specialist suppliers. It is possible to purchase sheets of glass coated with a lower layer of resistor and an upper layer of conductor; the user then defines the patterns required.

Thick *vs* Thin Film

Although the thick and thin film processes result in similar end products, as described, the manufacturing methods are totally different. The thickness of a thin film circuit is about $1 \, \mu m$ for a conductor and $10{-}100 \, nm$ for the resistors. The resolution of the tracks and resistors is usually in the region of $20{-}30 \, \mu m$. In the thick film process, the material thickness is in the region of $10{-}25 \, \mu m$ per layer with a consequent maximum resolution of $125{-}250 \, \mu m$.

Both these techniques can have extremely high reliability, provided proper packaging procedures are adopted. The difficulty is that packaging often costs more than the circuit, although some low cost conformal coatings produced by dipping the circuit in a plastic fluid are available for medium reliability circuits. The thick film technique is essentially a flow process, whilst thin film circuits require processing in batches. Hence the thick film process has a greater potential for automation with the resultant lower costs. However, thin film resistors have a higher performance, and one compromise often used is to incorporate thin film resistor circuits on a thick film hybrid substrate.

MICROWIRING

Effort has long been directed towards using copper clad laminate board (i.e. conventional printed circuit board or similar) and chip components as a means of reducing the size of circuitry without becoming involved in the complexities of thick or thin film techniques. As none of the components is an integral part of the substrate it is to be expected that the reliability of the assembly will be

less, since the number of connections is greater. Also, as the substrate
is to some extent flexible, there is a higher possibility of mechanical
fatigue of joints and wires. The process is rather less amenable to
automation than the thick or thin film techniques and requires that
standard, preferred value resistors are used.

However, the microwiring technique has been used successfully by
the Fisheries Laboratory, Ministry of Agriculture, Fisheries and
Food, at Lowestoft, to produce miniature acoustic and radio tags for
use on fish and terrestrial animals. The technique is particularly
suitable to small quantities of circuits (hundreds rather than
thousands) where, although reliability is important, it does not out-
weigh all other factors. It allows a high packing density of
components by using micro-miniature (μmin) packaged transistors
and resistors, but unpackaged capacitors (chips). All of these
components can be soldered to a flexible printed circuit board (PCB)
(Mitson & Young, 1975: plate 2) with the advantage that it can be
rolled into a cylindrical shape or folded around a rectangular bulky
component. If the board were to be kept in a flat form it would have
the dimensions of 21 × 11 mm, too large for most fish tagging
requirements (left-hand side of Fig. 6). The flexible board contains
32 of the total 38 circuit elements for an acoustic transponder and

FIG. 6. A microwired telemetry circuit. (Reproduced with the permission of the Ministry
of Agriculture, Fisheries and Food, Crown copyright.)

has an area of about $7\,mm^2$ for each component. It represents about the highest packing density to be obtained by use of solder and PCB assembly.

By using a wire bonding technique, which calls for a much greater outlay on capital equipment and level of operating skill, the same transponder has been reduced to a disc of 7 mm diameter; an area reduction of six times and a mean area for each component of 1.2 mm^2 (bottom right of Fig. 6). No significant weight reduction of the transponder is evident. The other two boards in Fig. 6 belong to a telemetry tag used for monitoring the direction in which a fish is heading. These boards process the signals from a miniature magnetic compass and are wire-bonded. There is little advantage in reducing further the size of the electronic circuit as the battery and transducer, or aerial in the case of radio tags, are still the largest components by a considerable margin. Figure 7 illustrates another example of micro-wiring.

SUMMARY

Obviously the optimum electronics packaging solution for any telemetry system would be to employ a custom silicon integrated circuit on a specially designed lead frame. However, such a solution is

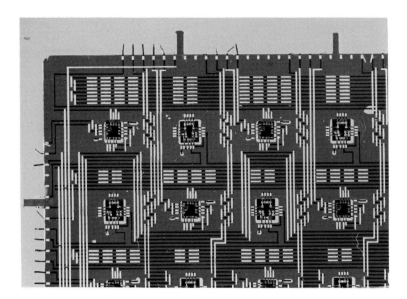

FIG. 7. An alternative microwiring system.

normally far too expensive for all but military and aerospace applications. If sufficient circuits are required to allow the use of an
uncommitted gate array, then this may well be a useful compromise,
but in general it is necessary to resort to a hybrid approach or to
microwiring. For "reasonable" quantities or when reliability is
important (particularly under adverse conditions) then either a thick
or thin film assembly is probably optimum. For small quantities, one
of the microwiring technologies could well prove useful.

However, the problem of employing unencapsulated complex
integrated circuits, or even simple graded devices, has not yet been
solved. The difficulty of testing the unpackaged device has meant
that either the integrated circuit manufacturer will not sell the
devices, or that if he does there is no guarantee of performance, particularly as many devices are dependent on the correct wire bonding
procedure being followed. One solution that seems to be emerging is
the use of leadless chip carriers (LCC), a picture of which is shown in
Fig. 8. These devices are about twice the size of the integrated circuit
contained in them. They are packaged and tested by the integrated
circuit manufacturer. Those currently available are of solid ceramic
construction and are attached to the circuit by reflow soldering.
Since they are a solid non-flexible construction it is preferable to
assemble them on a substrate of a similar material, which dictates
a thick film or ceramic substrate. However, work is in progress to
provide flexible miniature leaded chip carriers which would be
assembled on a conventional copper clad laminate. The circuits
in these carriers can be tested after encapsulation and thus are a
near panacea for the hybrid circuit manufacturer (and also for the
telemetry circuit designer). Interestingly, though, the motivation
for their development was to provide a viable method of increasing

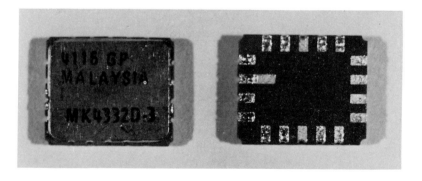

FIG. 8. Leadless chip carriers.

the number of leads on the integrated circuit package since the more familiar dual-in-line package can only really be used with up to about 40 leads, although larger packages have been produced and the more complex devices require more leads than this.

ACKNOWLEDGEMENTS

I should like to thank the Fisheries Laboratory of the Ministry of Agriculture, Fisheries and Food, Lowestoft, for the use of Fig. 6, and K. Bingham of I.C.L. for the use of Figs 4 and 7.

REFERENCES

Fitchen, F. (1970). *Electronic integrated circuits and systems*. New York: Van Reinhold.

Meyer, C. S., Lynn, C. & Hamilton, D. (1968). *Analysis and design of integrated circuits*. New York: McGraw Hill.

Mitson, R. B. & Young, A. H. (1975). A survey of the engineering problems of developing small acoustic fish tags. *Proc. Instrument. Oceanogr., Br. IRE* No. 32: 163–173.

FURTHER READING

(a) *Thick film technology*

Harper, C. A. (1974). *Handbook of thick film hybrid microelectronics*. New York: McGraw Hill.

Holmes, P. & Loasby, R. (1976). *Handbook of thick film technology*. Ayr: Electrochemical Publications Ltd.

(b) *Thin film technology*

Vossen, J. L. & Kern, W. (1978). *Thin film processes*. New York and London: Academic Press.

Symp. zool. Soc. Lond. (1982) No. 49, 61—73

Signal-to-Noise Ratio Enhancement in Receivers for Use in Radio Location

M. G. SLADE

E and EE Department, Royal Military College of Science, Shrivenham, Wilts., England

SYNOPSIS

This chapter discusses in general terms the limitations imposed on a radio-based animal tracking system by the propagation medium and examines how an integrate-and-dump matched filter can be used to improve the tracking-receiver output signal-to-noise ratio. A multi-transmitter tracking system is outlined and the problem of multichannel interference discussed.

INTRODUCTION

In the studies of relatively wild animals in the field, the biologist often requires physiological information for which short-range telemetry techniques are well suited. Often, however, the only parameter of interest is location. In such cases the animal, bird or fish is "tagged" with a small transmitter and radio direction-finding techniques are used to locate the subject.

The transmitter must be small and light in order not to hinder unduly the normal activities of the subject and yet be powerful enough to communicate with the locating receivers for the duration of the observations. The use of a high peak-power pulsed transmitter with a short duty cycle will help to minimize the power drain of the transmitter primary power source. An important factor in such circumstances is efficiency, and it is well established that aerials that are small compared with a wavelength are very inefficient (Harrison & King, 1961). There have been many investigations of methods to improve this figure but results indicate that efficiencies of less than 10% have been achieved. Substantial improvements to the transmitter power or life-time can only be achieved at the cost of increasing the overall load that the subject will have to carry. The propagation path is rarely line-of-sight and hence propagation is usually by surface wave. Approximate calculations in Appendix 1, show that, at a frequency of 100 MHz, attenuation over a range of 5 km is about 30 dB greater than for a line-of-sight path. The pre-detector signal-to-noise ratio (SNR), in a bandwidth of 25 kHz for a well engineered receiver with a noise figure of 10 dB, is shown to be 13.4 dB for a peak transmitted power of 1 mW. This result is based on the assumption that the ground is flat. In practice the addition of trees, buildings and ground undulations will further degrade the signal-to-noise ratio. As the subject is free to roam, the problems associated with multipath propagation will arise, the major one being signal fading. It is fairly obvious that a technique is needed to improve the receiver signal-to-noise ratio at extreme tracking ranges.

An improvement to the signal-to-noise ratio can be achieved by using a low-noise first-stage radio frequency amplifier connected to the receiver input and it is assumed that this measure has been taken. What other techniques can be utilized that do not involve major modifications to the tracking receivers?

At the receiver the direction-finding process consists of deciding whether, when the beam is pointed in a particular direction, the output is due to signal plus noise or noise alone. A high SNR at the output of the receiver will result in a good probability of detection and identification. An improvement in the receiver SNR could give an increased tracking range, minimize the weight of the radio tag, or improve the lifetime of the transmitter.

SNR enhancement techniques are widely used in pulse-radar systems, the improvement being achieved by discriminating against the noise in a narrow-band filter. A matched filter is one designed to maximize the output peak signal to average noise (power) ratio. The output of a matched filter receiver is the cross correlation between the received waveform and the impulse response of the filter, hence the shape of the input waveform is not preserved, though from the point of view of detecting signals this is of no importance.

THE MATCHED FILTER

The ability of a receiver to detect a weak signal is limited by the noise energy that occupies the same portion of the frequency spectrum as the signal energy. The weakest signal that can be detected is called the minimum detectable signal and its level is some-times difficult to define because of its statistical nature. Detection is based on establishing a threshold level at the receiver output; if the signal plus noise exceeds this level, a signal is assumed to be present.

The realization of a matched filter in practical cases is not always simple, but for the case of rectangular bursts of a sinewave a close approximation can be made using high-Q, RLC resonant circuits. It can be shown that for a pulse duration of time length τ, the maximum occurs when $B\tau \simeq 0.4$ where B is the half-power bandwidth. This value of $B\tau$ causes a degradation in the output signal-to-noise ratio of 0.88 dB compared with an ideal matched filter (Craig, 1960).

A closer approximation to the matched filter can be obtained with more complicated circuitry (Craig, 1960), for example, with rectangular bursts of a sinewave, a cascade of five single-tuned stages with a value of $B\tau = 0.672$, the signal-to-noise loss is 0.5 dB compared with an ideal matched filter.

A practical example in which tuned RLC circuits were used as matched filters is the Piccolo 32-Tone Telegraph System (Robin *et al.*, 1963). Each of 32 audio frequency tones is used to represent a character of the Murray code alphabet. A single tone is transmitted for the full period of the character. Piccolo was designed to operate with 75 band teleprinters at 10 characters per second, therefore each tone lasts for one tenth of a second. The tone frequencies range from 330 Hz (letter A) through 580 Hz (letter Z) to 640 Hz (all space) with 650 Hz additional for a stand-by function. The tones are separated in frequency by 10 Hz; this feature is important and will be discussed on pp. 65 and 67. Thirty-two matched filters are used to isolate the different audio signalling frequencies. Practical results show that a SNR of 9.2 dB gives a character error rate of 10%.

System Realization

The Piccolo system achieved its high-Q filters by means of LC resonant circuits to which positive feedback was applied. For the animal tracking system it was decided to use an operational amplifier with a Wein Bridge network providing the positive feedback path; the basic circuit is shown in Fig. 1. The gain of the amplifier may be changed by adjustment of R_4. Reduction of the gain to a value slightly below that required for oscillation results in a filter with a very high Q. The analysis for the design of this circuit is well established (Williams, 1970) and it may be shown that the resonant frequency f_0 is given by

$$f_0 = \frac{1}{2\pi\sqrt{C_1 C_2 R_1 R_2}}$$

The minimum gain for oscillation is achieved by having $R_4 = 2R_3$. Component values C_1, C_2, R_1, R_2 were chosen as in Fig. 2 to give an oscillation frequency of 5 kHz. To achieve good stability when

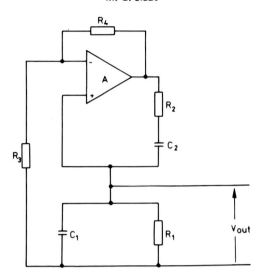

FIG. 1. Basic circuit for matched filter.

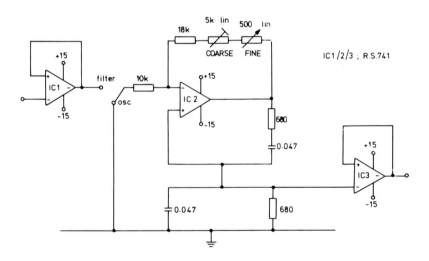

FIG. 2. Practical circuit for matched filter.

operating as a high Q-filter, input and output voltage-follower stages have been included, the output being taken from the junction of C_1 R_2 because of impedance switching requirements.

The performance of this circuit as a filter is such that the band-width may be as small as 10 Hz, which represents a Q = 500, without self-oscillation commencing. The Q required for a matched filter is 125; gain adjustment by the value of R_4 will make this circuit very suitable for use as a matched filter for pulses of sine waves.

An analysis of the response of the filter to a burst of sine waves is shown in detail by Robin *et al.* (1963). Sketches showing the envelope of the output response voltage waveform of the high-Q filter to bursts of sine waves are shown in Fig. 3.

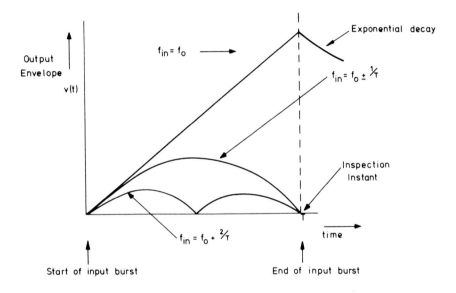

FIG. 3. Response of matched filter to orthogonal signals.

When the circuit is driven at its resonant frequency $(f_{in} = f_0)$ the amplitude of the response can be seen to rise very nearly linearly for the duration of the 10 ms input pulse. After this it decays exponentially, and the need to "dump" this oscillation, or ringing, before the next input pulse is apparent. When the circuit is excited by an input signal at a frequency offset by n/T Hz (n is an integer), there is a zero output at the time T. Thus if two tones are separated by $1/T$ Hz, or any multiple thereof, then detection at time T will result, ideally, in zero in one channel. Such a system is said to use orthogonal signalling. The implication is that a system using matched filtering of orthogonal signals in this way can have a large number of channels, separated in frequency by n/T Hz, the principle governing the operation of Piccolo. For the proposed tracking system the pulse duration of $T = 10$ ms means that the spacing of the individual transmission frequencies would be 100 Hz and in principle a large number of transmitters with the same pulse-repetition frequency would seem possible within the restricted very high frequency (VHF) band.

The orthogonality of the proposed system appears to be a very useful property if the monitoring of several transmitters is required, but the main feature of the matched filter is the maximization of the SNR for single channel operation. Measurement of SNR shows that the maximum improvement is obtained when B is chosen to satisfy $B\tau = 0.4$ but that the maximum improvement is approximately 2 dB below the theoretical value of 16 dB.

SINGLE-CHANNEL TRACKING SYSTEM

The system diagram (Fig. 4) shows a beat-frequency oscillator (BFO) which is adjusted so that the frequency of the resultant output signal is identical to the resonant frequency of the matched filter. Upon reception of an input signal the envelope of the filter output rises linearly with time and represents the integral of the signal. After the cessation of the signal pulse the filter continues to ring and must be dumped or shorted to zero before the reception of the next pulse.

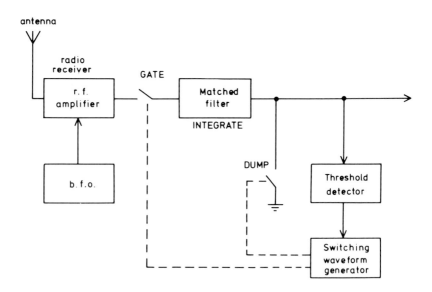

FIG. 4. Single channel tracking receiver including a matched filter.

This technique is often called "integrate and dump", and requires suitably synchronized gating pulses. Operation of the receiver in this mode requires that the integrate gate be open when a signal is expected and that the dump should operate from the cessation of the signal pulse until just before the next signal is due. Additionally, until a correct signal has been received the integrate gate should remain continuously open and the dump gate should remain inoperative. At the output of the matched filter the signal is amplitude detected and compared with a suitable reference voltage such that a decision can be made as to the existence of a correct signal. If the amplitude of the detected signal exceeds the decision threshold, a rectangular pulse will be produced, indicating the presence of a detected signal at the resonant frequency of the matched filter. This pulse is used as the timing reference to activate the necessary pulses for the integrate and dump sequences. Thus the receiver is self synchronizing.

Phase Comparison

If the direction finding technique is based on a null-signal method (Watson & Wright, 1971), the signal-to-noise enhancement will increase the accuracy of the bearing obtained. A more sophisticated technique which is based on measurement of phase requires two identical receivers (Winn, 1979). This technique usually involves a four-element rectangular array in which two channels, a sum and difference channel, are required. The sum channel indicates the presence of a signal and the difference channel is used to assess the direction of arrival of the signal by a phase comparison technique. This is a well established method of direction-finding for radar and is well documented. This SNR enhancement technique may be used provided any phase shift introduced by the matched filters is calibrated out of the bearing measurement (Skolnik, 1962).

Because of the roaming nature of the subjects, the transmitter locations may be nearby (100 m) or at extreme nominal line-of-sight range (5 km). There are thus wide variations in signal power at the receiver (see Appendix 1). Regulation of the signal strength within the receiver is provided by the automatic gain control (AGC) to avoid overloading of the receiver.

MULTI-CHANNEL TRACKING SYSTEM

Because of the orthogonality properties of signals that are separated in frequency by n/T Hz, it would seem possible to use several integrate and dump matched filters in parallel and track many transmitters simultaneously. In considering such a multi-tracking system there are several points that need to be carefully considered. The subjects to be tracked may be considered to be randomly distributed over a region which extends to the limit of the line-of-sight. Further, the transmitters are not synchronized and pulses will be received at differing times. Ideally each received pulse will identify the originating subject by its frequency.

Although the identifying signals are orthogonal the effects of different times of arrival, relative strengths of received signals and errors in the transmitted frequency need to be considered further.

Consider the example of a distant subject (5 km), transmitting a frequency of f_0 Hz, and a nearby subject (100 m), radiating an adjacent orthogonal frequency of $(f_0 + 1/T)$ Hz. If at the receiver the two signal pulses did not overlap in time, the dump grating would ensure the interchannel interference (crosstalk) was minimized. However, the AGC of the receiver will be fully operative for the strong signal, whereas for weak signals the AGC must allow the maximum gain of the receiver to be utilized. In the presence of a strong signal the AGC may well decrease the gain of the receiver so much that the weak signals from the distant subject are reduced below the decision threshold. AGC is used to overcome slow changes in signal strength and the time constant of the circuit is many times greater than that of the modulation. The problem is increased by considering time-overlapping orthogonal signals of widely differing strength. A strong signal controls the AGC and the presence of the weak signal may be missed completely. This is a problem often met in radar and a logarithmic amplifier is sometimes used to limit the strength of the largest signal. For the example quoted above, a difference in signal strength of at least 74 dB can be expected for the ranges quoted assuming that the mode of propagation is by surface wave, at a frequency of 100 MHz. If the pulse-repetition frequencies of the two transmitted signals are chosen to be significantly different, the time of overlap can be made to average out to a smaller quantity. Time lost owing to overlap will not then significantly affect the tracking accuracy.

For multi-channel tracking, direction finding by rotation of direction aerials is cumbersome and a phase comparison technique is more appropriate. Two such stations are required for triangulation, and accurate timing is required at both stations for co-ordination of results. The bearing is given virtually instantaneously, thus frequency, time and bearing from each station are required for the movements of each subject to be followed.

The transmitter tags are subjected to extreme environmental hazards in temperature and humidity and must obviously be mechanically robust. Such hazards will cause changes in frequency and in the short term these may be significant. Crystal oscillators are often quoted as having tolerances in frequency of a few parts per million. With good temperature and humidity control several orders better than this can be achieved. For an animal tag no such control can be implemented and the associated receiver must be tolerant to changes in transmitter frequency.

SUMMARY

The use of a Wein bridge oscillator circuit adjusted just below the threshold for oscillation gives a good approximation to a matched filter. This results in approximately a 14 dB increase in the pre-detector signal-to-noise ratio. Additions to the tracking receiver are minor in that they may be achieved by an appliqué unit attached to the output stage.

Orthogonality can readily be achieved by suitable choice of transmitter frequency and this property can be used to extend a single-subject tracking system into one in which multi-subject tracking is possible. In such a mode multichannel interference is a problem which can be alleviated by the use of a logarithmic amplifier. Undoubtedly multi-subject tracking requires real-time tracking and frequency monitoring so a direction finding (DF) technique, based on direct phase measurement, should be used.

REFERENCES

Craig, J. W. (1960). Optimum approximation to a matched filter response. *I.R.E. Trans. Inf. Theory* **IT6**, No. 3: 409–410.

Harrison, C. W. & King, R. W. P. (1961). Folded dipoles and loops. *I.R.E. Trans. Antenna Propag.* **AP-9**: 171–187.

Jordan, E. C. (1968). *Electromagnetic waves and radiating systems.* New York: Prentice Hall.

Robin, H. K., Bayley, D., Murray, T. L. & Ralphs, J. D. (1963). Multitone signalling system employing quenched resonators for use on noisy radio-teleprinter circuits. *Proc. I.E.E.* **110**: 1554–1568.

Skolnik, M. I. (1962). *Introduction to radar systems.* New York: McGraw Hill.

Watson, D. W. & Wright, H. E. (1971). *Radio direction finding.* London: Van Nostrand.

Williams, P. (1970). Bandpass filters using Wein Bridge. *Electron. Lett.* **16**, No. 6: 188.

Winn, R. T. E. (1979). New developments in communications electronic warfare. *Racalex Lect.* 1979: 85–90. Bracknell: Racal Electronic Ltd.

APPENDIX

This section shows the calculation that may be done to estimate the signal loss over a line-of-sight path when the mode of propagation is by surface wave. The calculation uses the formulation given by Jordan (1968: 635–654).

Magnitude of the space wave

$$E_{\text{sp}} = \frac{30\,\beta\,I\,l}{d}$$

where:

$\beta = 2\pi/\lambda$,
I = aerial current,
l = effective aerial length,
d = line-of-sight distance,
λ = wavelength of electromagnetic wave.

Magnitude of the surface wave

$$E_{\text{su}} = \frac{60\,\pi\,I\,l}{d}\,\frac{(1-R_{\text{v}})}{2\,p}$$

where

R_{v} = the ground reflection coefficient for vertically polarized waves. The ratio of loss to free space loss = $(1-R_{\text{v}})/2\,p$
$R_{\text{v}} \simeq -1$, therefore loss $\simeq 1/p$
$p = (\pi R/\lambda x)\cos b; b = \text{Tan}^{-1}\,(\epsilon_{\text{r}}+1)/x, x = (18\times 10^3\,\sigma)/f_{\text{MHz}}$
σ = conductivity of the ground,
ϵ_{r} = relative permittivity of the ground.

For good ground, $\sigma = 12$, $\epsilon_{\text{r}} = 15$ and a working frequency of 100 MHz gives

$$b = 82.3°, x = 2.16, p = 324.$$

For a 1 watt transmitter with a short dipole of gain $G_{\text{t}} = 1.5$, the power density P at the receiver 5 km distant, for free space propagation is given by

$$P = \frac{P_{\text{t}}G_{\text{t}}}{4\pi d^2} = 0.478 \times 10^{-8} \text{ W}$$

At a receiver with a two-element dipole array, $G_{\text{r}} = 3.28$, the received power is given by

$$\left(PG_r \frac{\lambda^2}{\pi}\right) = 1.25 \times 10^{-8} \text{ W}$$

The surface wave component is

$$\frac{1.25 \times 10^{-8}}{(324)^2} = 1.3 \times 10^{-11} \text{ W}$$

The signal-to-noise ratio is the important parameter and the contributions to the overall noisiness of the system are made up from the noise produced by the aerial and the receiver. At 100 MHz the aerial sees the sky noise at 4 K and the ground noise at 290 K; assuming that the receiving aerial is a dipole the total aerial noise temperature T_A is given by

$$T_A = 2\pi \int_0^\pi 294 \sin \theta \, d\theta = 3800 \text{ K}$$

The noise figure F of a well-engineered receiver will be in the region of 10 dB, and its contribution to the noise temperature will be

$$T_E = (F-1) \, 290 = 2610 \text{ K}$$

Noise power/Hertz $= k \, (T_E + T_A) = 8.8 \times 10^{-20}$ W/Hz, k is Boltzmann's constant, 1.37×10^{-23} joules, degree K^{-1}.

The predetector noise bandwidth is typically 25 kHz and thus the noise at this stage is $25 \times 10^3 \times 8.8 \times 10^{-20} = 2.2 \times 10^{-15}$ W. The predetector signal-to-noise ratio $= 1.3 \times 10^{-11}/2.2 \times 10^{-15} = 5.9 \times 10^3$ for 1 watt transmitted power.

The post-detector signal-to-noise ratio depends on the type of detector used. At best, a coherent detector will produce a 3 dB improvement giving a receiver output signal-to-noise ratio of 11.8×10^3.

The various assumptions made throughout this analysis make the final figure very approximate but undoubtedly the conclusion must be reached that with a 1 mW transmitted signal a signal-to-noise ratio of at most 12.0 dB at the receiver is not high enough to achieve confidence in the ability to track extreme line-of-sight subjects.

For a subject at 100 m, the loss of the surface wave component is decreased by $(50)^4 = 74$ dB.

DISCUSSION

R. Mitson (MAFF) — Did you test the system for use as an animal tracking method?

M. G. Slade — We tracked a student in the college grounds quite successfully! The 16 dB increase in the SNR should be very useful. The range would be increased by approximately twice, a small but significant amount.

J. W. R. Griffiths (Loughborough University) — Where does the improvement in SNR come from? What is the bandwidth of the input noise?

Slade — The bandwidth of the input noise will be reduced by the narrowness of the filter itself.

Griffiths — Any normal narrow bandwidth filter will give you such an improvement of this sytem over a normal narrow-band system?

Slade — The input response of an ordinary narrow bandwidth filter would not necessarily correlate with the incoming signal. Matching the impulse response is important.

Griffiths — A narrow-band filter with the appropriate bandwidth is an approximation to the correct matched filter. Use of the ideal gives a small further improvement.

Slade — One can make many types of filter with different impulse responses.

V. B. Kuechle (Minnesota) — With a matched filter a prior knowledge of frequency and pulse is required. Could you comment on tolerances?

Slade — A frequency tolerance of a few parts per million would probably be needed for an animal tracking system.

GENERAL DISCUSSION

E. Walton — I should like to make a general point not covered by any of today's speakers. It is my experience that commercially available equipment, in particular that from North American manufacturers, is inadequately specified. For example, the spectral output of transmitters is not given with the result that if one attempts to use a scanning receiver to identify several different transmitters, the frequency drift due to temperature will make individual transmitter identification difficult. These sort of problems are rarely referred to by the manufacturers and have not been aired today.

V. B. Kuechle (Minnesota) — I cannot speak for the North American manufacturers but would point out that the commercial supply of telemetry equipment in North America is largely a cottage industry. The firms usually have neither the technical expertise nor facilities to provide adequate equipment specification. I would suggest that it behoves the buyer, who also may lack the appropriate knowledge, to clearly specify the equipment he orders. In the example mentioned, for instance, one could specify that the frequency remains within a certain tolerance for a certain temperature range. We normally test all our transmitters for operation under field conditions. The problem of cross-over of spectral output is a receiver rather than a transmitter problem.

H. G. Lloyd (MAFF) — Perhaps the user should be allowed to field test the equipment before he decides to buy it.

T. Woakes (Birmingham University) — Could I ask whether the American systems being used in the UK have been passed by the Home Office?

R. M. Skiffins (Home Office) — The simple answer is that if equipment meets Home Office regulations the problems mentioned should not arise. Clearly much of the equipment being used in this country does not meet the regulations and buyers are not aware of this. I would suggest that prospective buyers submit a sample of the equipment they wish to use. The Home Office will test it free of charge and check for stability, harmonic radiations etc.

E. Walton — I raised this matter for the benefit of biologists new to the field of telemetry. I would recommend that biologists intending to use commercially available telemetry equipment should seek advice from people who have already used it. Manufacturers certainly will not tell customers about the potential drawbacks of their equipment.

Symp. zool. Soc. Lond. (1982) No. 49, 75–93

Acoustic Telemetry and the Marine Fisheries

F. R. HARDEN JONES and G. P. ARNOLD

Ministry of Agriculture, Fisheries and Food, Fisheries Laboratory, Lowestoft, Suffolk, England

SYNOPSIS

Special acoustic techniques have been applied to some problems in marine fisheries science at the Fisheries Laboratory, Lowestoft. Transponding acoustic tags have been developed for use with a high resolution sector scanning sonar to study the efficiency of the demersal otter trawl and the behaviour of migratory fish. A transponding compass tag has been developed which gives on-line information of the heading of a fish which can be assigned to one or other of eight 45° sectors. Plaice appear to maintain a surprisingly consistent course when in midwater, at night, and so without the usual rheotropic clues. The use of acoustic telemetry techniques has enabled some problems to be tackled in the open sea in a way which would have otherwise been impossible.

INTRODUCTION

This chapter is not a technical review of acoustic telemetry of which there have been several in recent years (Stasko & Pincock, 1977; Mitson, 1978; Nelson, 1978; Ireland & Kanwisher, 1978). Here we will show how special acoustic techniques have been applied to some problems in marine fisheries science. We will be dealing with work carried out at the Fisheries Laboratory, Lowestoft, which is supported by public funds. Not unexpectedly our Research and Development objectives have been defined, and

> research on marine fish and shellfish stocks and their environment is designed to provide the Fisheries Ministers with the scientific information needed to discharge their responsibilities to the fishing industry and consumers. It provides the basis for the formulation of domestic policy and the discharge of international obligations, particularly those related to conservation and environmental protection (Anon., 1974).

Within this framework the broad aims of research are seen as being directed towards the rational exploitation and conservation of the stocks of fish and shellfish; to increasing the efficiency of the industry by technological development; and to protecting the marine environment. We will show how acoustic telemetry techniques have been used in research projects relevant to two of the three broad aims: the efficiency of the otter trawl (relevant to technological development); and the migratory behaviour of fish (a problem of fish biology relevant to the rational exploitation and conservation of stocks).

OUR NEED FOR TELEMETRY

The Lowestoft approach to acoustic telemetry is somewhat different to that followed elsewhere and the difference arose because in 1969 we were able to install in our Research Vessel *Clione* a high resolution sector scanning sonar invented by Dr G. M. Voglis of the Admiralty Research Laboratory, Teddington. Cushing & Harden Jones (1979) give an account of the development of the Lowestoft Laboratory's interest in sector scanning sonar. The Voglis sonar — originally known as Bifocal and subsequently, in a modified form, as the ARL sector scanner — operated at 300 kHz and scanned a 30° sector out to ranges of 360 m with bearing and nominal range resolutions 0.33° and 8 cm respectively. Mitson & Cook (1971) have described the shipborne installation and stabilization system which allowed the sonar to be used in the horizontal or vertical mode and so give a plan or an elevation picture on a B-scan display of the area or target under surveillance. The facility for horizontal and vertical scanning allowed the range, bearing and depth of a target to be determined while the high resolution of the system gave a very detailed picture on the display. The original equipment was returned to the Admiralty in 1974 and replaced by a solid-state counterpart with improved reliability and performance (Holley, Mitson & Pratt, 1975).

When we first used the Voglis equipment on P.A.S. *Gossamer* in 1963 and 1964, we were particularly impressed by its picture-forming qualities, whether of fish shoals, bottom topography, underwater structures, or wrecks. The essential details of a midwater trawl could be clearly identified out to ranges of 180 m and we hoped that we would also be able to detect a demersal trawl against the bottom reverberation (Cushing & Harden Jones, 1966). Our expectations were fully realized and in further trials in 1969 an otter trawl towed by one research vessel was detected by sector scanning sonar on R.V. *Clione* out to ranges of 180 m: details of the main warps, otter boards, eddy trails, bridles, dan lenos, headline and floats, footrope, selvedges, and cod end of the net were all clearly visible.

Our particular interest in the otter trawl arose from the importance of this gear in the United Kingdom fisheries. In 1969 the otter trawl accounted for 82% of the value of the demersal fish landed by British vessels. Although the pattern of our fisheries has changed over the last ten years, the gear still accounted for 58% by value of the demersal catch in 1979. We were interested in the efficiency of the otter trawl: for example, did it catch 40, 60 or 80% of the fish that came within its influence? While it was easy to frame this simple question, obtaining an answer was a different matter. Our line of approach will be easier to understand when considering the history and development of the otter trawl (Fig. 1).

In the 1870s sailing smacks and steam trawlers fished with a net whose mouth was kept open by a wooden beam supported by two iron shoes which rode along the bottom. In the otter trawl the opening of the net was maintained by the spreading action of two otter boards, or doors, fixed to each wing end. This net was invented and patented in 1894 by an engineer, James Robert Scott of Granton, near Edinburgh, in Scotland. Scott's father was the manager of the General Steam Fishing Company of Granton which, at the time, was facing a steady decline in earnings. The Company survived as the new

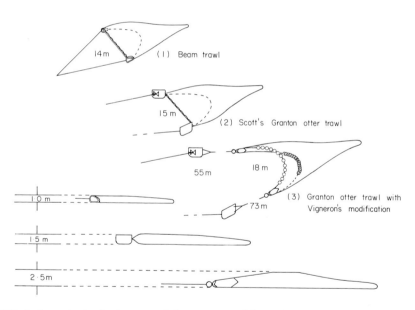

FIG. 1. Stages in the development of the otter trawl. (1) The beam trawl as towed by sailing smacks and steamers. (2) Scott's otter trawl which was introduced in 1894. The wooden beam was replaced by otter boards attached directly to the wing ends of the net. (3) In 1920 sweeps or bridles (Vigeron's modification) were placed between the otter boards and wing ends. Later, spherical dan lenos and angled spreading bars, headline floats, and ground-rope bobbins were added to the gear. When fishing for plaice, lengths of heavy chain (the so-called ticklers) are rigged between the boards, the wings of the net, and in the middle of the ground rope.

gear caught 50% more roundfish such as cod and haddock. Further-more it fished as well by day as by night, in contrast to the beam trawl whose catches were low during the day. Vessels licensed to work Scott's gear outfished all others. The length of voyages was cut with a consequent improvement in the quality of the catch and its sale price. By 1896 the new trawl was generally adopted by British steam trawlers and with later additions and modifications — such as bridles, floats, bobbins and special footropes — Scott's original invention provided the basic design for many of the bottom trawls now in use.

The spread of Scott's otter trawl from wing to wing was similar to that of the larger beam trawls of the day — 14 to 15 m — so the area of bottom swept clear by both gears was probably the same. It was never claimed that the otter trawl caught more flatfish (plaice, sole, turbot) than the beam trawl. The increased catch of roundfish (cod, haddock, whiting) was attributed to a higher headline while the otter trawl's ability to catch fish by day was thought to be due to the absence of the beam which could have scared fish when they were still in a position to avoid the net.

In the 1920s a French shipowner, J.-A. Baptiste Vigneron, intro-duced the use of bridles, or sweeps, between the otter boards and the wing ends of the trawl. One of the reasons for this measure was to allow silt and stones disturbed by the doors to pass outside the wing ends and so reduce damage to fish in the cod end. But the introduction of bridles had an unexpected gain: there was a marked increase in catch proportional to the extra spread between the boards which suggested that the net was now catching fish which were originally positioned outside the wings. These fish would have to move into the path of the net to be caught and they may have done so because they were disturbed by the boards or the bridles themselves. These gains were welcomed by the trawling industry at a time when catches were dropping following the short-lived boom after the 1914—18 war.

A similar increase and subsequent decline in catch-rate followed the 1939—45 war. This stimulated a substantial United Kingdom effort[*] in the 1950s to design, on hydrodynamic principles, a trawl with a greater spread, a higher headline, but similar hydrodynamic drag, when compared with the gear then in general use. The engineer-ing objectives were largely achieved. The mouth area of the new trawl was 2.5 times that of the standard trawl and the drag, although some

---

[*] The so-called SARO project, the acronym being derived from Saunders-Roe Ltd, the agent for the development work.

10% greater, was acceptable. But the larger trawl did not lead to any substantial increase in catch.

When trying to account for this unhappy result, it should be remembered that the fishing power of a trawl has two components. First, there is the area or volume swept by the gear within which fish may be caught, and secondly, there is the proportion of the fish within the area or volume swept clear that are caught. The distinction is between fishing intensity and efficiency. Fishing intensity may be measured in terms of area or volume swept clear per unit time. This may be increased by towing a larger trawl, or the same trawl at a faster speed. For example, among beam trawlers the relatively slow sailing smacks were outfished by the faster steamers; steamers using Scott's new otter trawl outfished those using beam trawls; and otter trawls rigged with bridles outfished those without. The advantage of one class of vessel or gear over another is probably attributable to differences in fishing intensity. However, the efficiency of the gear must also be taken into consideration. If the fishing intensity is increased by towing at a very much faster speed, or by towing a very much larger trawl, efficiency may be reduced to such an extent that the catches are not increased and may even fall. This is a possible explanation of the apparent failure of the United Kingdom SARO trawl.

It is difficult to measure the efficiency of a trawl. The fish caught in the cod end can be counted. But what about the fish between the otter boards? Trawls are worked in depths greater than 30 m, where visibility may be restricted to a few metres while the boards themselves may be more than 50 m apart. Under these conditions divers, television or photographic techniques cannot provide quantitative data, and trials with an acoustic arch spread over the boards to count the fish between them were never entirely satisfactory. We hoped that the high resolution sector scanning sonar, with its unique picture-forming qualities, would help to provide the answer to the problem. The sector scanner can detect 30—40 cm roundfish out to ranges of 150 m. But we found that it was impossible, when working in horizontal mode, to make a distinction between the echoes returned by single fish in the path of the gear and the echoes from small rocks or stones on the sea bed. This difficulty was overcome by the development of an acoustic transponding tag which returned a powerful and unambiguous signal when insonified by the outgoing signal of the sonar. The tag was invented and described by Mitson & Storeton-West (1971) and when used in conjunction with the sector scanner provided a new approach to the problem of gear efficiency.

THE EFFICIENCY OF THE OTTER TRAWL

We found that a plaice fitted with an acoustic transponding tag and released from R.V. *Clione* went to the bottom where it could be kept under surveillance with sector scanning sonar for a 2–4 h settling-down period. A second research vessel, the R.V. *Corella*, towing a Granton trawl, was then directed by VHF radio to catch this particular fish, the complete fish-catching process being followed by sonar. The capture or escape of the fish could then be related to its position with respect to the gear, while details of its behaviour were subsequently analysed from video or 16 mm film records of the display. A more detailed account of this work has been published elsewhere (Harden Jones, Margetts *et al.*, 1977; Harden Jones, 1980a).

Statistically acceptable estimates of the efficiency of the gear required the release of several hundred individual fish to obtain a sufficient number of valid attacks, and the work was spread over seven years. The results are summarized in Table I. They show that R.V. *Corella*'s Granton trawl, which was rigged for general purposes rather than specially for plaice fishing,* caught about 70% of the plaice that lay within the otter boards. We consider this to be a high and acceptable level of efficiency. Although the estimate only applies to one gear under the particular conditions of our experiment, the results suggest that it might be profitable to pay some attention to other aspects of trawling, such as how to reduce drag (and thus fuel costs) or vulnerability to damage (and thus loss of fishing time in mending and replacement). The importance of drag will be appreciated from the breakdown in Table II. To what extent could the drag attributable to various components of the gear be reduced while maintaining the volume swept clear and the present efficiency? Our results have helped us to see the research and development problems in a more realistic context.

TRACKING PLAICE IN THE OPEN SEA

We gained a lot of experience with sector scanning sonar and acoustic transponding tags in tackling a practical problem of direct interest to the fishing industry, and the technique that emerged proved to be

* R.V. *Corella*'s trawl had longer bridles, fewer and shorter tickler chains, less groundrope chain and more headline floats when compared with the usual commercial rig for plaice fishing on good grounds.

TABLE I

The efficiency of the Granton otter trawl for plaice as determined by the capture of fish fitted with acoustic transponding tags. The data relating to the efficiency of the gear without a board-to-board tickler chain has been published (Harden Jones, Margetts et al., 1977) while that relating to the gear with such a chain is new and should be regarded as preliminary only (Harden Jones, Arnold & Greer Walker, unpublished data)

Position of fish	Board-to-board chain not fitted			Board-to-board chain fitted		
	Valid attacks	Number of fish caught	Efficiency (%)	Valid attacks	Number of fish caught	Efficiency (%)
Within boards	166	73	44	48[a]	32	67
Between boards and wing ends	72	16	22	21	10	48
In path of net	94	57	61	24	19	79

[a] The position of three of these fish with respect to the wing ends of the net was uncertain.

TABLE II

The breakdown of the drag of a Granton trawl (24 m headline, 37 m ground-rope) used in the Humber fleet in the early 1960s. Towing speed 1.75 m/s, warp out 198 m. Data from Dickson (1964)

Component of gear	Drag (tonnes)	Percentage of total
Net	3.4	48
Bobbins, groundrope, headline and floats	0.7	10
Bridles and dan lenos	0.2	3
Otter boards	2.0	29
Warps	0.7	10
Totals	7.0	100

useful for studying the movements of fish in the open sea. We soon found that it was possible to keep individual fish fitted with acoustic transponding tags under continual surveillance for several days. The facility for both horizontal (plan) and vertical (depth) scanning — the change from one mode to the other being made in a few seconds literally at the flick of a switch — allowed us to position the fish with reference to the ship to within a few metres, with a similar accuracy in depth. Furthermore when the fish were at ranges of 100 to 150 m, it was possible to make the distinction between the depth categories *on the bottom, close to the bottom,* (and thus likely to be occasionally touching the bottom) *just off the bottom,* and *in midwater.**

Using this technique, we completed a series of tracking exercises with plaice in the Southern Bight of the North Sea, which showed that fish which covered a substantial distance (more than 10 to 15 km) during the period of surveillance did so when they left the bottom and moved into midwater. The track of Plaice 7 provides an example (Fig. 2). This fish was released off Southwold at 10:09 on 12 December, 1971 and reached the Cross Sand, a sandbank more than 40 km to the north, by noon the following day. The plaice moved rapidly to the north on a north-going tide and Fig. 3 shows that on this tide the fish was in midwater and at times over 20 m above the bottom. The relation between speed over the ground, tide, and the fish's position in the water column is summarized in Table III.

This behaviour, in which the vertical movements are related to the change in the direction of the tidal stream rather than to a change in light intensity, leads to a pattern of movement which we have called

---

* These distinctions are important when considering the tactile and visual clues likely to be used by a fish when orientating to water currents.

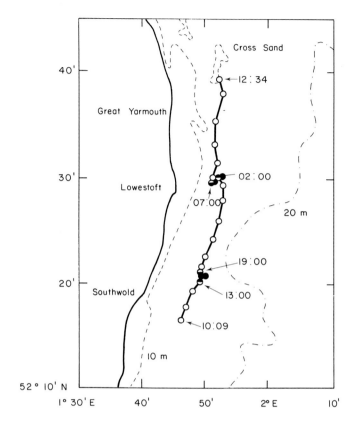

FIG. 2. The track of a plaice (Plaice 7) fitted with an acoustic transponding tag released off Southwold at 10:09 on 12 December 1971. Hourly positions of the fish are indicated and the times of slackwater are shown. The fish was abandoned at 12:34 on 13 December when it swam on to the Cross Sand, a sandbank to the north of Great Yarmouth. ○, North-going tide; ●, south-going tide; ◓, low-water slack; ◒, high-water slack. The 10 m and 20 m depth contours are shown.

selective tidal stream transport (Greer Walker, Harden Jones & Arnold, 1978). It would seem to be an economical transport system for migrants on passage between their feeding and spawning areas as full or ripening fish, and on the return journey between their spawning and feeding areas as spent or recovering fish.

The next step was to show that tidal stream transport occurred in a natural wild population. The plaice of the Southern Bight of the North Sea provide a convenient stock on which to work from Lowe-stoft. These plaice have a summer feeding area between the Norfolk

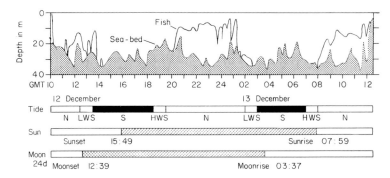

FIG. 3. The depth of Plaice 7 in relation to the direction of the tide and other environmental factors. The track was abandoned at 12:34 on 13 December when the fish went into shallow water at the Cross Sand (see Fig. 2).

Banks and the Outer Silver Pit. Conventional tagging experiments have shown that in November and December the ripening fish leave the feeding area and migrate towards their spawning area some 150 km to the south, the centre of egg production (52° 00′N, 02° 30′ E) being midway between England and the Netherlands. Spawning reaches its peak in the middle of January and subsequently the spent fish return to the north in February and March.

We argued that if the plaice were using tidal stream transport when on migration, in November and December ripening fish would be expected to be more abundant in midwater on south-going tides, and in February spent fish would be expected to be more abundant in midwater on north-going tides. A series of fishing experiments using a large midwater trawl showed that this was indeed true (Table IV), the difference between the tides being clear cut and consistent with the gonad maturation stages characteristic of ripening and spent fish.

TABLE III

Speed over the ground, direction of tide, and position in the water column for Plaice 7 (see Figs 2 and 3)

Time GMT	Speed over the ground (cm/s)	Direction of tide	Position in water column
10:00—13:00	80	North	Midwater
13:00—19:00	9	South	On bottom
19:00—02:00	89	North	Midwater
02:00—07:00	8	South	On bottom
07:00—12:30	115	North	Midwater

TABLE IV

Paired tows made with a midwater Engel trawl on consecutive north-going and south-going tides in the Southern Bight of the North Sea: a comparison of catches for those pairs in which plaice were caught in one or both tows. Data from Harden Jones, Arnold et al. *(1979)*

Months	No. of paired tows	Comparison of catches More plaice caught on		Catches equal
		North-going tide	South-going tide	
Nov., Dec.	39	5	33	1
Feb.	31	20	1	0

These results suggest that we may have identified the transport system used by the migrants on passage. But this cannot be the end of the story. If plaice do use the tide for transport when on migration, they must join the appropriate tide at the start of the journey; stay with the correct tide when on passage; and finally leave the transport system on arrival. While we have no satisfactory answers to these three problems, it seems clear that the original choice or selection of tide is critical and that the would-be migrant needs to make a clear distinction between low-water slack which is followed by a south-going tide and high-water slack which is followed by a north-going tide. It is not uncommon for fish to leave the bottom at sunset and an unambiguous cue to join a south-going tide could be provided by the coincidence of low-water slack, sunset, and full moon: such a conjunction of events might entrain a plaice to a semi-diurnal (12 h) rhythm of vertical movement characteristic of tidal transport (Harden Jones, 1980b). If this hypothesis were correct, no directional information would be necessary for the migrant either at the start or during the course of the journey.

THE ACOUSTIC TRANSPONDING COMPASS TAG

The question of direction, and thus orientation, brings us to a recent development in telemetry at the Lowestoft Laboratory. The use of the tidal stream by migrating plaice would appear to absolve them from any need to maintain a particular course or direction through the water. But when simple vector analysis was applied to those midwater tracks for which current meter data were available, the results showed that some fish maintained a surprisingly consistent course for periods of up to two hours (Greer Walker *et al.*, 1978: figs 5, 11, and 17).

We wanted to take this matter further and accordingly pressed our colleagues at the Fisheries Laboratory to produce a tag which would let us know the direction in which the fish was heading. They invented an acoustic transponding compass tag so that we can now obtain a continuous record of the position, depth and heading of a fish in real time. The new tag is attached to a plaice by two Petersen disks and carefully aligned along the body axis on the upper (right) side. The compass reference is provided by a magnet attached to a slotted disk, which is mounted on a jewelled pivot. Changes in bearing are detected optically using a circular array of eight infrared detectors to determine the bearing of the slot — and so the heading of the fish — which can be assigned to one or other of eight 45° sectors. When the tag is interrogated at 300 kHz, it responds with a reference pulse. From this signal the range, depth, and bearing of the fish can be determined by sector scanning sonar in the usual way. The heading of the fish is indicated by a second pulse, the delay between the two signals — ranging in 6-ms steps from 24 to 66 ms — serving to identify the relevant sector. The two pulses are displayed on a B-scan sonar display and, from selected channels of the receiver beam, on an Alden paper recorder (Fig. 4). The heading of the fish can also be read directly from a digital decoder. A full description of the compass tag will be given elsewhere by Pearson & Storeton-West (in preparation) and details of the proving trials by Mitson, Storeton-West & Pearson (in preparation).

Although the compass tags are somewhat larger than was hoped for, we — as users — are well pleased. The tags are reliable, and the life of about 50 h is sufficient to extend the tracking team. To date we have tracked seven plaice fitted with compass tags and our colleague Dr Greer Walker has similarly tracked two salmon. A preliminary account of one of the plaice tracks has already been published (Harden Jones, 1981).

We were particularly interested in the orientation of plaice at night and in midwater when they would be unlikely to receive visual or tactile clues from the bottom. Under these circumstances a plaice 5 m to 20 m above the bottom (total depth of water 25 m to 30 m), and on a night when neither the moon nor stars were visible, would maintain a consistent heading and remain within two 45° sectors for up to an hour. These results were quite unexpected. They may underestimate the ability of a plaice to maintain a steady course as there were instances where the compass signals were alternating between two adjacent sectors which suggested that the fish were keeping to a narrow range of bearings towards the edge of either sector.

These results raise two new problems: first, what is the reference

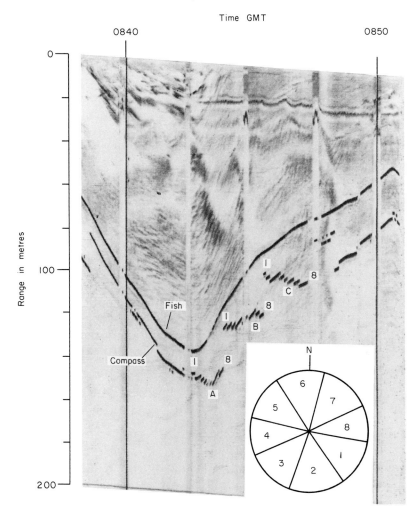

FIG. 4. A section of an Alden recorder chart (28 May 1980) showing the signals returned from a plaice fitted with a transponding acoustic compass tag. The plaice is in midwater, well above the bottom. The first signal gives the range from the ship which varies from 45 m to 127 m. The second signal is from the compass, the delay between the two signals identifying the compass sector in which the fish was heading. Between 08:42:30 and 08:47:30 the fish completed three clockwise circles labelled A, B, and C. Sectors 1 and 8 are numbered (the shortest and longest delays respectively), and the compass bearings associated with the sectors are indicated in the insert. The interruptions to the record at 08:42:30, 08:45, and 08:47:30 mark the times when the transducer was turned from the horizontal to the vertical mode to determine the depth of the fish.

used to select the heading that the fish adopts when in midwater; and secondly, how is the course subsequently maintained?

One observation has suggested a possible answer to the first question. One plaice was seen to be swimming very close to the bottom along the crest of a sand wave before "taking off". When in midwater the fish maintained the same heading: the comparison with an aircraft and its runway was irresistible. This led to the suggestion that for plaice, the heading maintained in midwater might be related to that held on the bottom shortly before "take-off". The data that have been analysed are consistent with this hypothesis. We suspect that some topographical feature could provide the bottom reference: sand waves, whose crests are aligned at 90° across the tidal axis, would be suitable.

We do not know how a plaice maintains a consistent heading in midwater. The height of the fish above the bottom, on nights when neither the moon nor stars were visible, would seem to rule out the usual rheotropic clues. It is possible that the heading is maintained by a form of inertial guidance through the labyrinth or from on-line information of the earth's magnetic field detected through an unidentified sensor: such matters are for future study.

While we have been making a point of the plaice's ability to maintain a consistent heading in midwater, a no less remarkable observation is that they sometimes go round in circles. Figure 4 shows a section of an Alden chart recording three consecutive clockwise circles completed by a midwater plaice. Another plaice completed nine consecutive circles over a period of 80 min. We do not know the significance of this pattern of movement.

A NEW REQUIREMENT

There are several environmental and biological parameters which have been considered as candidates for telemetry and the list includes salinity, water flow past the fish, water and body temperature, and the rate of heart beat. But when money is short priorities decide the order of battle and at present we are particularly interested in a pitch and roll indicator. Such a device would tell us something of the attitude of a fish in terms of pitch about the transverse (y) axis, and roll about the longitudinal (x) axis. This information is needed in connection with the use of acoustic methods to obtain estimates of the abundance of fish stocks. When using such methods it is important to know the target strength of the species being surveyed. The target strength of a fish depends critically on its attitude, tilt angles of a few degrees being sufficient to introduce errors of up to 60% into the estimates of stock abundance (Foote, 1980). It is for this reason that studies on the attitude of fish are thought to be important. In the past we have been well served by our colleagues who have invented the acoustic tags for which we asked: we suspect that the call for a tilt-tag will not go unanswered.

OTHER TELEMETRY SYSTEMS

The Lowestoft shipborne sector scanning sonar system is a sophisticated tool which requires extensive engineering skills and logistic support. However, it is reliable (Cushing & Harden Jones, 1979), and has three particular advantages. First, the picture-forming qualities enable the movements of fish to be observed in relation not only to fishing gear but also to topographical features such as sand waves and wrecks. Secondly, the facility to scan in the vertical as well as the horizontal mode enables the position of the fish to be very rapidly determined in all three dimensions: tracking studies have shown that depth is of the greatest importance. Thirdly, the research vessel is a mobile platform which can readily follow the unrestricted movements of fish in the open sea.

There is a worldwide interest in tracking fish continuously for long periods and several simpler systems have been developed which are suitable for other applications. These involve both radio and acoustic telemetry. Radio systems have certain advantages in fresh water and these are discussed by our colleague Dr Solomon in the next chapter. But in the sea high conductivity renders such systems useless and under these circumstances acoustics are essential.

The simplest acoustic tracking system employs a hand-steered directional hydrophone used in conjunction with a tag transmitting continuous trains of pulses. But the range of the transmitter from the receiver is not known and there is some uncertainty in determining the position of the fish. In small enclosed bodies of water this problem may be reduced by the use of fixed systems of location. Single hydrophones have been used in fish passes (Pincock *et al.*, 1974), while in a small Scottish loch two or more directional hydrophones have been used for triangulation (Young, Tytler & Holliday, 1975). More precise location has been achieved in a large sea loch, using the principle of hyperbolic navigation and an array of omnidirectional hydrophones (Hawkins, Urquhart & Smith, 1980).

Acoustic pressure waves are generated with miniature piezo-electric transducers, usually made from lead-zirconate-titanate, and operating at frequencies between 20 and 300 kHz. Within this band the lower frequencies give much greater range, but only at the expense of much larger transducers which then determine the size of the acoustic tag. A cylindrical transducer resonant at 20 kHz is approximately 4 cm in diameter compared with 0.3 cm for one resonant at 300 kHz. At the higher frequencies battery size becomes limiting. At 300 kHz the maximum working range is approximately 0.5 km; at 30 kHz ranges up to 5 km have been achieved under ideal conditions (Sciarrotta & Nelson, 1977). Low frequencies and large tags are only applicable to large fish such as sharks and tunas.

With the Lowestoft sector scanner the depth of a fish can be determined from the tilt angle θ of the transducer when in vertical mode. The depth of a target in the middle of the receiver beam is then equal to $R \sin \theta + C$, where R is the range in metres, θ the tilt angle, and C the depth in metres of the transducer below the surface (here 4 m). Simpler tracking systems require a depth telemetry tag. Several pressure-sensitive tags have been developed enabling, for example, the depth of the fish to be determined to ± 0.3 m in 30 m of water over a temperature range of 5° to 20°C (Pincock & Luke, 1975). But such devices have had only a limited biological application (Stasko & Rommel, 1974; Stasko, 1975). Other tags have been developed, which telemeter such environmental and physiological variables as internal and/or external temperature (Lawson & Carey, 1972; Coutant & Carroll, 1980), heart rate (Priede & Young, 1977; Storeton-West, Mitson & Greer-Walker, 1978), electrocardiogram (Kanwisher, Lawson & Sundnes, 1974) and ventilation rate (Eriksson & Ulveland, 1977; Oswald, 1978). Swimming speed can be deduced from tail-beat frequency (Young et al., 1972; Harden Jones, 1973; Stasko & Horrall, 1976), or measured directly (Voegeli & Pincock, 1980). The most sophisticated system is that devised for Pacific sharks by which on-line information on swimming speed and direction can be telemetered in various combinations with depth, light intensity and water temperature (Sciarrotta & Nelson, 1977; Standora & Nelson, 1977). But with the obvious exception of temperature, none of these systems has advanced to the point where it can be deployed routinely: biological applications have, accordingly, been limited.

Work at sea is expensive. Our conventional 300 kHz transponding acoustic tag costs about £100. Tags are returned by fishermen and one in four of those left in the sea is subsquently recovered. The compass tags bear a substantial charge for research and development and are currently costed at £1700 each. But here again some are recovered. The greatest single recurrent cost is undoubtedly that of a research vessel which at today's (1981) prices is charged at a rate of up to £3000 a day. There is no obvious and practical way to dispense with the research vessel and still meet the general principle that for any particular application the mobility of the telemetry system should match that of the target. Financial considerations have brought with them the need to define the objectives, to set the scientific priorities, and to assess the results. We have found the discipline helpful.

REFERENCES

Anon. (1974). *1973 Annual Report, Fisheries Research and Development Board.* London: HMSO.

Coutant, C. C. & Carroll, D. S. (1980). Temperatures occupied by ten ultrasonic-tagged striped bass in freshwater lakes. *Trans. Am. Fish. Soc.* 109: 195–202.

Cushing, D. H. & Harden Jones, F. R. (1966). Sea trials with modulation sector scanning sonar. *J. Cons. perm. int. Explor. Mer* 30: 324–345.

Cushing, D. H. & Harden Jones, F. R. (1979). Sector scanning sonar and the fisheries since 1969. In *Proc. Inst. Acoustics, Underwater Acoustics Group*: 1–10. Meeting on Progress in Sector Scanning Sonar, 18-19 December 1979. Printed at the University of Bath. (mimeo.).

Dickson, W. (1964). Performance of the Granton trawl. In *Modern fishing gear of the world* 2: 521–525. London: Fishing News (Books) Ltd.

Eriksson, L.-O. & Ulveland, S. (1977). Long-term telemetric system for gill-strokes of fish. *Aquilo* (Zool.) 17: 61–64.

Foote, K. G. (1980). Effect of fish behaviour on echo energy: the need for measurements of orientation distributions. *J. Cons. perm. int. Explor. Mer* 39: 193–201.

Greer Walker, M., Harden Jones, F. R. & Arnold, G. P. (1978). The movements of plaice (*Pleuronectes platessa* L.) tracked in the open sea. *J. Cons. perm. int. Explor. Mer* 38: 58–86.

Harden Jones, F. R. (1973). Tail beat frequency, amplitude, and swimming speed of a shark tracked by sector scanning sonar. *J. Cons. perm. int. Explor. Mer* 35: 95–97.

Harden Jones, F. R. (1980a). Acoustics and the fisheries: recent work with sector-scanning sonar at the Lowestoft Laboratory. In *Advanced concepts in ocean measurements for marine biology* 1: 409–421. Diemer, F. P., Vernberg, F. J. & Mirkes, D. Z. (Eds). Columbia: University of South Carolina Press.

Harden Jones, F. R. (1980b). The migration of plaice (*Pleuronectes platessa*) in relation to the environment. In *Fish behavior and its use in the capture and culture of fishes*: 383–399. Bardach, J. E., Magnuson, J. J., May, R. C. & Reinhart, J. M. (Eds). Manila: International Center for Living Aquatic Resources Management.

Harden Jones, F. R. (1981). Fish migration: strategy and tactics. In *Animal migration* XXX–XXX: 139–165. Aidley, D. J. (Ed.). [Soc. exp. Biol. Seminar Series, XX.] Cambridge: University Press.

Harden Jones, F. R., Arnold, G. P., Greer Walker, M. & Scholes, P. (1979). Selective tidal stream transport and the migration of plaice (*Pleuronectes platessa* L.) in the southern North Sea. *J. Cons. perm. int. Explor. Mer* 38: 331–337.

Harden Jones, F. R., Margetts, A. R., Greer Walker, M. & Arnold, G. P. (1977). The efficiency of the Granton otter trawl determined by sector-scanning sonar and acoustic transponding tags. *Rapp. P.-v. Réun. Cons. int. Explor. Mer* 170: 45–51.

Hawkins, A. D., Urquhart, G. G. & Smith, G. W. (1980). Ultrasonic tracking of juvenile cod by means of a large spaced hydrophone array. In *A handbook on biotelemetry and radio-tracking*: 461–470. Amlaner, C. J. & Macdonald, D. W. (Eds). Oxford: Pergamon Press.

Holley, M. L., Mitson, R. B. & Pratt, A. R. (1975). Developments in sector scanning sonar. *I.E.R.E. Conf. Proc.* 32: 139–153.

Ireland, L. C. & Kanwisher, J. W. (1978). Underwater acoustic biotelemetry: procedures for obtaining information on the behavior and physiology of free-swimming aquatic animals in their natural environments. In: *The behavior of fish and other aquatic animals*: 341–379. Mostofsky, D. I. (Ed.). London and New York: Academic Press.

Kanwisher, J., Lawson, K. & Sundnes, G. (1974). Acoustic telemetry from fish. *Fish. Bull., U.S.* 72: 251–255.

Lawson, K. D. & Carey, F. G. (1972). *An acoustic telemetry system for transmitting body and water temperature from free swimming fish.* Woods Hole Oceanographic Institution Technical Report, WHOI-71-67, 21 pp (mimeo.).

Mitson, R. B. (1978). A review of biotelemetry techniques using acoustic tags. In *Rhythmic activity of fishes*: 269–283. Thorpe, J. E. (Ed.). London and New York: Academic Press.

Mitson, R. B. & Cook, J. C. (1971). Shipboard installation and trials of an electronic sector-scanning sonar. *Radio electron. Engr.* 41: 339–350.

Mitson, R. B. & Storeton-West, T. J. (1971). A transponding acoustic fish tag. *Radio electron. Engr.* 41: 483–489.

Mitson, R. B., Storeton-West, T. J. & Pearson, N. D. (In preparation). *Trials of an acoustic transponding fish tag compass.*

Nelson, D. R. (1978). Telemetering techniques for the study of free-ranging sharks. In *Sensory biology of sharks, skates, and rays*: 419–482. Hodgson, E. S. & Mathewson, R. F. (Eds). Arlington, Va.: U.S. Office of Naval Research, Department of the Navy.

Oswald, R. L. (1978). The use of telemetry to study light synchronization with feeding and gill ventilation rates in *Salmo trutta. J. Fish. Biol.* 13: 729–739.

Pearson, N. D. & Storeton-West, T. J. (In preparation). *Design of an acoustic transponding compass tag for fish.*

Pincock, D. G. & Luke, D. McG. (1975). Systems for telemetry from free-swimming fish. *I.E.R.E. Conf. Proc.* 32: 175–186.

Pincock, D. G., Luke, D. McG., Church, D. W. & Stasko, A. B. (1974). An automatic monitor for detecting and recording passage of transmitter-fitted fish. *Can. Tech. Rep. Fish. Aquat. Sci.* No. 499: 1–59.

Priede, I. G. & Young, A. H. (1977). The ultrasonic telemetry of cardiac rhythms of wild brown trout (*Salmo trutta* L.) as an indicator of bioenergetics and behaviour. *J. Fish. Biol.* 10: 299–318.

Sciarrotta, T. C. & Nelson, D. R. (1977). Diel behavior of the blue shark, *Prionace glauca*, near Santa Catalina Island, California. *Fish. Bull., U.S.* 75: 519–528.

Standora, E. A. & Nelson, D. R. (1977). A telemetric study of the behavior of free-swimming Pacific Angel sharks, *Squatina californica. Bull. Sth. Calif. Acad. Sci.* 76: 193–201.

Stasko, A. B. (1975). Progress of migrating Atlantic salmon (*Salmo salar*) along an estuary, observed by ultrasonic tracking. *J. Fish. Biol.* 7: 329–338.

Stasko, A. B. & Horrall, R. M. (1976). Method of counting tailbeats of free-swimming fish by ultrasonic telemetry techniques. *J. Fish. Res. Bd Can.* 33: 2596—2598.

Stasko, A. B. & Pincock, D. G. (1977). Review of underwater biotelemetry, with emphasis on ultrasonic techniques. *J. Fish. Res. Bd Can.* 34: 1261—1285.

Stasko, A. B. & Rommel, S. A. (1974). Swimming depth of adult American eels (*Anguilla rostrata*) in a saltwater bay as determined by ultrasonic tracking. *J. Fish. Res. Bd Can.* 31: 1148—1150.

Storeton-West, T. J., Mitson, R. B. & Greer Walker, M. (1978). Fish heart rate telemetry in the open sea using sector scanning sonar. *Biotelemetry* 5: 149—153.

Voegeli, F. A. & Pincock, D. G. (1980). Determination of fish swimming speed by ultrasonic telemetry. *Biotelem. Patient Monit.* 7: 215—220.

Young, A. H. Tytler, P. & Holliday, F. G. T. (1975). New developments in ultrasonic fish tracking. *Proc. R. Soc. Edinb.* 75: 145—155.

Young, A. H., Tytler, P., Holliday, F. G. T. & MacFarlane, A. (1972). A small sonic tag for measurement of locomotor behaviour in fish. *J. Fish. Biol.* 4: 57—65.

DISCUSSION

J. R. Chandler (Southern Water Authority, Worthing) — If plaice and other tidally transported fish have no need for a directional response how can they determine what direction the tide is running in?

F. R. Harden Jones — Some form of synchrony is required with the state of the tide. In the spawning season full moon, slack water and sunset are in conjunction which would provide an entraining clue to turn a diurnal into a semi-diurnal rhythm. This would be the cue to start swimming. An olfactory clue might indicate when to stop swimming. I would stress that these are hypotheses.

Symp. zool. Soc. Lond. (1982) No. 49, 95–105

Tracking Fish with Radio Tags

D. J. SOLOMON

Ministry of Agriculture, Fisheries and Food, Fisheries Laboratory, Lowestoft, Suffolk, England

SYNOPSIS

Where they can be used, radio tags have several advantages over acoustic tags for fish tracking in fresh water. Attenuation of radio signals in sea water or brackish water is so high that radio tags are ineffective in these situations, and even in fresh water attenuation is such that their use in deep lakes is limited. However, in rivers where the fish is unlikely to be further than a few metres from the surface, radio tags excel; with the equipment currently in use by the Lowestoft laboratory, tags in shallow water can be detected at ranges up to 1.5 km. Triangulation using directional receiver aerials, or a riverside maximization of signal strength, allows location of tagged fish within 10–20 m. Use of an underwater aerial allows location to within a few centimetres.

Tags can be externally, gut or body cavity mounted, with integral or trailing aerials; those used in the study described in this chapter are inserted in the gut of salmon and sea trout and have integral aerials. The study involves monitoring the reactions of migratory fish to natural and regulated fluctuations in river flow.

INTRODUCTION

Approach

The aim of this chapter is to discuss the strategies and tactics of radio tracking experiments with fish, rather than to deal with equipment design *per se*. It is written from the viewpoint of a user rather than a designer of tracking gear. It is based largely on the writer's experience in developing and using a practical system of long-term tracking of freshwater fish in rivers. However, at several stages in the development the user's experience and changing requirements were fed back to the designers and builders of the equipment, thus allowing the rapid evolution of an effective system. Many of the equipment design criteria and experimental procedures may be of interest to others entering the field from a similar standpoint of ignorance.

Advantages of Radio Tracking

The advantages of radio over acoustic tracking in shallow fresh waters are so overwhelming that Stasko & Pincock (1977) suggested that the use of acoustic systems is becoming limited to situations where radio cannot be used. The main limitation of radio signals is their high attenuation in water. At the frequency allocated for radio telemetry in the UK (104.60–105.00 MHz) salt water is, for all practical purposes, opaque. Even fresh water of low conductivity causes such high attenuation that remote sensing depends on the signal emerging from the water directly above the fish and travelling most of the intervening distance in air. However, this is not a problem in many situations, e.g. streams and shallow lakes, where the fish is unlikely to be more than a very few metres from the surface. The main advantages of radio over acoustic telemetry for freshwater fish are:

(a) Lower power consumption, allowing longer tag life. One type of tag used in our study is powered by two 450 mAh lithium cells, and has a design life of six months. An equivalent 76 kHz acoustic tag has a life of about two weeks.

(b) Freedom from interference from ambient noise and entrained air bubbles. This is particularly important in rocky streams, with much broken water. Such conditions predominate in many rivers containing migratory salmonid fish, one of the main groups studied in freshwater tracking.

(c) No "line of sight" requirement underwater. Again, a major advantage in rocky rivers, or smaller streams with a "riffle and pool" configuration.

(d) Detection possible over considerable distances in air. This allows tags to be located periodically without any requirement for continual tracking. This gives the system tremendous flexibility, and allows the simultaneous use of many tags.

The River Fowey Experiment

Throughout this chapter reference will be made to the project on which we are using radio tags in this laboratory. Such illustration by example is considered to be valid, particularly as many of the problems of operation became apparent only after the project was under way, and required empirical solution. The experiment aims to elucidate the influence of natural and regulated river flows (discharge) on upstream migration of adult Atlantic salmon (*Salmo salar* L.) and sea trout (*S. trutta* L.). Migration is well known to be correlated with natural increases in discharge following rain (freshets), but information on the effects of reservoir releases (artificial freshets) is conflicting. The project is being undertaken on the River Fowey, in south-west England (Fig. 1). A major regulating reservoir is to be built on one of the tributaries, and the few years before its commission are being used to develop the tracking system, and to study the natural situation. Other aspects of the scheme, including observations on fish movements using electronic counters and conventional tagging methods, are being investigated by the South West Water Authority.

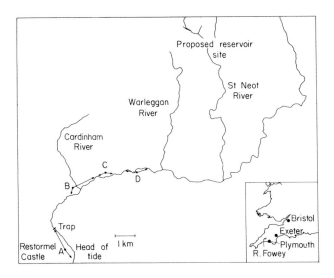

FIG. 1. The River Fowey system. Most fish tracked are caught and tagged at the trap of Restormel. Points A to D are examples of vantage points from which considerable lengths of river can be monitored.

PERFORMANCE

Attenuation of radio signals in water is predominantly dependent on conductivity, increasing rapidly as conductivity increases (Fig. 2). The only practical way that radio telemetry can be used for marine animals is by locating the transmitter on a float, attached to the animal by a line. Such methods are very limited in application, and are effectively limited to pelagic organisms; Priede (1980) used such a system for basking sharks. In fresh waters, the distance travelled in water by a signal to be received in air is effectively the depth of water over the fish, as only signals hitting the surface within 6° of normal penetrate, or are re-radiated, into the air. Attenuation in air is low, so providing a reasonable signal reaches the water surface, detection from considerable distances is possible. With the equipment we use, ranges of up to 1.5 km have been achieved where the tagged fish has been lying in shallow water. Overhanging banks, deep water and intervening land masses or dense vegetation all reduce the range from which the tag is detectable, but in practice a range of a few hundred metres is almost always achieved. Mammal biologists using tags of similar specification have achieved ranges in air under test conditions of several kilometres.

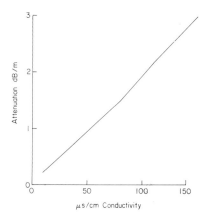

FIG. 2. Relationship between attenuation of radio signals at 100 MHz and conductivity of fresh water. The relationship holds over the range of frequencies 3 MHz to 200 MHz. Data from Stasko & Pincock (1977).

Tags can be located under water using an immersible aerial, but a combination of long wavelength and reflection of signals makes remote directional sensors impractical. We have developed a system using signal strength as an indicator of proximity for exactly locating tags – this is described on p. 102.

The tags we use transmit pulsed signals which are used only to locate and identify the tag – different tags have different pulse rates and radio frequencies. However, there is scope for the signal to carry other information in the same manner as acoustic tags. Possibilities include temperature, pressure (depth), heartbeat, opercular beat or other physiological or environmental variables.

The long life and long range of radio tags make locating tags, not followed continually, fairly straightforward. This is particularly useful when continual monitoring is not necessary, for example when spot checks on location will suffice, or a number of short-term continuous observations are required. Many tagged animals can be at large at the same time (up to 40 in our experiments) making the achievement of statistically significant results a simpler task. There is the possibility of using passive listening stations, which record the passage of tags past a fixed point; these would be particularly suitable for migratory fish in a river.

Use of radio tags may be limited where a combination of depth of water and high conductivity leads to high attenuation. In deep lakes it may be possible to locate tagged fish only while they are near the surface. A trailing wire acting as an underwater antenna, 75 m in length, was used by Knight, Marancik & Layzer (1977) for detecting tags at depths exceeding 45 m. The detectable range was about 10 m. At depths less than 45 m the tags could be detected from the surface. The radio frequency they used was 30 MHz, but no values for water conductivity were quoted.

TAGS AND TAGGING

Tags

Circuit design is outside the scope of this chapter. Generally, tags used for terrestrial tracking are effective for fish tracking, with suitable encapsulation or casings. They may be externally mounted, or inserted into the stomach or body cavity. Both integral aerials and trailing antennas (led through the oesophagus and gill cover with stomach-mounted tags) have been used successfully; the tags we have used so far have been inserted into the stomach and have integral aerials. They are enclosed inside polystyrene tubes with hemispherical ends. The cap is sealed on to the tube using model-makers' polystyrene cement, after the circuit is switched on and an hour or more before the tag is to be inserted. Many tags have been recovered after several weeks inside fish or under water, and none has shown any signs of leaking.

Two sizes of tag cases are used. The larger, 54 mm × 17 mm diameter, encloses the tag with two 450 mAh lithium cells, which has a maximum life of six months. The other, 36 mm × 15 mm diameter, has two 220 mAh mercury cells with a maximum life of three months. The signal is pulsed, the pulses being of about 50 ms duration with a repetition rate of between 40 and 100 per minute. Up to four tags with different repetition rates can be used at once on a single nominal radio frequency. The radio frequencies are separated nominally by 5 kHz intervals. Problems with ensuring that the transmitted frequency is close to the nominal frequency, and gradual drift in both pulse repetition rate and apparent radio frequency, make closer spacing of radio frequencies and pulse repetition rates inadvisable.

Insertion of Tags

With fish of less than about 2 kg body weight, the tag is simply pushed down the oesophagus with a shaped dowel rod. The rod diameter is slightly less than that of the tags, and the end is hollowed so that the rounded end of the tag fits into it. With larger fish, the tag can twist in the oesophagus so that the rod slips past; for these fish, a slightly more complex inserter is used. This comprises a shaped plastic tube with an internal diameter slightly larger than the tag. This is inserted into the stomach via the oesophagus, and the tag is pushed down the tube using a plunger. The fish are anaesthetized for this operation using either insertion method.

For the first experiments fish were captured by electric fishing in the lower reaches of the river. Latterly, the fish have been taken in a temporary trap near the tidal limit operated by the South West Water Authority on selected nights during the migration season (June to September). A problem arose with a high proportion of the salmon dying within a week of tagging. At least some of these mortalities were due to damage done to the gut by the tag during insertion or *in situ*. Much of the problem was removed by improving the methods of insertion and the shape of the tag casing — the cases are now completely smooth and round-ended. However, a proportion of the fish are still dying, and others are seen to be unhealthy, with attacks of fungus. This is likely to be caused by handling the fish during a time of physiological stress, as they are adapting to the change from salt to fresh water. It is hoped that refinements in trapping and handling methods will ameliorate this situation.

The sea trout have presented different problems. None has at any time appeared to have been adversely affected by trapping, handling and tagging. They appear to resume their upstream migration within 24 h of release, and some have been caught at traps further up the river. However, most sea trout regurgitate their tags after between five and ten days of tagging. During this time they may migrate many kilometres upstream and provide useful results, but this situation is of course using only a fraction of the potential of the tag system. The time delay between insertion and regurgitation may be linked to normal digestion rates, and the fish becoming aware that the bulk in its stomach is not being digested. The two obvious solutions to this problem are external mounting or body cavity mounting of tags. External mounting would be used with reluctance, as there is potential for the tag to catch on rocks or debris, and for the wounds to become infected or to be made worse by the drag and buffetting of the tag in fast water. Body cavity insertion involves surgery under field conditions, but may provide the answer; it is being investigated for possible future use.

TRACKING

Approach and Methods

Fish may be tracked continuously, or spot checks taken at regular intervals on their position. The latter approach still leaves the possibility of continuous tracking during events of particular interest, for example freshets in our study. Low water conditions may prevail for days or weeks after the fish have been tagged, but because tags have a long life and can be easily located, when conditions of interest do occur there is a good population of marked fish to observe. This approach has the added advantage that the fish should have overcome the trauma of capture and tagging and so be behaving naturally by the time that the critical observations are made.

Several methods have been used for coarse location of radio tags in rivers. Use of non-directional aerials from light aircraft allows location within a few hundred metres and long stretches of river can be covered. Power & McCleave (1980) completed a successful study of movements of Atlantic salmon in the Penobscot River, Maine, using almost exclusively such airborne observation. They were able to monitor several hundred kilometres of river twice weekly during the periods of study. Where such long distances are involved, location within a few hundred metres is adequate.

More accurate remote locating can be undertaken using directional aerials from such vantage points as high ground and buildings. Usually a single bearing is all that is required; in most situations such a line will only intersect the river once. On the River Fowey, several

of the tracking spots can cover 2 km of river; some are shown in Fig.
1. There are other reaches which require more intensive cover, from
lower vantage points or bridges. There are several stretches where
roads run close to the river. Here, spot checks from lay-bys or continuous listening using a non-directional aerial from a moving vehicle
can cover considerable distances fairly quickly. A few stretches,
particularly where the river runs in a deep gorge, are so inaccessible
that a search on foot is the only alternative to an airborne survey.

If a more accurate fix is required, once a tag is located by coarse
methods, walking the bank until signal strength is maximized can
indicate location within about 10–20 m. More precise still, an underwater aerial we have developed allows the tag to be pinpointed
within a few centimetres. This has proven particularly useful in
checking whether or not the tag is still inside a fish, and to allow
recovery of regurgitated tags or those inside dead fish. Details of the
underwater aerial are to be published elsewhere (Solomon &
Storeton-West, in preparation); it consists of a simple loop of coaxial
cable about 12 cm in diameter, mounted at right angles to the end of
a pole. The aerial is probed underwater, and the signal strength
increases exponentially as the tag is approached. The signal is
maximized when the tag is actually within the loop. This gear has
allowed the recovery of many regurgitated tags, several in dirty water
conditions where the tag could not be seen until it was lifted from
the water.

Equipment

The receiver used in our studies is manufactured to our specification
by Mariner Radar Ltd (Mariner M57). It is continually tunable over
the range 104.70–104.75 KHz, and allows identification of transmitter frequency within 100 Hz.

Non-directional aerials used in fish tracking include loops
(McCleave, Power & Rommel, 1978), single dipole (our study, for
aircraft tracking) and whip (our study, for road vehicle tracking).
Directional aerials include Yagi and H-Adcock arrays. Although the
Yagi has a higher gain, we have predominantly used H-Adcock aerials
as they can be folded simply to allow on-foot tracking through bank-
side vegetation.

The underwater aerial has been described above, where its use is
discussed.

INTERPRETATION OF RESULTS

A major problem with mammal tracking has been the difficulty in recording, presenting and analysing observations on movements in two or even three dimensions. Fish in a river are effectively limited to movements in one dimension in gross terms, and this makes experimental design and interpretation of results much simpler. With migratory fish the situation is simplified even further as the movements are likely to be in one direction only at any stage of the migratory cycle. However, even then it is important to have the aims of the experiment clear, and to establish how the results obtained can be used to test a hypothesis. With the Fowey experiment, the hypothesis would be that artificial freshets are not as effective as similar sized natural freshets at stimulating upstream movements. Significance tests could be done on the proportion of fish stimulated to migrate upstream by each freshet type, or on the average distance travelled.

Although a clear objective is the primary justification for the cost and effort of a tracking programme, the experimenter should be aware of the potential for other, perhaps unexpected, facets of behaviour to be revealed. Examples could be diurnal variations in behaviour, more limited ranges of migration than expected, tendency for cohesive groups to be maintained, and the unexpected influence of such environmental variables as barometric pressure or phases of the moon.

Because of the problems experienced so far on the Fowey project (pp. 100–101) there are as yet few results to report. We have some information on rates of travel of sea trout, including the observation that they appear to move fairly short distances each night, even in low flow conditions. Observations on the lies occupied by fish are very interesting — they are basically the pools well known to fishermen, plus some much smaller holes under overhanging banks. There are some stretches of river which appear to offer no holding lies to fish, which could simplify searching for tagged fish. Locations of some sea trout or their regurgitated tags are shown in Fig. 3, with the time in days after tagging.

Problems remain, but potential solutions for those which are known are in hand; many problems have already been solved. If we can overcome them, we will have an effective research tool with which to approach a long-standing problem of salmonid fishery management.

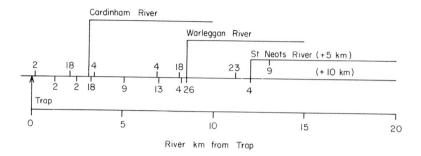

FIG. 3. Time (days) taken by sea trout to reach various points in the Fowey system, June—September 1980. Times shown are between tagging at the trap and relocation, so they are maximum times for movement. At least a further 10 km of the main river and 5 km of the St Neots River are accessible to sea trout.

ACKNOWLEDGEMENTS

I am grateful to the Electronics Section at the Fisheries Laboratory for supplying the equipment required for this study, and for their prompt and effective response to changes in specifications and requirements.

Particular thanks are due to Hugh Sambrook, South West Water Authority, for making available the trapping facilities and his invaluable help with several aspects of the project. I also wish to thank Mr Roger Furniss and his staff, South West Water Authority, for their help and support.

REFERENCES

Knight, A. E., Marancik, G. & Layzer, J. B. (1977). Monitoring movements of juvenile anadramous fish by radio telemetry. *Progve Fish Cult.* **39**: 148—150.

McCleave, J. D., Power, J. H. & Rommel, S. A. (1978). Use of radio telemetry for studying upriver migration of adult Atlantic salmon (*Salmo salar*). *J. Fish Biol.* **12**: 549—558.

Power, J. H. & McCleave, J. D. (1980). Riverine movements of hatchery-reared Atlantic salmon (*Salmo salar*) upon return as adults. *Envir. Biol. Fish* **5**: 3—14.

Priede, A. G. (1980). An analysis of objectives in telemetry studies of fish in the natural environment. In *A handbook on biotelemetry and radio tracking*: 105—118. Amlaner, C. J. & Macdonald, D. W. (Eds). Oxford and New York: Pergamon Press.

Stasko, A. B. & Pincock, D. G. (1977). Review of underwater biotelemetry, with emphasis on ultrasonic techniques. *J. Fish. Res. Bd Can.* **34**: 1261—1285.

DISCUSSION

J. White (Loughborough University) — (i) Why did you use an RF system in preference to a sonar system? (ii) What are the relative costs of both systems?

D. J. Solomon — (i) Battery size is more limiting with sonar systems. But as a general rule sonar is only suitable for freshwater use in deep water such as lakes and radio is more suitable for shallow rivers. (ii) Receivers for sonar and radio systems are approximately the same at £400–£500 each. Transmitter costs are also approximately equal.

V. B. Kuechle (Minnesota) — Consideration of which system should be used depends on the application. Costs are approximately equal. If the conductivity of water in the study area is over 50 mS per metre and the water is deep, sonar is more suitable. If the water is turbulent and contains many bubbles which would cause noise in the sonar system then radio has to be used.

H. G. Lloyd (MAFF) — Do salmon migrate at night in all rivers or only in smaller rivers?

Solomon — We do not know, but electronic counts of fish suggest that in clear water conditions there is a tendency to migrate at night while in dirty water conditions migration takes place around the clock.

Symp. zool. Soc. Lond. (1982) No. 49, 107–128

Telemetry of Physiological Variables from Diving and Flying Birds

P. J. BUTLER and A. J. WOAKES

Department of Zoology and Comparative Physiology, University of Birmingham, Birmingham, England

SYNOPSIS

Completely implantable transmitters (a two-channel FM, and a single-channel PIM) have been built specifically to study changes in the heart rate and respiratory frequency of free range birds.

The FM system confirmed and extended earlier observations, made with an externally mounted, commercially available transmitter, that when pochards or tufted ducks dive naturally there is no prolonged reduction in heart rate (bradycardia) below the level recorded when the animals are swimming on the water. This is contrary to the accepted view, based largely on experiments performed on restrained, forcibly submerged domesticated ducks, that bradycardia and selective peripheral vasoconstriction occur during submersion as part of an oxygen conserving mechanism. Underperfused tissues are thought to metabolize anaerobically, while normally or hyperperfused tissues (brain and heart) metabolize aerobically. The results from naturally diving ducks, together with approximate calculations of oxygen stores and metabolic rate while submerged, indicate that the ducks may be completely aerobic during natural dives of normal duration. Ciné photography has allowed us to relate the behaviour of the ducks to changes in heart rate.

Long-term (2–3 months) recordings of heart rate in tufted ducks have been made, using the PIM transmitter, before and after denervation of the carotid bodies. In distinct contrast to the situation during enforced submersion of restrained domestic ducks, these organs do not appear to play a major role in controlling heart rate during natural dives.

The two-channel transmitter has been implanted into barnacle geese that had previously been trained, by imprinting, to follow a human standing in the back of a truck travelling around an oval track. Heart rate, respiratory frequency and wing beat frequency (from ciné film) do not change with flight velocity, but there is a more or less constant 3:1 correspondence between wing beat frequency and respiratory frequency. During periods of gliding, both heart rate and respiratory frequency decrease.

INTRODUCTION

According to Schmidt-Nielsen (1977), a "turning point in modern zoological science" has been the movement away from studies on anaesthetized or restrained animals in the laboratory towards investigations into how animals function in nature. Studies of some natural activities, such as diving, have for many years been performed on restrained animals, but in cases where this was not possible, e.g. flight of birds, information on physiological variables was scant until methods for recording from the performing animal were devised. Recently, radio telemetry has been used in studies of diving and flying and it is already clear that it has an important part to play in our understanding of the animal in its natural environment.

DIVING

Historical Background

When diving or flying, a bird is exercising, but when diving it has only the oxygen contained within its body at its disposal. On the basis of the total amount of oxygen contained within the body and the survival times of restrained animals with their heads held under water, it was suggested by a number of workers at the end of the last century (see Scholander, 1940 for references) that the oxygen stores are insufficient to enable aerobic metabolism to continue at its normal level during diving. For example, Scholander quotes Richet (1899) who calculated that a duck weighing 1.5 kg has an oxygen store of 90 ml. Thus at a resting consumption of 30 ml min^{-1} the store would last for 3 min, but the duck could "dive" (i.e. survive with its head held under water) for 20 min. It was concluded that oxygen consumption must be greatly decreased during a dive.

Perhaps the first indication as to how this could be achieved in diving animals was given by a Fellow of the Zoological Society of London, R. H. Burne, when in 1909 he suggested that in the walrus, *Odobenus rosmarus*, blood flow to "the vegetative as opposed to the voluntary organs" may be reduced or cease altogether during a dive. If this were the case, total oxygen consumption would decline as blood flow would be maintained only to a few organs, e.g. CNS and skeletal muscle. A similar idea was proposed by Irving (1939) and the broad details of the adjustments in the cardiovascular system and in the metabolizing cells during enforced dives of birds and mammals, were contained in a series of papers published by Scholander and Irving in the 1940s.

The fact that oxygen consumption at the end of an enforced dive was less than would be expected if aerobic metabolism had continued at normal levels during the periods of submersion, supported the idea that oxygen usage was reduced during diving (Scholander,

1940). The fact that during enforced submersion the concentration of lactic acid increased to a considerably greater level in skeletal muscle than it did in the blood, whereas upon surfacing it increased substantially in the blood, indicated that perfusion of the muscles had been reduced during submersion and that they had metabolized anaerobically (Scholander, 1940; Scholander, Irving & Grinnell, 1942). Central arterial blood pressure remained close to normal during enforced submersion, despite the reduced perfusion of certain tissues, as there was an accompanying reduction in heart rate (brady-cardia) and hence in cardiac output (Irving, Scholander & Grinnell, 1942; Murdaugh *et al.*, 1966). Recent work on restrained mammals and birds has generally confirmed and extended these observations. In mallard and domestic ducks, *Anas platyrhynchos*, blood flow during submersion is reduced to almost all tissues with the notable exception of the CNS, heart and adrenal glands, to which it increases (Johansen, 1964; Jones *et al.*, 1979). In the Weddell seal, *Leptony-chotes weddelli*, blood flow to the brain remains unchanged during submersion whereas coronary perfusion decreases in proportion to the reduction in myocardial work (Zapol *et al.*, 1979). The efficacy of these adjustments in conserving oxygen can be illustrated by the fact that in young harbour seals, *Phoca vitulina*, depletion of oxygen in the blood and lungs during submersion was slower in animals with lower heart rates (Irving, Scholander & Grinnell, 1941) and that blood oxygen tension in submerged domestic ducks declined faster after abolition of the bradycardia or peripheral vasoconstriction (Butler & Jones, 1971).

Thus, observations on restrained birds and mammals with their heads or whole bodies submerged into water have led to the generally accepted idea that, when diving, their metabolic rate and in particular their usage of oxygen, decreases as blood flow to most of the body is reduced. These underperfused areas undergo anaerobiosis. Blood flow to the brain is at least maintained and if coronary blood flow does decrease it is in direct proportion to the decline in cardiac work. Blood pressure is maintained at near normal levels despite the selective vasoconstriction as the result of an accompanying reduction in heart rate and cardiac output. The bradycardia is, in fact, often taken as an indication of the occurrence of the other adjustments to diving (Påsche & Krog, 1980).

Forcible submersion of seals, *Phoca vitulina, Halichoerus grypus, Cystophora cristata* (Scholander, 1940; Elsner, 1965), muskrats, *Ondatra zibethica* (Drummond & Jones, 1979) and coypus, *Myocas-tor coypus* (Folkow, Lisandar & Öberg, 1971) causes an immediate, substantial reduction in heart rate. On the other hand in the mallard duck or its domesticated varieties, there is a gradual decline in heart rate to $25-40$ beats min^{-1} over $30-60 s$ (Butler & Jones, 1971; Butler & Taylor, 1973). The progressive nature of the bradycardia (and other cardiovascular adjustments) in birds is related to the progressive changes in blood oxygen and carbon dioxide tensions during the period of apnoea (Butler & Taylor, 1973). Sectioning of the nerves to the arterial chemoreceptors (carotid bodies) of domestic ducks abolishes the majority of the bradycardia that occurs during enforced submersion (Jones & Purves, 1970). It might be expected that to be maximally effective, any oxygen conserving mechanism would come into operation before there were any changes in blood gas tensions and not in response to them. In order to determine whether or not bradycardia was immediate during natural dives, it was decided to use radio telemetry to record heart rate from free diving ducks. There were, therefore, two new requirements; a telemetry system and some ducks that would dive spontaneously.

Telemetry Systems

At first a commercially-available FM transmitter was attached to an Australian white eye duck, *Aythya australis*. It was placed in a watertight box which was held in position on the back of the duck by passing straps underneath the base of the wings. This arrangement was not ideal, as the birds never accepted the presence of the box and spent much of their time pecking at it, but it did lead to the discovery that there is not a maintained bradycardia when these birds dive naturally (Butler & Woakes, 1976). This prompted the designing and building of an implantable, two-channel FM transmitter for monitoring heart rate and respiratory frequency from free range birds (Woakes & Butler, 1975). After compromising between minimum size, maximum transmission range and maximum battery life, the transmitter could be operated continuously for six days.

A bipolar electrode lay close to the heart underneath the sternum and the e.c.g. was transmitted directly. A small thermistor placed in the lumen of the trachea monitored the variation in temperature of the inspired and expired air and modulated the frequency of a subcarrier oscillator. The lead from the thermistor was taken under the skin and the transmitter itself lay in the abdominal cavity. Initial problems during use included breakage of the leads and difficulties with reed switches which were used to extend the life of the RM625 mercury cell. These were overcome by using 4-start-coiled stainless steel wires for the leads and a mercury film switch to replace the reed switch. The former have proved to be totally reliable whereas the latter continued to present some problems. It was designed for coil operation and had to be modified by adding a small magnet to polarize the moving shuttle in order to increase its sensitivity to a permanent magnet. Even so, it did not always operate successfully, it was the most expensive single component and the birds had to be caught in order to switch the transmitter on and off, before and after a series of recordings. Nevertheless it was an essential component because the birds seldom started to dive until six to seven days after the implantation of the transmitter and it was not desirable to increase the bulk or mass of the transmitter by using a larger battery (see Woakes, 1979). An unavoidable limitation to the duration of recording was the eventual coating of the respiratory thermistor with tracheal secretions, which increased its response time.

The two-channel FM transmitter was used during an initial series of observations on diving ducks and with flying geese. In the latter case it was possible to replace the single RM625 mercury cell with a separate battery pack containing two RM625s. This doubled the continuous operating life of the system and allowed us to dispense with the mercury switch. Recently, when a single channel e.c.g. transmitter was required for experiments lasting up to three months, a transmitter using pulse interval modulation was built. A design by Fryer (1970) was modified to suit the available components and the receiver (Sony CRF 5090) being used. As the e.c.g. can contain major frequency components of up to 100 Hz, a pulse repetition frequency (effectively the sampling rate) of about 400 Hz was chosen. The pulse duration is approximately $50 \mu s$, which is the shortest value that the receiver can follow without undue distortion. The transmitter has a range of 15 m, a current consumption of approximately $40 \mu A$ and a continuous operating life of at least 14 weeks. Its dimensions are 21.5 x 11.5 x 5.5 mm with a mass of 3.9 g, including the battery but without leads and encapsulation. The construction techniques were similar to those outlined for the two-channel FM transmitter (Woakes & Butler, 1975; Woakes, 1979). A special decoder was designed to demodulate the output from this pulse interval transmitter, and details are given in Woakes (1980).

Naturally Diving Birds

The two-channel FM transmitter was implanted into pochards, *Aythya ferina*, or tufted ducks, *A. fuligula*, which were released initially onto a shallow pond (0.65 m depth) and latterly onto a deeper pool (2 m depth). Although the average durations of the dives were different, being longer on the deeper pool (*cf.* Dewar,

1924), the changes in heart rate and respiratory frequency were similar in each case (Butler & Woakes, 1979; Butler, 1979). Spontaneous dives often occurred in a series and there could be in excess of 40 dives in quick succession, particularly on the shallower pool Typically, the first dive in a series was heralded by increases in heart rate (see Figs 1, 2 and 4) and respiratory frequency. Immediately upon submersion there was an instantaneous reduction in heart rate below the level that was recorded before the pre-dive increase. Heart rate then increased during the early stages of the dive and more or less stabilized at a level which, 20 s after submersion, was similar to the value recorded during high levels of swimming activity at the surface (Fig. 1). The mean heart rate of 237 beats min^{-1} 20 s after natural submersion compares with a rate of 47 beats min^{-1} 20 s after forcible submersion of restrained tufted ducks and 110 beats

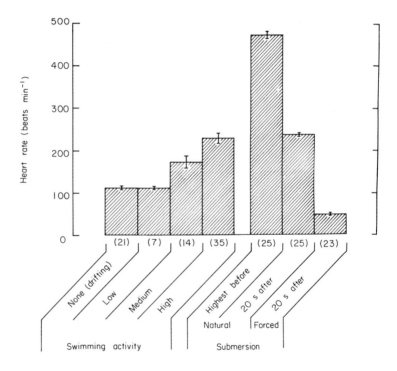

FIG. 1. Mean values of heart rate in tufted ducks, *Aythya fuligula*, during various levels of swimming activity (*cf.* Butler & Woakes, 1979), just before and 20 s into spontaneous submersions which were in excess of 30 s duration (Butler, 1979) and 20 s into forced submersions of the head of restrained ducks. Vertical lines associated with each column are ± S.E. of mean. The number in parentheses beneath each column indicates the number of events contributing to the mean value. The value for high swimming activity is for pochards. In the other categories of swimming activity there was no significant difference in heart rate between pochards and tufted ducks (Butler & Woakes, 1979).

min^{-1} during low-level swimming activity of free range ducks (Fig. 1). Upon surfacing there was an immediate increase in heart rate and rapid breathing. The following dive often occurred while these variables were still elevated.

By taking ciné films of tufted ducks on a glass sided tank (1.55 m deep) it has been possible to relate the behaviour of the animal to changes in heart rate (Fig. 2). At the end of the period of cardiac acceleration before submersion, the duck extended the web of both feet and placed them beneath the sternum. They were then quickly thrust backwards as the neck extended and the whole body arched itself into the water. During this period, before the head was submerged, there was an extension of the interval between heart beats, i.e. the heart began to slow before submersion of the nostrils. This confirms the earlier observation that heart rate slows before the cessation of ventilation (Butler & Woakes, 1976). Also, just before submersion, the bird compressed its feathers close to its body and while it was swimming down, a stream of bubbles was left in its wake (Fig. 2). During the period of descent, which for five dives lasted an average of 2.7 s and was at a mean velocity of 0.57 m s^{-1}, the legs beat in phase at an average frequency of 274 beats min^{-1} and heart rate was at its lowest. Once at the bottom, the body was held somewhere between 45° and the vertical and the mean frequency of the leg beat decreased to 188 beats min^{-1}. Heart rate increased slightly during the dive, so that after 12 s it was, on average, 193 beats min^{-1}

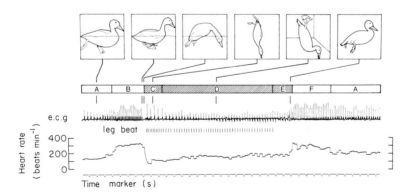

FIG. 2. Relationship between the behaviour of a tufted duck, *Aythya fuligula*, and changes in heart rate during a spontaneous dive on a glass-sided tank, 1.55 m deep. From above, downwards: tracings of duck from ciné film showing from left to right, swimming, preparing to dive, moment of submersion, descending, feeding on bottom, surfacing (10 cm from surface); time periods of A-swimming on surface, B-cardiac acceleration before submersion, C-descent, D-feeding on bottom, E-surfacing, F-cardiac acceleration following surfacing; e.c.g; end of power stroke of each leg beat cycle; heart rate; time marker(s). The lines between the pictures of the ducks and the time boxes join coincident points in time.

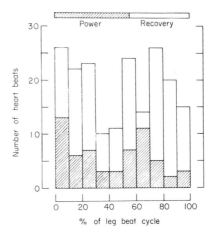

FIG. 3. Histogram showing the position in the leg beat cycle at which the QRS wave of the e.c.g. signal occurred during the period that a tufted duck was feeding on the bottom of a tank of water 1.55 m deep. Open columns refer to the total period at the bottom, cross-hatched columns refer to the period 12 s after submersion. The horizontal bar at the top shows the proportion of the power (55.4%) and recovery phases of the leg beat cycle.

and there was a clear 1:1 correspondence between leg beat and heart beat (Fig. 2).

During the whole time that the ducks were at the bottom, the power stroke occupied, on average, 55.4% of the leg beat cycle and the heart did not beat at any particular phase of the leg cycle. However, after the ducks had been at the bottom for 12 s, when the 1:1 relationship between leg beat and heart beat was apparent, the heart tended to beat more often at the beginning of the power stroke or at the beginning of the recovery stroke (Fig. 3). The significance of this is not clear. Although one group of leg muscles was in a relaxed state when the heart beat and thus their perfusion was not impeded by occlusion of the capillary beds, the other group of muscles was contracting. With a 1:1 frequency relationship, it might have been expected that the heart would beat when the muscles involved with the power stroke were relaxing, particularly at the end of a dive when arterial oxygen tension might have been reduced. To surface the bird stopped beating its legs and rose passively by its own buoyancy (*cf.* Dewar, 1940; Brooks, 1945). Ascent time was, on average, 2.84 s at a mean velocity of $0.55 \, \text{m s}^{-1}$. Heart rate began to increase before the bird penetrated the water surface, thus confirming the observations of Butler & Woakes (1979) and Butler (1979). On one or two occasions, when "fun diving," the birds used their wings to increase their under water velocity.

Both on the outdoor ponds and on the indoor tank it was noted that the ducks never appeared to be exhausted after a series of natural dives, no matter how many there were in a sequence. This is not consistent with their having metabolized anaerobically and having accumulated lactic acid during submersion. This subjective assessment, together with the lack of bradycardia during natural dives, even when they exceeded 30 s duration (Butler, 1979), indicates that the oxygen conserving adjustments that have been shown to occur in forcibly submerged ducks may not occur during natural dives. The implication is that metabolism is completely aerobic during normal dives in nature. One way to investigate the possibility of such a proposal is to estimate the available oxygen present in the body before submersion and the oxygen utilization during the period spent under water. Unfortunately there are no relevant data on the diving ducks used in the present studies, but information from other birds (mostly ducks) can be usefully adapted.

Using the exponent of 0.9 from the allometric equation for the volume of the respiratory system in birds (Lasiewski & Calder, 1971), and the measured functional volume for a 2.5 kg domestic duck (Scheid, Slama & Willmer, 1974), a 0.8 kg tufted duck (or pochard) would have a functional respiratory volume of 190 ml. As the ducks hyperventilate before a dive, the oxygen concentrations measured in the air sacs of starlings (*Sturnus vulgaris*) at the beginning of a flight (Torre-Bueno, 1978) may be taken as representative of those present in ducks just before diving. If the posterior sacs represent 45% of the total respiratory volume (Scheid *et al.*, 1974) then the gaseous O_2 stores at the start of a dive are 33.6 ml. Taking the air sac values in a mallard duck when breathing oxygen at 45 mmHg (Colacino, Hector & Schmidt-Nielsen, 1977), the P_{O_2} in the air sacs of the tufted duck at surfacing might be approximately 30 mmHg, leaving 7.4 ml of oxygen in the respiratory system. The available oxygen in the respiratory system would then be 26.2 ml. Assuming that the hyperventilation and tachycardia preceding the dive serve to raise the oxygen stores, arterial blood may reach 100% saturation and venous blood may be 70% saturated, which in a domestic duck is approximately equivalent to $P_{v}O_2$ of 65 mmHg at pH 7.43 (Scheipers, Kawashiro & Scheid, 1975). If $P_{a}O_2$ at the end of a dive is close to that in the air sacs (say 27 mmHg), then arterial blood will be approximately 25% saturated at a pH of 7.38 (Scheipers *et al.*, 1975). Venous blood can be assumed to be almost completely deoxygenated and to constitute 70% of the blood volume. Taking values from mallard ducks of 20 vol% for oxygen capacity (Black & Tenney, 1980) and 11% for blood volume as a proportion of body weight (Bond & Gilbert, 1958), then oxygen stores available during a dive would be 4.3 ml in arterial blood and 9.9 ml in the venous blood. Oxygen bound to myoglobin is probably only going to be of use in the active muscles themselves and assuming a leg muscle mass of approximately 60 g (Prange & Schmidt-Nielsen, 1970) and an oxygen carrying capacity of 30 ml kg^{-1} muscle (Schmidt-Nielsen, 1979), this will amount to an oxygen store of 1.8 ml. Thus the total, usable oxygen store is calculated to be 42.2 ml.

Prange & Schmidt-Nielsen (1970) found that at its maximum sustainable swimming speed, the mallard duck had an oxygen uptake of 3.9 l kg^{-1} h^{-1} (approximately four times the resting values). Using the allometric power coefficient of 0.73 for oxygen uptake during swimming or running in birds (Butler, 1981), the 0.8 kg tufted duck under similar conditions would consume 0.92 ml O_2 s^{-1}. Assuming this to be its maximum rate of oxygen usage when submerged, its stores would last at least for 45.7 s. Using similar calculations, the oxygen stores of a tufted, or pochard, duck weighing 0.6 kg would last 43.8 s. These calculations, although very approximate, indicate that all of the natural dives of pochards and tufted ducks that we have observed could have been completely aerobic. The maximum dive duration that Dewar (1924) recorded for these ducks was 40 s for a tufted duck. It is illuminating to note that Richet's (1899) duck with its oxygen store of 90 ml and a resting consumption of 30 ml min^{-1}, could have metabolized aerobically for 45 s at four times its resting rate when diving naturally. That is, if it ever did dive naturally; it was most probably a domesticated variety of the mallard duck, which rarely dives.

As well as obtaining the physiological data from freely diving ducks, we were also faced with the fact that when left to their own choice, ducks (and most birds for that matter) do not normally dive for the prolonged periods that have been used under experimental conditions. This had been appreciated by other workers, in particular by Eliassen (1960). He refuted the idea that vasoconstriction and anaerobiosis occur in the muscles of birds during natural diving. He based his conclusion on his own laboratory experiments, his interpretation of previous experiments by other workers and calculations of oxygen stores and oxygen usage during dives of normal duration. His proposals were largely rejected and the idea of major areas of vasoconstriction and anaerobiosis during diving persisted (Andersen, 1966).

As well as our own recordings from ducks, the use of radio telemetry with other birds has indicated differences in heart rate during natural and forced submersion. Cormorants, *Phalacrocorax auritus*, do not show a bradycardia during natural dives (Kanwisher, Gabrielsen

& Kanwisher, 1981) and although there is a bradycardia during natural diving in the penguins, *Pygoscelis papua* and *P. adeliae*, it is not as severe as during forced submersion (Millard, Johansen & Milsom, 1973). Even this mild bradycardia in penguins may not have been associated with anaerobiosis. Certainly Millard *et al.* (1973) found little reduction in arterial oxygen tension at the end of natural dives,

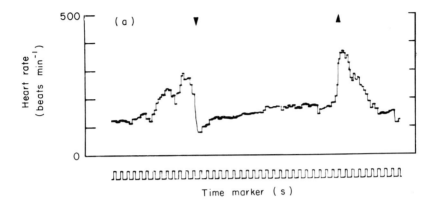

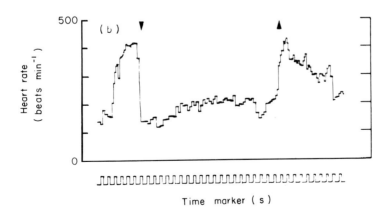

FIG. 4. Changes in heart rate of a tufted duck during spontaneous submersion on an outside pond (2 m deep). (a) Before, (b) after, bilateral denervation of the carotid bodies. Arrows indicate time of submersion and surfacing.

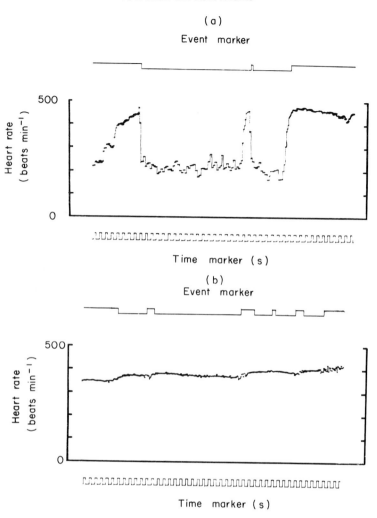

(a)

Event marker

Time marker (s)

(b)

Event marker

Time marker (s)

FIG. 5. Changes in heart rate during escape dives of two different tufted ducks. (a) Untreated, (b) after i.v. injection of atropine ($2\,mg\,kg^{-1}$). Downward deflections of event marker indicate periods of submersion.

and the Weddell seal exhibits mild bradycardia during relatively short dives (Kooyman & Campbell, 1972) but any dive of less than 20 min duration is completely aerobic in nature (Kooyman *et al.*, 1980).

The single-channel PIM transmitter was implanted into tufted ducks so that heart rate during natural dives could be recorded before and after denervation of the carotid bodies. Preliminary indications are that bilateral carotid body denervation does not affect the occurrence or duration of dives, neither does it have any dramatic effect on heart rate during submersion. Fairly typical changes in heart rate during spontaneous dives before and after denervation are shown in Fig. 4. Heart rate is higher before and during the dive following denervation and it takes longer to return to normal after surfacing. A consistent, but inexplicable feature of the denervated animals is the prominent, transient reduction in heart rate just before surfacing. Following injection of atropine $(2\,mg\,kg^{-1})$ into a tufted duck, the bird was placed on water in the glass sided tank and induced to dive by chasing it with a net (*cf.* Butler & Woakes, 1979). This was merely to ensure that it dived before the effect of the atropine had worn off. The reduction in heart rate, from its elevated level, upon submersion was abolished after the injection of atropine (Fig. 5). If anything, heart rate increased with successive dives. Thus, although naturally diving ducks do not exhibit bradycardia in the classical sense, the reduction in heart rate from the pre-dive tachycardia is mediated by the vagus nerves. The carotid bodies are not involved in this part of the response, but they may exert a slight negative influence before and during submersion.

FLIGHT

Historical Background

Few birds dive and compared with marine mammals their perform-ance under water is not outstanding. Most birds fly and some of them perform most impressive feats of physical endurance. The Pacific golden plover, *Pluvialis dominica fulva*, may perform nonstop flights for 3800 km (Johnston & McFarlane, 1967), while a flock of bar headed geese was observed flying over the summit of Mt. Everest, an altitude of approximately 9000 m (Swan, 1961). It is not sur-prising, therefore, that avian flight has been of interest to man for many centuries. Because of enormous technical problems, however, direct measurements of physiological variables from flying birds have only occurred on a fairly routine basis during the last 20 years or so. Basically, two techniques have been used, radio telemetry and wind tunnels.

One of the earliest accounts of recordings from a free-flying bird is that of Lord, Bellrose & Cochran (1962). They accidentally recorded respiratory frequency from a free-flying mallard duck, *A. platyrhynchos*, when they were tracking it. The transmitter had a mass of 38 g, was mounted externally and had a range of 150 m. It was found that when the duck had been flying at an apparent ground velocity of approximately $18 \, m \, s^{-1}$ for more than 5 min its respiratory frequency was 96 breaths min^{-1}. Heart rate and blood pressure were recorded by Eliassen (1963a,b). He mounted a transmitter with a mass of 40 g on the back of mallards or great backed gulls, *Larus marinus*, and found for the latter that there was an increase in arterial pulse pressure, but surprisingly, no change in heart rate during a flight over 100 m distance. This latter observation has not been confirmed by other workers.

Roy & Hart (1963, 1966) designed and built transmitters which were capable of transmitting e.c.g., temperature, respiratory frequency and tidal volume. They had a mass of 36 g and 30 g and were mounted externally. When measuring tidal volume, a mask was also attached to the birds (Hart & Roy, 1966; Berger, Hart & Roy, 1970). In addition to the excessive payload, the flights were of short duration ($< 20 \, s$) as the bird was prevented from flying away by attaching it to a nylon line. A number of birds were studied, e.g. pigeon, evening grosbeak, *Hesperiphona vespertina*, ring billed gull, *Larus delawarensis*, and black duck, *Anas rubripes*. It was found that, during flight, heart rate increased to 2–4 times its resting value whereas oxygen uptake increased by 12–14 times resting value. With the exception of the pigeon, minute ventilation volume increased by a similar proportion to oxygen uptake. Radio telemetry has been used to record heart rate from free-flying herring gulls, *Larus argentatus* (Kanwisher, Williams, Teal & Lawson, 1978). A high frequency (432 MHz) transmitter with a mass of 50 g was mounted externally. The transmitter itself had a 17 cm whip aerial and high gain receiving antennae were used, giving a maximum range of 80 km. Recording over such large distances is very useful, but the technique does not allow any control over the duration of the flight or of the air velocity of the bird.

Torre-Bueno (1976) was rather critical of experiments in which bulky transmitters were mounted externally and were thus likely to affect the animal's flying performance. He implanted a small transmitter (Southwick, 1973) into starlings (*Sturnus vulgaris*) which had been trained to fly in a wind tunnel and recorded their core and skin temperatures during flights of 0.5–2 h duration. As with pigeons (Hart & Roy, 1967), core temperature was independent of ambient temperature although skin and subcutaneous temperatures were variable. Birds seem to be able to vary their insulation and Torre-Bueno (1976) suggests that a high body temperature is maintained at all environmental temperatures in order to increase the efficiency of the flight muscles.

The combined use of implanted transmitters and wind tunnels has great potential. There is no doubt that since wind tunnels were first used to record oxygen uptake in budgerigars, *Melopsittacus undulatus* (Tucker, 1966), and to study the power requirements of flight in pigeons (Pennycuick, 1968), they have provided the most useful data on the physiology of bird flight. They have enabled birds to be flown for long periods of time at known air velocities. Although, theoretically, the relationship between power output and flight velocity is "U"-shaped (Pennycuick, 1968; Rayner, 1979), power input (i.e. oxygen uptake) rises with increasing velocity in the laughing gull, *Larus atricilla* (Tucker, 1972) and the fish crow, *Corvus ossifragus* (Bernstein, Thomas & Schmidt-Nielsen, 1973) and is independent of velocity in the starling, *Sturnus vulgaris* (Torre-Bueno & Larochelle, 1978). Of the birds studied so far, only the budgerigar has a U-shaped relationship between power input and flight velocity (Tucker, 1968). Respiratory frequency and tidal volume are also constant over a wide range of velocities in the fish crow (Bernstein, 1976). Again theoretically, the power requirement of flight (i.e. power output) is proportional to (body mass)$^{1.166}$ (Wilkie, 1959; Pennycuick, 1975), whereas data from flights in wind tunnels indicate that power input for birds is proportional to (body mass)$^{0.75}$ (Butler, 1981).

Despite the major advances that have been made using wind tunnels, most of the studies have involved the attachments of masks on the head for measurements of oxygen uptake, or leads and cannulae for recording of electrical signals or withdrawal of blood samples (Tucker, 1968; Butler, West & Jones, 1977). It has been suggested that wind tunnels may adversely affect the flying performance of birds (Greenewalt, 1975), and Butler *et al.* (1977) demonstrated that the flight pattern of pigeons in a wind tunnel is different from that of free-flying pigeons. Because of size restrictions, it is also noticeable that data have not been obtained from birds of greater than 0.5 kg mass when performing flapping flight in a wind tunnel, although there is a clear need for data from larger birds (see Butler, 1981).

Free-flying Birds

Barnacle geese, *Branta leucopsis*, were raised from eggs and imprinted on a human, so that they would follow him for long periods. They were eventually taken to an airfield where their willingness was tested to fly behind a moving truck containing the foster parent. The two-channel FM transmitter was completely implanted into those birds that successfully flew behind the truck and allowed themselves to be recaptured. By using ciné photography and recording from the transmitter by way of a receiver mounted in the truck, it was possible to record wing beat frequency, heart rate and respiratory frequency from the free-flying birds (Butler & Woakes, 1980). Although the investigators varied the ground speed of the truck, the bird was under no physical obligation whatsoever to fly in a particular direction or at a particular velocity. It voluntarily flew close enough to the truck to enable film to be shot and information from the transmitter to be received. There was no artefact on the e.c.g. signal at all during the flights, thus enabling an instantaneous rate counter to be used (Fig. 6).

The mean resting values of heart rate (72 beats min^{-1}) and respiratory frequency (8.5 breaths min^{-1}) in the barnacle geese were approximately 50% lower than the values for birds of similar mass as calculated by the allometric formulae of Calder (1968) and Berger, Hart & Roy (1970). During flight, however, there was less of a discrepancy so that the increase in heart rate above the resting level was greater in the geese (7.2 times) than the 2–4 times increase that had been reported for other species (Berger & Hart, 1974). Using an implanted telemetry system allowed values closer to those

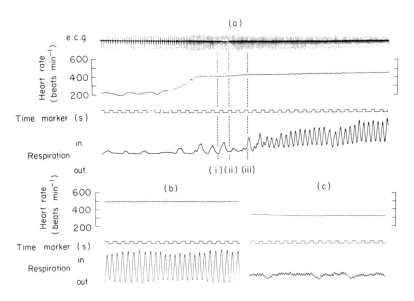

FIG. 6. Traces from ♀ barnacle goose showing heart rate and respiratory frequency before, during and after a flight of 11 min 52 s duration. (a) Take off (note clean e.c.g. signal during flight); (b) steady flapping flight at 22 ms^{-1}, 5 min after take off; (c) 3 min after landing. The vertical dashed lines in (a) indicate when (i) the truck started to move; (ii) the bird began to run and flap its wings; (iii) the bird was airborne. (From Butler & Woakes, 1980.)

of the truly resting animal to be obtained. None of the measured variables changed appreciably with variations in flight velocity from $15-26\,m\,s^{-1}$. Mean heart rate during flight was 512 beats min^{-1}, mean respiratory frequency was 99 breaths min^{-1} and mean wing beat frequency was 287 beats min^{-1}. Flights lasted for an average of 14.4 min at a mean air velocity of $18.7\,m\,s^{-1}$. On two occasions the birds slope-soared on the airstream rising over the front of the truck for longer than 20 s during which time heart rate declined appreciably to 261 beats min^{-1} and respiratory frequency was 68 breaths min^{-1}. During flapping flight there was a 3:1 correspondence between wing beat frequency and respiratory frequency.

By synchronizing the recording of respiratory air flow with a simultaneous film sequence it was possible to demonstrate that wing beat was tightly locked to fixed phases of the respiratory cycle (Fig. 7). A 1:1 correspondence between wing beating and ventilation has been demonstrated in pigeons (Hart & Roy, 1966; Butler *et al.*, 1977) whereas ratios as high as 5:1 have been reported for the black duck, *A. rubripes*, quail, *C. coturnix* and pheasant, *Phasianus colchicus*

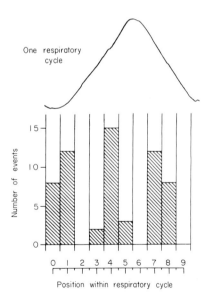

FIG. 7. Histogram showing the position during the respiratory cycle at which the wings were fully elevated (called "events") during the flight of a barnacle goose. Twenty respiratory cycles of equal duration were chosen and each was divided into ten equal intervals. Below the histogram is plotted the mean position of each group of events (o) and above the histogram is a trace of one of the respiratory cycles, inspiration-up. (From Butler & Woakes, 1980).

(Berger, Roy & Hart, 1970). The latter authors concluded that co-ordination was not obligatory during flights of a few seconds duration. It is quite clear that in the barnacle goose, the close phase relationship between the two activities is present during flights of relatively long duration, but the precise significance of such close co-ordination is not yet apparent.

If the necessary transducers can be developed for recording tidal volume (see Woakes & Butler, 1979) and oxygen uptake, it is felt that this method of imprinting larger birds (ducks, geese, swans) upon a human and using implanted transmitters to monitor such variables, could provide the most useful information on the energetics of free range flight in birds.

ACKNOWLEDGEMENTS

The authors' work was supported by the Science Research Council. They wish to thank Mr E. Keijer for his assistance.

REFERENCES

Andersen, H. T. (1966). Physiological adaptations in diving vertebrates. *Physiol. Rev.* 46: 212—243.

Berger, M. & Hart, J. S. (1974). Physiology and energetics of flight. In *Avian biology* 4: 415—477. Farner, D. S. & King, J. R. (Eds). New York and London: Academic Press.

Berger, M., Hart, J. S. & Roy, O. Z. (1970). Respiration, oxygen consumption and heart rate in some birds during rest and flight. *Z. vergl. Physiol.* 66: 201—214.

Berger, M., Roy, O. Z. & Hart, J. S. (1970). Co-ordination between respiration and wing beats in birds. *Z. vergl. Physiol.* 66: 190—200.

Bernstein, M. H. (1976). Ventilation and respiratory evaporation in the flying crow, *Corvus ossifragus*. *Respir. Physiol.* 26: 371—381.

Bernstein, M. H., Thomas, S. P. & Schmidt-Nielsen, K. (1973). Power input during flight of the fish crow, *Corvus ossifragus*. *J. exp. Biol.* 58: 401—410.

Black, C. P. & Tenney, S. M. (1980). Oxygen transport during progressive hypoxia in high-altitude and sea-level water fowl. *Respir. Physiol.* 39: 217—239.

Bond, C. F. & Gilbert, P. W. (1958). Comparative study of blood volume in representative aquatic and non aquatic birds. *Am. J. Physiol.* 194: 519—521.

Brooks, A. (1945). The underwater action of diving ducks. *Auk* 62: 517—512.

Burne, R. H. (1909). Notes on the viscera of a walrus (*Odobaenus rosmarus*). *Proc. zool. Soc. Lond.* 1909: 732—738.

Butler, P. J. (1979). The use of radio telemetry in the studies of diving and flying of birds. In *A handbook on biotelemetry and radio-tracking*: 567—

577. Amlaner, C. J. & Macdonald, D. W. (Eds). Oxford and New York: Pergamon Press.

Butler, P. J. (1981). Respiration during flight. In *Advances in physiological sciences:* 10: 155–164. Hutás, I. & Debreczeni, L. A. (Eds). Oxford and New York: Pergamon Press.

Butler, P. J. & Jones, D. R. (1971). The effect of variations in heart rate and regional distribution of blood flow on the normal pressor response to diving in ducks. *J. Physiol., Lond.* 214: 457–479.

Butler, P. J. & Taylor, E. W. (1973). The effect of hypercapnic hypoxia, accompanied by different levels of lung ventilation on heart rate in the duck. *Respir. Physiol.* 19: 176–187.

Butler, P. J., West, N. H. & Jones, D. R. (1977). Respiratory and cardiovascular responses of the pigeon to sustained, level flight in a wind tunnel. *J. exp. Biol.* 71: 7–26.

Butler, P. J. & Woakes, A. J. (1976). Changes in heart rate and respiratory frequency associated with spontaneous submersion of ducks. In *Biotelemetry* 3: 215–218. Fryer, T. B., Miller, H. A. & Sandler, H. (Eds). New York and London: Academic Press.

Butler, P. J. & Woakes, A. J. (1979). Changes in heart rate and respiratory frequency during natural behaviour of ducks with particular reference to diving. *J. exp. Biol.* 79: 283–300.

Butler, P. J. & Woakes, A. J. (1980). Heart rate, respiratory frequency and wing beat frequency of free flying barnacle geese, *Branta leucopsis. J. exp. Biol.* 85: 213–226.

Calder, W. A. (1968). Respiratory and heart rates of birds at rest. *Condor* 70: 358–365.

Colacino, J. M., Hector, D. H. & Schmidt-Nielsen, K. (1977). Respiratory responses of ducks to simulated altitude. *Respir. Physiol.* 29: 265–281.

Dewar, J. M. (1924). *The bird as a diver.* London: H. F. & G. Witherby.

Dewar, J. M. (1940). Timing the under-water activities of diving birds. *Br. Birds* 33: 58–61.

Drummond, P. C. & Jones, D. R. (1979). The initiation and maintenance of bradycardia in a diving mammal, the muskrat, *Ondatra zibethica. J. Physiol., Lond.* 290: 253–271.

Eliassen, E. (1960). Cardiovascular responses to submersion asphyxia in avian divers. *Årbok Univ. Bergen* 2: 1–100.

Eliassen, E. (1963a). Preliminary results from new methods of investigating the physiology of birds during flight. *Ibis* 105: 234–237.

Eliassen, E. (1963b). Telemetric registering of physiological data in birds in normal flight. In *Bio-telemetry*: 257–265. Slater, L. E. (Ed.). Oxford: Pergamon.

Elsner, R (1965). Heart rate response in forced versus trained experimental dives in pinnipeds. *Hvalråd. Skrift.* 48: 24–29.

Folkow, B., Lisander, B. & Öberg, B. (1971). Aspects of the cardiovascular control in a mammalian diver (*Myocastor coypus*). *Acta physiol. scand.* 82: 439–446.

Fryer, T. B. (1970). *Implantable biotelemetry systems. NASA SP-5094.* Washington: NASA Technology Utilization Publications.

Greenewalt, C. H. (1975). The flight of birds. *Trans. Am. phil. Soc.* NS 65: 5–67.

Hart, J. S. & Roy, O. Z. (1966). Respiratory and cardiac responses to flight in pigeons. *Physiol. Zool.* 39: 291–306.

Hart, J. S. & Roy, O. Z. (1967). Temperature regulation during flight in pigeons. *Am. J. Physiol.* 213: 1311–1316.

Irving, L. (1939). Respiration in diving mammals. *Physiol. Rev.* 19: 112–134.

Irving, L., Scholander, P. F. & Grinnell, S. W. (1941). Significance of the heart rate to the diving ability of seals. *J. cell. comp. Physiol.* 18: 283–297.

Irving, L., Scholander, P. F. & Grinnell, S. W. (1942). The regulation of arterial blood pressure in the seal during diving. *Am. J. Physiol.* 135: 557–566.

Johansen, K. (1964). Regional distribution of circulating blood during submersion asphyxia in the duck. *Acta physiol. Scand.* 62: 1–9.

Johnston, D. W. & McFarlane, R. W. (1967). Migration and bioenergetics of the Pacific golden plover, *Condor* 69: 156–168.

Jones, D. R. Bryan, R. M., West, N. H., Lord, R. H., & Clark, B. (1979). Regional distribution of blood flow during diving in the duck (*Anas platyrhynchos*). *Can. J. Zool.* 57: 995–1002.

Jones, D. R. & Purves, M. J. (1970). The carotid body in the duck and the consequences of its denervation upon the cardiac responses to immersion. *J. Physiol., Lond.* 211: 279–294.

Kanwisher, J. W., Gabrielsen, G. & Kanwisher, N. (1981). Free and forced diving in birds. *Science, N.Y.* 211: 717–719.

Kanwisher, J. W., Williams, T. C., Teal, J. M. & Lawson, K. O. (1978). Radiotelemetry of heart rates from free-ranging gulls. *Auk* 95: 288–293.

Kooyman, G. L. & Campbell, W. B. (1972). Heart rates in freely diving Weddell seals, *Leptonychotes weddelli. Comp. Biochem. Physiol.* 43A: 31–36.

Kooyman, G. L., Wahrenbrock, E. A., Castellini, M. A., Davis, R. W. & Sinnett, E. E. (1980). Aerobic and anaerobic metabolism during voluntary diving in Weddell seals: evidence of preferred pathways from blood chemistry and behaviour. *J. comp. Physiol.* 138: 335–346.

Lasiewski, R. C. & Calder, W. A. (1971). A preliminary allometric analysis of respiratory variables in resting birds. *Respir. Physiol.* 11: 152–166.

Lord, R. D., Bellrose, F. C. & Cochran, W. W. (1962). Radio telemetry of the respiration of a flying duck. *Science, N.Y.* 137: 39–40.

Millard, R. W., Johansen, K. & Milsom, W. K. (1973). Radio telemetry of cardiovascular responses to exercise in diving penguins. *Comp. Biochem. Physiol.* 46A: 227–240.

Murdaugh, H. V., Robin, E. D., Millen, J. E., Drewry, W. F. & Weiss, E. (1966). Adaptations to diving in the harbor seal: cardiac output during diving. *Am. J. Physiol.* 210: 176–180.

Påsche, A. & Krog, J. (1980). Heart rate in resting seals on land and in water. *Comp. Biochem. Physiol.* 67A: 77–83.

Pennycuick, C. J. (1968). Power requirements for horizontal flight in the pigeon *Columba livia. J. exp. Biol.* 49: 527–555.

Pennycuick, C. J. (1975). Mechanics of flight. in *Avian biology* 5: 1–75. Farner, D. S. & King, J. R. (Eds). New York and London: Academic Press.

Prange, H. D. & Schmidt-Nielsen, K. (1970). The metabolic cost of swimming in ducks. *J. exp. Biol.* 53: 763–777.

Rayner, J. M. V. (1979). A new approach to animal flight mechanics. *J. exp. Biol.* 80: 17–54.

Richet, C. (1899). De la résistance des canards à l'asphyxie. *J. Physiol. Path. gén.* 1. [As quoted by Scholander, P. F. (1940).]

Roy, O. Z. & Hart, J. S. (1963). Transmitter for telemetry of biological data from birds in flight. *Bio-Med. Electron. BME IEEE Trans.* 10: 114–116.

Roy, O. Z. & Hart, J. S. (1966). A multichannel transmitter for the physiological study of birds in flight. *Med. Biol. Engng* 4: 457—466.

Scheid, P., Slama, H. & Willmer, H. (1974). Volume and ventilation of air sacs in ducks studied by inert gas wash-out. *Respir. Physiol.* 21: 19—36.

Scheipers, G., Kawashiro, T. & Scheid, P. (1975). Oxygen and carbon dioxide dissociation of duck blood. *Respir. Physiol.* 24: 1—14.

Schmidt-Nielsen, K. (1977). The physiology of wild animals. *Proc. R. Soc. Lond.* (B) 199: 345—360.

Schmidt-Nielsen, K. (1979). *Animal physiology*. 2nd edn. Cambridge; University Press.

Scholander, P. F. (1940). Experimental investigations on the respiratory function in diving mammals and birds. *Hvalråd. Skrift.* 22: 1—131.

Scholander, P. F., Irving, L. & Grinnell, S. W. (1942). Aerobic and anaerobic changes in seal muscles during diving. *J. biol. Chem.* 142: 431—440.

Southwick, E. E. (1973). Remote sensing of body temperature in a captive 25 g bird. *Condor* 75: 464—466.

Swan, L. W. (1961). The ecology of the high Himalayas. *Scient. Am.* 205: 68—78.

Torre-Bueno, J. R. (1976). Temperature regulation and heat dissipation during flight in birds *J. exp. Biol.* 65: 471—482.

Torre-Bueno, J. R. (1978). Respiration during flight in birds. in *Respiratory function in birds*: 89—94. Piiper, J. (Ed.). Berlin: Springer—Verlag.

Torre-Bueno, J. R. & Larochelle, J. (1978). The metabolic cost of flight in unrestrained birds. *J. exp. Biol.* 75: 223—229.

Tucker, V. A. (1966). Oxygen consumption of a flying bird. *Science, N.Y.* 154: 150—151.

Tucker, V. A. (1968). Respiratory exchange and respiratory water loss in the flying budgerigar. *J. exp. Biol.* 48: 67—87.

Tucker, V. A. (1972). Metabolism during flight in the laughing gull, *Larus atricilla. Am. J. Physiol.* 222: 237—245.

Wilkie, D. R. (1959). The work output of animals: flight by birds and by manpower. *Nature, Lond.* 183: 1515—1516.

Woakes, A. J. (1979). Construction techniques using subminiature discrete components. In *A handbook on biotelemetry and radio tracking*: 213—215. Amlaner, C. J. & Macdonald, D. W. (Eds). Oxford: Pergamon Press.

Woakes, A. J. (1980). *Biotelemetry and its application to the study of avian physiology*. Ph.D. thesis: Birmingham University.

Woakes, A. J. & Butler, P. J. (1975). An implantable transmitter for monitoring heart rate and respiratory frequency in diving ducks. *Biotelemetry* 2: 153—160.

Woakes, A. J. & Butler, P. J. (1979). An implantable transducer for the measurement of respiratory airflow. In *A handbook on biotelemetry and radio tracking*: 287—292. Amlaner, C. J. & Macdonald, D. W. (Eds). Oxford: Pergamon Press.

Zapol. W. M., Liggins, G. C., Schneider, R. C., Qvist, J., Snider, M. T., Creasy, R. K. & Hochachka, P. W. (1979). Regional blood flow during simulated diving in the conscious Weddell seal. *J. appl. Physiol.: Respirat. Environ. Exercise Physiol.* 47: 968—973.

DISCUSSION

R. Kenward (Institute of Terrestrial Ecology) — If you used an externally mounted transmitter with implanted electrodes would it be possible to achieve a greater range than with the implanted transmitter which necessitates following the bird with a truck?

P. J. Butler — We tried externally mounted transmitters at the outset. The transmitter was mounted by straps passing under the wings and ducks were very successful at removing them. Perhaps our methods of attachment could have been improved but this problem just does not arise with implanted transmitters. Also, there is no problem with drag using implanted transmitters. This is important in both flying and diving. An advantage of having an imprinted bird flying close to a vehicle is that a high degree of control over the experimental conditions can be achieved. This is obviously important when telemetering physiological variables.

Symp. zool. Soc. Lond. (1982) No. 49, 129–137

Devices for Telemetering the Behaviour of Free-living Birds

R. E. KENWARD

Institute of Terrestrial Ecology, Monks Wood Experimental Station, Abbots Ripton, Huntingdon, England

G. J. M. HIRONS

The Game Conservancy, Fordingbridge, Hampshire, England

and

F. ZIESEMER

Staatliche Vogelschutzwarte Schleswig-Holstein, Olshausenstr. 40-60, D-2300 Kiel, West Germany

SYNOPSIS

The measurement of bird activity using transmitters with variable pulse rates is described. In one transmitter the current, and hence the pulse rate, is reduced during flight by the cooling of a thermistor strapped under the wing. The other tail-mounted tag uses mercury switching of a capacitive sub-circuit to give a slow pulse rate when the tail is vertical in a resting bird, a fast rate when the tail is horizontal during flight, and an alternating rate during feeding. These tags have been used with portable receivers in behaviour studies of woodcock and goshawks, and could be of great value in automatic activity monitoring.

INTRODUCTION

Biotelemetry, the "measurement of biological parameters from afar", is a wide field which includes wire, acoustic, light, radar and storage telemetry as well as radio telemetry (review in Kimmich, 1980). This chapter is concerned with radio telemetry using transmitters carried by the subject.

It was U.S. Navy personnel, aware of early success in radio telemetry of physiological data from "flying mammals" (jet aircraft pilots, in Barr, 1954), who developed the first small transmitter for animals (Le Munyan, White, Nybert & Christian, 1959). This 120 g transmitter package, with a six-month life and 20 m range, was

implanted in chipmunk (*Marmota monax*) and intended as a telemetry device, although its potential for radio tracking was also
realized and further developed in smaller, longer range systems
(Cochran & Lord, 1963; Craighead, Craighead & Davies, 1963; Verts,
1963). Some of the earliest papers described telemetry of bird wing-
beat and respiration frequency (Eliassen, 1960; Lord, Bellrose &
Cochran, 1962) rather than radio tracking. Since that time, transmitters with implanted sensors have been used to monitor temperature, heart rate, blood flow, respiration rate (reviews in Amlaner,
1978; Butler, 1980) and digestive activity (Fuller, Duke & Maxfield,
in preparation). Very few studies have obtained more than two or
three weeks of data from transmitters, for two main reasons. First,
most experiments have relied on frequency modulation (FM) or
amplitude modulation (AM) of continuous signals, which reduce
battery life by taking more power than pulse-interval or pulse-width
modulated (PIM, PWN) signals. When the latter methods are used
pulses can be initiated or otherwise modulated by the physiological
events (review in Macdonald & Amlaner, 1980), although this
increases weight and cost of circuits. Secondly, if packages are
implanted they require more power for the same range as external
transmitters; if the sensors alone are implanted, these are difficult to
keep correctly positioned and connected to the remote transmitter
for long. There may also be welfare (and hence legal) restrictions on
implantation in free-living birds. This chapter describes two ways of
transmitting flight activity by PIM from free-living birds, based on
simple externally-mounted transmitters without implanted sensors.

METHODS

Both transmitters described here have the circuit shown in Fig. 1,
from Cochran (1967) and Amlaner (1978) (further details in
Kenward, this volume, p. 175). In the thermistor flight-monitoring
package, which was harness-mounted (on woodcock, *Scolopax
rusticola*, see Hirons & Owen, this volume, p. 139), the load resistor
(R_1) was replaced by a miniature bead thermistor giving 470 kΩ at
25°C (Radiospares 151–164). Fine PTFE-insulated multistrand
copper wires (BICC B7/0.08) were soldered to the thermistor leads as
close to the bead as possible, and the joints covered with rapid drying
Araldite epoxy adhesive. The transmitter and battery were also
encapsulated in Araldite, and all lead and aerial emergence points
were sealed with silicone rubber (General Motors Silicone Glue and
Seal). The woodcock harness consisted of four 15 cm straps of 3.5 ×

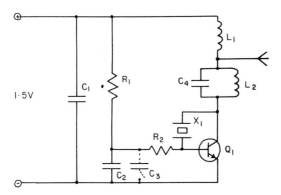

FIG. 1. Transmitter circuit, showing position of optional thermistor and of capacitive switching circuit. $L_1 = 6-7$ turns 40 SWG on 3 mm diameter; $L_2 = 12-14$ turns 40 SWG on 3 mm diameter; $R_1 = 470$ k or thermistor; $R_2 = 8.2$ k; $C_1 = 1$ n; $C_2 = 2.2$ μ (or 1 μ when C_3 present); $C_3 = 1.5$ μ; $C_4 =$ selected $15-22$ p; $Q_1 =$ MMT 74; $X_1 =$ third overtone crystal.

0.8 mm four-core clothing elastic, which passed on either side of the body anterior and posterior to the wings and knotted at a single point under the breast. The 3.5 cm long wires to the thermistor were fastened along one harness strap, which passed behind the wing and to which the thermistor was bound and glued so that the bead projected about 5 mm away from the bird's side under the wing (Fig. 2). In this position the thermistor was warmed by the bird while insulated from the exterior by the closed wing, but cooled by passing air in flight. Containing a (Mallory) RM625 mercury cell and a 25 cm long 0.4 mm diameter stainless steel whip aerial, this package weighed 8.5 g complete with harness.

The posture-monitoring package used a 12.7 mm long mercury switch (Gordos HG 520LO) to connect a second capacitor in parallel with C_3 (Fig. 1), thus decreasing the pulse rate. This package was designed for tail-mounting on goshawks (*Accipiter gentilis*), to indicate feeding and flight behaviour, and had 470 kΩ at R_1 with parallel 1.5 μF capacitors at C_3. The mercury switch was positioned in a vertical plane at 45° to the longitudinal axis of the tail (Fig. 3), connected so that C_3 capacitance was low with the tail horizontal in flight, and high with the tail close to vertical when the hawk perched. The pulse rate was thus fast in flight and slow at rest. When the hawk arched its back to pull at food, and then relaxed to swallow, the tail inclined at more, and then less, than 45° from the horizontal, producing a characteristic irregular pulse rate. The package and mounting was as described in Kenward (1978), except that a 1.5 V lithium/ copper oxide battery weighing 7.5 g (Saft-Sogea LCO2) was used giving 1400 mAh, instead of the earlier 1.35 V mercury cells, which

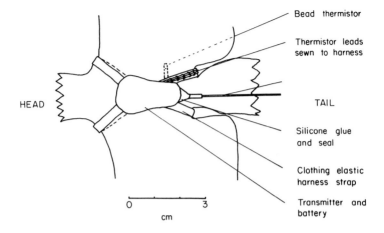

FIG. 2. Flight monitoring package on woodcock back, with thermistor mounted on harness under one wing.

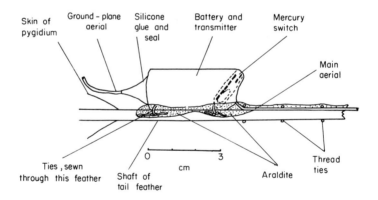

FIG. 3. Posture monitoring package mounted on goshawk tail feather. Ties on the other side of the package are knotted round the adjacent feather, so that two feathers support the package but can moult independently.

gave 575 mAh at 8 g (RM640) or 1000 mAh at 12 g (RM1). A further improvement was that nylon-coated multi-strand stainless steel angling trace wire was more durable for aerials than the PTFE-coated copper wire used in the earlier package. Package weight was 14 g, 1–2% of goshawk body weight.

RESULTS AND DISCUSSION

Thermistor Package

Pulse rates from the thermistor flight monitoring package were much slower in flight than at rest (Table I). The flight rates, and to a smaller extent the rest rates, varied with air temperature, but the flight rates never approached rest rates for any transmitter at air temperatures of up to 20°C. The thermistors responded very rapidly to temperature change, so that a fall in pulse rate was apparent within a second or two of birds being seen to take off. It was the change in pulse rate between consecutive observations that indicated initiation, or cessation, of flight.

TABLE I

Performance data for woodcock radio packages with thermistors for flight monitoring

Bird	Monitoring period[a]		Range of pulse rates (per minute)		Air temperature range (°C)
	Dates	Duration in days	In flight	At rest	
A	31/3−5/5	36	76−134	172−180	−
B	29/3−6/6	68	102−120	152−180	11−18
C	19/4−22/5	33	96	196−224	6−11
D	10/5−11/6	32	120	175−180	11−20
E	30/3−6/6	67	72−110	222	10−18

[a] End of period may correspond with bird leaving area rather than end of signal life.

Although the battery life of up to 68 days was adequate for recording flight activity through much of the woodcock roding (display) period, it was lower than the field life expected from this package without a thermistor. With a fixed value of 470 kΩ at R_1, the 350 mAh cell drains at about 4 mAh per day for about 90 days. Life could be increased slightly by incorporating the thermistor in a bridge circuit, with one parallel and one series resistance, to reduce the temperature sensitivity. Use of a positive temperature coefficient thermistor is less desirable in the flight monitor than in squirrel thermistor collars (Kenward, this volume, p. 175) because reduced R_1 in resting birds increases transmitter power when it is most needed: with the bird on the ground.

Posture-sensing Package

In this package, capacitance is varied instead of load resistance, so current remains constant (Macdonald & Amlaner, 1980) and field life is the same as without the sensor. The previous 13–15 g goshawk tail package gave about five months' life with the 575 mAh mercury cell. Lithium cells have a much higher power: weight ratio than mercury cells (Ko, 1980), so that the present 14 g package (12 g without the mercury switch) would operate for about one year on the 1400 mAh battery with the same circuit current. This package has been in use for less than a year, so no data on field life are yet available. However, the discharge curve for lithium cells differs from that for mercury cells (Fig. 4), so that at 2.2 V the initial current is higher than when the voltage stabilizes at the nominal 1.5 V. This will probably reduce field life somewhat below that expected at constant 1.5 V, and has other implications. With an elevated initial voltage and current, signal strength and hence transmission range is greatest when it is most needed: for the first few days of radio tracking. After a time, tagged individuals become easier to track because their usual haunts are known. The disadvantage of the large initial voltage drop in lithium cells is that the pulse rates and the radio frequency drop too. The radio frequency change may be as much as 5 kHz, so that a transmitter cannot be detected at the initial receiver frequency setting after the first few days. These changes can be anticipated by noting frequency and pulse rate(s) with each transmitter coupled to a 1.5 V source (e.g. a zinc/carbon cell) before connecting the lithium cell. A related problem is that the high initial voltage may overload

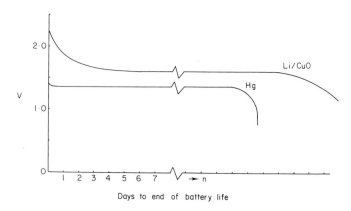

Days to end of battery life

FIG. 4. Voltage characteristics during discharge of mercury and lithium/copper oxide batteries used in transmitters.

the transistor. Overloading can be avoided by coupling a 150–390 Ω resistance in series with the battery. The wastage of power in this resistance is minimized if the resistor can be connected across external contacts to switch on the transmitter for 48 h before attachment to the bird. After 48 h the voltage has dropped to about 1.7 V, and the resistor may be removed before soldering together and sealing the external contacts; at this time there is also little further change in frequency and pulse rates.

Application

For attaching transmitters to birds, tail-mounting is preferable to harnesses, which are not moulted after use and can adversely affect flight activity, social status, feeding and survival (review in Kenward, 1980). However, there are many bird species for which tail posture is not a good indicator of flight. For some of these it might be possible to position flight-cooled thermistors from tail-mounted or other non-harness packages (e.g. on the undersurface of primaries in very large birds), but where packages must operate through a moult or are too heavy for tail-mounting, then harnesses may be unavoidable.

The thermistor flight-monitor has been used in sequence-sampling of up to six male woodcock throughout each evening roding session, counting the pulses received during each short time interval and recording whether or not the birds were flying, to link roding activity with mating success (Hirons & Owen, this volume, p. 139, and in preparation). The goshawk packages have been used mainly for tracking and recording predation. The feeding signal not only avoids wasting time in approaching birds which are perched low without kills, but also helps in recording small, rapidly eaten kills. A number of kills which were eaten in trees would not have been detected without this signal. Both these package types can also serve as mortality indicators. Thermistor packages indicate body cooling associated with death (Stoddart, 1970), and posture packages will indicate death (without the power reduction of a cooled thermistor package) if the body posture becomes atypical, as when a hawk is continuously horizontal.

Although the packages have not yet been used in this way, they would be ideal for automatic monitoring of flight in energetics or behaviour studies. Previous automatic flight monitoring has depended on recording movement from one location to another by triangulation from two costly remote-controlled directional receiving aerials (e.g. Forbes & Warner, 1974), or by interpreting chart-recorded signal amplitude variations (Widén, this volume, p. 153), which can be ambiguous and hence time-consuming to analyse. On the other hand, the posture-sensing and the thermistor packages produce unambiguous PIM flight signals, which can be received at one fixed station with an inexpensive omnidirectional aerial, and automatically recorded and counted to give a digital record for computer graphics and analysis. Avoiding the need for recording and decoding by hand is an important consideration at a time of high labour costs and financial stringency. It is likely that ingenious development of novel sensor mechanisms for automatic radio telemetry of free-living animals will provide many new opportunities in behavioural and ecological studies.

ACKNOWLEDGEMENTS

We thank W. Muschiol for providing his trained goshawk for transmitter trials, J. P. Dempster, R. Green and A. Village for ideas and comments and M. B. Haas for typing the manuscript.

REFERENCES

Amlaner, C. J. (1978). Biotelemetry from free-ranging animals. In *Animal marking − recognition marking of animals in research*: 205−228. Stonehouse, B. (Ed.). London: Macmillan.

Barr, N. L. (1954). Radio transmission of physiological information. *Military Med.* 114: 79.

Butler, P. J. (1980). The use of radio telemetry in the studies of diving and flying of birds. In *A handbook on biotelemetry and radio tracking*: 569−577. Amlaner, C. J. & Macdonald, D. W. (Eds). Oxford: Pergamon.

Cochran, W. W. (1967). *154−160 MHz beacon (tag) transmitter for small animals*. American Institute of Biological Sciences Bioinstrumentation Advisory Council Information Module 15.

Cochran, W. W. & Lord, R. D. (1963). A radio-tracking system for wild animals. *J. Wildl. Mgmt* 27: 9−24.

Craighead, F. C., Craighead, J. J. & Davies, R. S. (1963). Radio-tracking of grizzly bears. In *Biotelemetry: the use of telemetry in animal behaviour and physiology in relation to ecological problems*: 133−148. Slater, L. E. (Ed.). New York & London: Macmillan.

Eliassen, E. (1960). A method for measuring the heart rate and stroke/pulse pressures of birds in normal flight. *Arbok Univ. Bergen, Mat.-Nat.-Sci.* 12: 1−22.

Forbes, J. E. & Warner, D. W. (1974). Behaviour of a radio-tagged saw-whet owl. *Auk* 91: 783−795.

Fuller, M. R., Duke, G. E. & Maxfield, L. (In preparation). *Sensing, telemetry and interpretation of gastric motility in owls.*

Kenward, R. E. (1978). Radio transmitters tail-mounted on hawks. *Ornis scand.* 9: 220–223.

Kenward, R. E. (1980). Radio monitoring birds of prey. In *A handbook on biotelemetry and radio tracking*: 97–104. Amlaner, C. J. & Macdonald, D. W. (Eds). Oxford: Pergamon.

Kimmich, H. P. (1980). Artifact free measurement of biological parameters: biotelemetry, a historical review and layout of modern experiments. In *A handbook on biotelemetry and radio tracking*: 3–20. Amlaner, C. J. & Macdonald, D. W. (Eds). Oxford: Pergamon.

Ko, W. H. (1980). Power sources for implant telemetry and stimulation systems. In *A handbook on biotelemetry and radio tracking*: 225–245. Amlaner, C. J. & Macdonald, D. W. (Eds). Oxford: Pergamon.

Le Munyan, C. D., White, W., Nybert, E. & Christian, J. J. (1959). Design of a miniature radio transmitter for use in animal studies. *J. Wildl. Mgmt* 23: 107–110.

Lord, R. D., Bellrose, F. C. & Cochran, W. W. (1962). Radio telemetry of the respiration of a flying duck. *Science, N.Y.* 137: 39–40.

Macdonald, D. W. & Amlaner, C. J. (1980). A practical guide to radio tracking. In *A handbook on biotelemetry and radio tracking*: 143–159. Amlaner, C. J. & Macdonald, D. W. (Eds). Oxford: Pergamon.

Stoddart, L. C. (1970). A telemetric method for detecting jackrabbit mortality. *J. Wildl. Mgmt* 34: 501–507.

Verts, B. J. (1963). Equipment and techniques for radio-tracking striped skunks. *J. Wildl. Mgmt* 27: 325–339.

Symp. zool. Soc. Lond. (1982) No. 49, 139—152

Radio Tagging as an Aid to the Study of Woodcock

G. J. M. HIRONS

The Game Conservancy, Fordingbridge, Hampshire, England

and

R. B. OWEN, JR

School of Forest Resources, University of Maine, Orono, Maine, USA

SYNOPSIS

Cryptic plumage, crepuscular habits, and a woodland existence make the wood-cock *Scolopax rusticola* a challenging species to study, and without radio tagging many aspects of its basic biology would remain unknown or misinterpreted. Conventional radio tracking alone has provided new information on habitat preferences, home range size and breeding biology. However, it was necessary to develop a more specialized technique, involving temperature-sensitive transmitters and a programmable automatic scanning receiver to investigate one fundamental aspect of the woodcock's complex polygamous mating system — the mating success of individual males in relation to the timing and duration of their display flights. Daily feeding, incubation and brooding rhythms have been monitored with a strip-chart recorder. Details of these techniques and examples of the information obtained are presented.

INTRODUCTION

The Problem

The combination of perfect camouflage, lack of sexual dimorphism, a solitary woodland existence and crepuscular habits make the wood-cock *Scolopax rusticola* a challenging species to study. Except for the males' conspicuous dusk and dawn display flights, direct observation of woodcock away from the nest is extremely difficult making worthless conventional techniques of marking for individual recognition. For this reason some of the most elementary facts have been either missing or misinterpreted in accounts of the woodcock's life history, ecology and behaviour (Shorten, 1974).

This chapter describes the contribution that radio tagging has made to a study of the woodcock's breeding biology, food and habitat requirements. Detailed results are not presented. Instead we outline the basic techniques of radio location employed in the study, and then demonstrate by the use of examples how these have provided information additional to the mere cataloguing of individual movements which characterized many early radio tagging studies. Finally, we describe how these basic techniques have been developed to investigate particular aspects of the woodcock's behaviour — methods which may also be applicable to studies of other species.

Background

The woodcock, although related to waders (Scolopacidae), is adapted for life in woodland. Its most distinctive features are a long beak, large eyes set high on a round head, a stout body (average weight c. 325 g) with short legs, and the "dead-leaf" camouflage provided by its marbled, brown, grey and black plumage. There are no obvious sex or age differences in appearance.

The presence of woodcock in breeding areas is revealed by the characteristic display or advertisement flight of the males, termed roding. At dusk and dawn throughout the long breeding season (in Britain, late-February to mid-July) roding woodcock make extensive flights above the woodland canopy, calling repeatedly. Originally these flights were regarded as displays defining territory boundaries although the nature of the pair bond was unknown (Hirons, 1980). Similarly, published information was inconclusive (Shorten, 1974) on the number of broods produced per female per season, the age of first-breeding, and whether woodcock feed by day or by night.

Study Area

Woodcock have been equipped with transmitters in a 171 ha mainly deciduous wood in north-east Derbyshire (Whitwell Wood) from March to July each year from 1978 to 1980. The wood is surrounded by arable farmland interspersed with isolated pastures and leys, and numerous small copses. Two 15 ha blocks lying one kilometre away across open fields constitute the nearest woodland to Whitwell of any consequence. The whole area is made accessible for radio tracking by numerous public roads, and Whitwell Wood has a comprehensive network of rides, some of which are negotiable by vehicle. The topography of the area is gently undulating but some of the most useful vantage points for radio tracking are the spoil heaps associated with the mining of coal.

METHODS

Radio Equipment

Transmitters

So far 70 woodcock have been fitted with transmitters. The transmitters have been of two types, both operating in the 173.2–173.5 MHz waveband: standard one-stage transmitters (Cochran & Lord, 1963) weighing 2 g, obtained from several commercial outlets and powered by mercury batteries (either RM-675 (2 g) or RM-625 (4.6 g)); or one-stage transmitters with a miniature bead thermistor sewn into the attachment harness beneath the wing. The signal pulse rate of this transmitter slows when the bird flies. The construction and use of this flight-sensing transmitter is described elsewhere in this volume (Kenward, Hirons & Ziesemer, this volume, pp. 129–137). Whip aerials 25 cm long made of 0.4 mm diameter stainless steel wire were attached to both types of transmitter. Battery life for the standard transmitters has averaged 50 (RM-675) and 75 (RM-625) days depending on battery type.

Transmitter attachment

Two methods of transmitter attachment have been used: saddles made of thin dental latex with wing holes cut out (Godfrey, 1970) or four narrow strands of elastic (four cord, nylon-coated clothing elastic) passed around the bird in a figure of eight and tied at the breast (Amlaner, Sibly & McCleery, 1978). In both cases the transmitter lies on the bird's back between the wings with the whip aerial parallel to the tail. The latex saddle was unsatisfactory because in every case it perished within 21–28 days of attachment and the birds lost their transmitters. The elastic harness was more durable and transmitters were recovered from all the four birds which were shot in the winter following marking (7–8 months after transmitter attachment). However only one of the nine radio tagged birds recaptured in the next breeding season (9–12 months after radio tagging) retained its transmitter (after 10 months) although another one still had its harness intact.

The complete radio packages weighed between 5 g (RM-675 battery) and 8 g (RM-625) or 2.0–3.5% of the body weight of the radio tagged woodcock.

Receiving equipment

Radio tagged birds were located with a portable receiver (Type LA-12, AVM Instrument Company, Champaign, Illinois) linked either to a hand-held, three-element unidirectional Yagi aerial or a rotatable six-element Yagi mounted 20 m above the ground at the top of a tree in the highest part of the wood. The range over which transmitted signals could be detected varied from less than 200 m with the three-element Yagi on wet days for a bird in thick cover (400 m with the six-element Yagi) to several kilometres for a flying bird. Line-of-sight range with both the bird and the observer on the ground averaged about 600 m. On several occasions birds were located from the air using a light aircraft with a three-element Yagi aerial attached to each wing strut.

In the third year of the study the timing and duration of the dawn and dusk roding flights of individual males equipped with flight-sensing transmitters were monitored throughout the long breeding season. An omni-directional colinear array (Type 7073, Jaybeam Ltd, Northampton) was fixed to the top of another 20 m tree. This was linked to a programmable automatic scanning receiver (supplied by Cedar Creek Bioelectronics Laboratory, University of Minnesota) which scanned sequentially each of the programmed frequencies (up to four birds per observation period) for 7.5 s. Whether the radio tagged birds were flying or not was determined by counting the number of pulses received in each 7.5 s interval scanned.

Activity Recording

Variation in the strength of received signals caused by the movement of the radio tagged birds can be used to determine the presence or absence of activity. Moreover certain components of the bird's behaviour can be recognized by the pattern of changes in signal strength as registered by an external strip-chart recorder. Provided the correlation betwen particular types of behaviour and the chart record can be confirmed by direct observation, it is possible to quantify activity and to monitor it remotely for long periods. In this study this method has been used to make seasonal comparisons of the daily activity rhythms of particular individuals and to monitor the duration of bouts of incubation and brooding. Full details of the equipment and field methods employed are appended.

Techniques of Radio location

The roding areas of individual radio tagged males were determined by accumulating through the season all observations of marked birds passing overhead. Observations were made throughout the wood, mainly at the intersections of rides which allowed the longest view of roding birds. Even so, they could seldom be observed for more than a few seconds from such places. Accordingly, additional direct observations of radio marked birds were made from the tops of trees and other vantage points. At high light intensities the flight paths of roding birds could be plotted accurately over considerable distances by reference to known landmarks. However, even when light levels were too low to see the roding birds some information on the extent of roding grounds could still be obtained by gauging the strength and direction of the radio signals alone.

The position of stationary birds was determined by triangulation with the hand-held Yagi after initial location with the tree-mounted six-element Yagi. Whenever possible the bird's position was determined from distances less than 50 m away as signal reflection and diffraction reduce the accuracy of triangulation at long range in woodland, particularly after the leafing of the tree canopy. The accuracy of position fixes was checked regularly by locating the radio tagged birds visually at very close range, but without flushing them. This method also enabled the positions of birds to be determined exactly whenever necessary, for example to locate nests (particularly late in the season when the incubating bird is usually completely hidden by vegetation), broods (in order to weigh chicks to study their growth) and feeding sites (identified by the presence of the woodcock's distinctive droppings) and to determine the diets of individual birds (from examination of faeces). All observations were plotted on large-scale maps of the study area (1:10560).

The Effect of Radio Packages on Woodcock

Woodcock fitted with transmitters have moulted, roded, mated, laid eggs and reared broods. However, this does not necessarily mean that the radio packages have no effect on their behaviour, although unfortunately by the very nature of the species it is impossible to make quantitative comparisons between radio tagged and unmarked woodcock. However, as some individuals have been equipped with transmitters more than once in the same (up to three per season) or successive breeding seasons (ten individuals) it has been possible to make some observations on the physical effects of their wearing harnesses.

Four individuals are known to have caught their beaks in the elastic harness (from one to ten weeks after attachment), presumably while preening, and in one instance this caused the bird's death. This problem probably results from a too loosely fitting harness. A further two birds had suffered extensive feather loss on the breast and flanks although their harnesses did not appear ill-fitting.

A high proportion of the birds marked early in the breeding season have disappeared within ten days. These may have been migrants from overseas (two known cases) or possibly they may have moved as a consequence of being disturbed too early in the breeding season.

After radio tagging most birds undergo a period of "adjustment" lasting from one day to a week, during which they may cease displaying (males) or fail to make the regular shifts in location at dusk and dawn, which are characteristic of radio marked birds. There is, however, much individual variation in the response to radio attachment and birds radio tagged previously have displayed the same day they were caught.

In our opinion any effects on behaviour are outweighed by the usefulness of the technique. Put simply, without radio tracking many aspects of the woodcock's behaviour and ecology would remain entirely unknown. Some effects of radio packages on behaviour recorded in studies of other species are briefly reviewed by Kenward (this volume, p. 175).

SOME RESULTS FROM RADIO TAGGING WOODCOCK

Breeding Biology

Females breed in their first year but many first-year males do not display or take part in breeding. Lost clutches and sometimes broods are rapidly replaced. Two females that lost broods (two and ten days after hatching respectively) began incubating replacement clutches only 12 days later. Males do not incubate or accompany broods.

Regularly locating the broods of radio tagged females has shown that chicks can fly when 21 days old but continue to be accompanied by their mother for about another 17—18 days. Whether any females are truly double-brooded has not yet been established but, given the long period in which eggs can be found (mid-March to mid-July), this is a definite possibility.

A high proportion of males return each year to display over the same area. Females are less site-faithful and almost invariably change woods between breeding attempts (a 10 km movement being the furthest recorded).

Mating System

Radio tracking has shown male woodcock to be successively polygynous. Contrary to earlier reports they do not defend either an exclusive or specific area to which females are attracted, and in which mating and/or nesting takes place. Instead males display solitarily over large areas, often over 100 ha in extent. These may overlap completely the areas over which other males rode. Individual males have been recorded displaying over points 3 km apart on the same evening.

Roding males are seldom continuously in the air for more than 20 min (maximum yet recorded 43 min) and usually a bird's evening display consists of two to four flights. Most birds display for about twice as long in the evening as in the morning and the maximum time spent roding by a radio tagged male in a 24 h period has been 64 min. Radio tagged birds have displayed throughout the entire breeding season (i.e. over four months).

When a male finds a receptive female he remains with her constantly (usually 3–5 days) before resuming display flights. The frequency of such non-roding periods during the breeding season can be used as a measure of each individual's ability to locate mates (from none to at least four females per season for males studied). The use of the programmable automatic scanning receiver enables the length of dawn and dusk roding flights of individual males equipped with flight-sensing transmitters to be monitored simultaneously through the breeding season. Variation in the success of individual males in locating (and presumably mating) with females can thus be related to differences in the timing and duration of their display periods (e.g. Fig. 1). The next step will be to remove the most successful males and see how this affects the subsequent display behaviour and mating success of the remaining males.

Feeding Behaviour

Radio tracking has established that in winter (all birds) and early in the breeding season (non-breeders, females prior to laying, roding males) woodcock fly out from woodland at dusk to pasture fields. Around Whitwell Wood arable outnumber pasture fields by 18:1 and the average distance flown to fields by ten individuals was about one kilometre (range 400–1800 m), twice that flown by woodcock wintering in predominantly stock-farming areas on the Lizard Peninsula in Cornwall (personal observation). In winter birds return to woodland at dawn and spend most of the day in heavy cover, but as breeding gets under way birds not associated with nests or broods

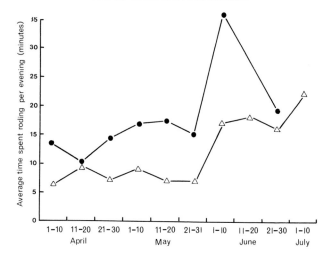

FIG. 1. The success of male woodcock in locating mates in relation to the length of time spent displaying over Whitwell Wood, Derbyshire. For each 10 day interval the average time spent displaying per evening by males which located a mate in that period (circles) is compared to the time spent displaying by those that did not (triangles). The figure is based on over 120 observations of 14 individual males and 13 recorded matings.

switch to spending the night within woodland, often in clearings, or on arable fields.

Examination of the stomach contents of birds shot at various times of day and night has shown that in winter woodcock feed mainly at night on earthworms. In 1980 the activity of radio tagged individuals was monitored continuously throughout the day and night on several occasions in the period March—July, by the use of a strip-chart recorder (see Appendix 1). This enabled seasonal shifts in patterns of activity to be monitored with great accuracy. It was found that in early spring when woodcock are spending the night on pasture, they are active, and presumably feeding, for a high proportion of the night (Table I, Fig. 2) and then inactive for most of

TABLE I

Activity patterns of radio tagged male woodcock in relation to season and habitat. Activity is expressed as the percentage of 5-min periods monitored in which the bird was active.

		Day			Night		
Bird	Date 1980	Number of 5-min periods monitored	% active	Habitat	Number of 5-min periods monitored	% active	Habitat
"Arthur"	1/2 April	138	46	Woodland	110	76	Pasture
"Noddy"	8/9 April	158	97	Woodland	93	18	Woodland
"R"	25/26 April	179	79	Woodland	83	13	Arable
"Arthur"	4/5 May	177	95	Woodland	64	3	Woodland

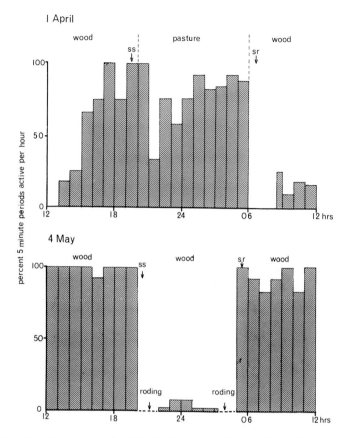

FIG. 2. The circadian rhythm of activity for the same radio tagged breeding male woodcock at two periods in the breeding season. In early April the bird spent the daytime largely inactive in thick cover within woodland, and flew out to a pasture field at dusk where it was active and so presumably feeding for most of the night. In May the bird spent the whole 24-h period within woodland, active for most of the day (and so presumably feeding) and mostly immobile (and presumably roosting) during the night.

the day which is spent in woodland. As nights get shorter, the grass longer, and the ground on pasture fields harder, the birds switch to feeding in woodland during the day and roosting at night (Table I, Fig. 2).

From systematic radio location of woodcock in the breeding season it was apparent that particular areas of the wood were frequented more often than others by nesting and feeding woodcock. Sampling earthworms at the exact sites where radio tagged woodcock were located and comparing these with random sites indicate that in summer woodcock select areas of the wood where earthworm densities are highest (Fig. 3).

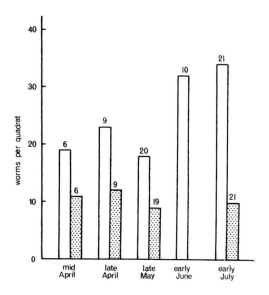

FIG. 3. The average number of earthworms per 0.25 m² quadrat extracted by the formalin method from sites in woodcock feeding areas compared with sites selected at random (shaded) in Whitwell Wood, Derbyshire, 1980. Numbers above the histograms refer to the number of quadrats sampled.

CONCLUSIONS

The technique of radio tagging often has its most immediate impact on studies of elusive species, like the woodcock, for which conventional marking techniques are of little use. For these species even the indiscriminate application of radio tracking and the collection of anecdotal information on movements can sometimes provide useful results. However, to be really effective in ecological or behavioural studies of any species, radio tracking must be applied to the solution of specific problems. With "difficult" species the problems may not even become apparent until after the initial application of radio tagging. However, this should not prevent the research worker from capitalizing on the advantages of the technique by modifying methods and data collection so as to provide answers to relevant questions. In the study described above radio telemetry is providing a unique opportunity to investigate the mating success of individual males, and female mate choice, within a complex polygamous mating system. However, the existence of this type of mating system in the woodcock was not known at the outset of the study.

In this chapter we have tried to show that radio tracking should involve more than just "following animals around"; too many radio tracking studies appear to us to fail to exploit the technique's full potential.

REFERENCES

Amlaner, C. J., Jr., Sibly, R. & McCleery, R. (1978). Effects of transmitter weight on breeding success in herring gulls. *Biot. Pat. Montn* 5: 154—163.
Cochran, W. W. & Lord, R. D., Jr. (1963). A radio-tracking system for wild animals. *J. Wildl. Mgmt* 27: 9—24.
Godfrey, G. A. (1970). A transmitter harness for small birds. *Inland Bird Band. Newsl.* 42: 3—5.
Hirons, G. (1980). The significance of roding by woodcock *Scolopax rusticola*: an alternative explanation based on observations of marked birds. *Ibis* 122: 350—354.
Shorten, M. (1974). *The European Woodcock*. Report on a search of the literature since 1940. Fordingbridge, Hampshire: The Game Conservancy.

APPENDIX

Activity Monitoring

In this study variations in received signal strength caused by movement of radio tagged woodcock have been used to determine the presence or absence of activity.

Most receivers employed in radio tracking studies allow the insertion of an external meter in series with the receiver's own internal signal strength meter which enables signal strength to be recorded on a strip-chart recorder. By this method the activity rhythms of radio tagged individuals can be monitored over long time periods.

We used a portable radio tracking receiver (a programmable automatic scanning receiver supplied by Cedar Creek Bioelectronics Laboratory, University of Minnesota) linked to a 0—1 mA movement Rustrak DC Recorder (Model 288, supplied by Gulton Europe Limited, Brighton). The recorder prints a series of dots through the impinging action of a stylus driven by a chopper bar against pressure sensitive chart paper. The resolution of the recorder is governed by its chart speed and strike rate; changes in signal strength over short time periods are easier to distinguish with fast strike rates and chart speeds. In this study a strike rate of 2/s (drive motor 8 r.p.m.) and a chart speed of 61 cm/h were used.

Movement of the radio tagged bird causes the strength of the received signal to vary as the orientation of the transmitter aerial and/or signal path changes. This variation in signal strength is reproduced on the chart record (Fig. 4). Moreover, certain types of activity may produce a characteristic pattern of signal fluctuations recorded on the chart allowing these activities to be quantified. Woodcock are good subjects for this type of study since they remain immobile for long intervals when not feeding or displaying, making it easy to distinguish periods of activity.

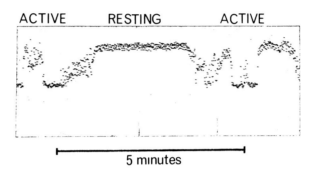

FIG. 4. A section of the chart produced by a Rustrak recorder used to monitor automatically the activity of radio tagged woodcock. Activity is indicated by fluctuations in signal strength.

The interpretation of the chart records is made easier if a capacitor is introduced between the receiver output and the recorder. At slow pulse rates the stylus of the recorder falls back between pulses, but the capacitor, by storing the impulses of current, has the effect of bridging the gap between the received signals making the band of dots on the chart narrower. This makes gross fluctuations in signal strength easier to detect. The value of capacitor used depends on the pulse rate of the transmitter and the strike rate of the recorder. We used no capacitor for pulse rates above 200 beats per minute, a 470 μ capacitor for pulse rates of 140–200 b.p.m. and a 2200 μ capacitor for pulse rates below 140 b.p.m. Switching between different capacitors was made easier by inserting them into the circuit via small crocodile clips.

Both the recorder and receiver were powered by a 12 V car battery. However, changes in voltage caused variation in the chart speed of the recorder which complicated the fitting of a time base onto the chart records. This problem was alleviated by programming a 7.5 s scan of a "nonsense" frequency into the receiver to give a time check on the chart every 2.5 min.

The receiver and the recorder were housed together in a plastic washing-up bowl. This, together with the battery, was enclosed in a large plastic sack. The receiver was linked either to a three-element Yagi or omni-directional ground plane aerial usually strapped to a tree about 100 m from the bird being monitored. Care was taken to ensure that movement of the vegetation between the receiving aerial and the bird was not itself causing fluctuations in received signal strength. The whole system was easily portable, and waterproof.

In this study, this method has been used to monitor the attentiveness of incubating and brooding females and to make seasonal and individual comparisons of activity rhythms.

DISCUSSION

J. White (Loughborough University) — Was it possible to determine whether matings were successful?

G. J. M. Hirons — We know that female woodcock call down roding males and it is possible to observe the pair together for between three and four days. It is not possible, however, to determine whether mating takes place.

R. Kenward (Institute of Terrestrial Ecology) — Do the males have the right of refusal?

Hirons — Females will form bonds with several males and sometimes an additional male may be observed close by a mated pair presumably on the off-chance of obtaining a mating. We have no evidence that males "refuse" the advances of females.

T. Woakes (Birmingham University) — What precautions did you take to ensure that transmitter attachment did not affect the behaviour you were investigating?

Hirons — Woodcock are very difficult to keep in captivity and because of their shy and secretive habits it is not possible to make comparative studies of marked and unmarked birds in the wild and so it is not possible to quantify the effects of transmitter attachment. However, the transmitter harness does cause some feather loss and one bird is known to have died as a result of getting its beak caught in the harness.

S. Pickering (Durham University) — Is there any physical difference between the males that achieve matings and those that do not?

Hirons — Firstly, most first year males do not rode but we do not have enough information to say whether age has any influence on the mating success of roding males. Secondly, there does seem to be a weight difference but it is difficult to differentiate between cause and effect. The lightest males seem to be the ones which achieve most matings and this may be due to weight loss caused by their extensive periods of roding. Roding males decline in weight from $c.$ 320 g to 250 g on average during the breeding season. However, this decline can be very rapid and it may even be an advantage for birds to be light, perhaps because they can fly more slowly!

H. V. Thompson (MAFF) — Did you have any problems in catching woodcock and were your methods selective?

Hirons — It was easy to catch nesting females on the ground. Males could be caught when roding by blanket mist netting. A specialized method involves throwing out a domestic hen in front of roding males which then fly into pre-positioned nets. This method may select the more eager males!

Symp. zool. Soc. Lond. (1982) No. 49, 153—160

Radio Monitoring the Activity of Goshawks

PER WIDÉN

The National Swedish Environment Protection Board, Grimsö Wildlife Research Station, S-770 31 Riddarhyttan, Sweden

SYNOPSIS

The daily activity of three goshawks during 10 January—3 March 1980 is described. The hawks were equipped with pulsed transmitters for radio tracking, and the signals were automatically registered on chart rolls, where it was possible to analyse the activity pattern. Thirty-six days of activity recording of usable quality were collected. Recording was separated into two categories: activity and inactivity. Activity means only locomotive activity, i.e. flying, whereas inactivity includes activity-in-place, e.g. eating and preening, as well as resting. The activity of the goshawks was confined to the daylight hours. The hawks were active in short periods, with an average length of 60 s, whereas the intervening periods of inactivity lasted for 9 min on average. In total, the goshawks were active 3.9% of the recorded time. To analyse the influence of hunger on the activity, the change in activity between observed kills was monitored. The total activity increased with the time after the last kill ($n = 14$ kills). On the same day as a kill, but after finishing the meal, total activity was 3.2%, on the second day 5.8%, and on the third day 8.3%. After this, in all cases but one, the goshawk had made a new kill. This increase in activity was achieved mainly by an increase in the frequency, rather than in the length, of activity periods.

INTRODUCTION

Radio telemetry is an excellent tool for activity studies of free-ranging animals. However, most telemetry studies of raptors have concentrated on locating the birds, rather than monitoring their activity. Studies of activity pattern, hunting techniques, time- and energy budget etc., have mainly been carried out on easily observed species, living in open habitats (e.g. Tarboton, 1978; Wakeley, 1978 a,b,c; Warner & Rudd, 1975), where visual observations are possible and binoculars the only instrument needed. For raptors of more secretive habits, such as the forest-living goshawk, there is little information. Radio tracking was used by Kenward (1976) to locate hawks for observation in deciduous woodland after leaf-fall, but these direct observations are impossible in coniferous forest. This chapter describes how automatic activity recording has been used in a study of goshawk (*Accipiter gentilis*) predation in a central Swedish coniferous forest area.

STUDY AREA

The study was carried out at Grimsö Wildlife Research Station, which is situated in south central Sweden (59°25'N, 15°25'E), in the southern boreal zone (Lindquist, 1966), the south taiga. The research area is fairly flat at 75—150 m above sea level. The vegetation is mainly coniferous forest (74% of the area), containing Scots pine (*Pinus sylvestris*) and spruce (*Picea abies*), sometimes mixed with deciduous tree species. Wetlands, both peat bogs and fens, comprise 18% of the area. Lakes and rivers form 5% of the area, and only 3% of the research area is farmland.

MATERIAL AND METHODS

Goshawks were equipped with radio transmitters, giving a pulsed signal on the 150 MHz band. The general type of transmitter used is described by Cederlund, Dreyfert & Lemnell (1979). The transmitters were mounted on the tail-feathers (Dunstan, 1973; Kenward, 1978). Kenward (1978) evaluated the effect of the transmitter package on behaviour and condition, concluding that tail-mounted packages did not affect the hawks. The transmitter package used in this study weighed approximately 20 g, which is 1.6—2.4% of the body weight of a goshawk, and had a lifetime of about 120 days. Signals were received by omnidirectional half-wave aerials, or by 4-element Yagi aerials, and recorded on chart rolls by a Gould Brush 220 two-channel analogue recorder, in combination with an AVM model LA 12 receiver. The whole equipment was placed in a 20-m tower on a hill, so that the receiving aerials were well above tree-top height of the surrounding area (Cederlund & Lemnell, in press).

A signal good enough for activity recording could be obtained at a distance of about 15 km, when the hawk was flying or perched in a tree. With the hawk on the ground, the activity recording range decreased to a few kilometres.

As the transmitter position or orientation altered, variations in signal strength indicated activity. Since the aerial was fixed to a tail feather, it was always kept straight, and the polarization of the transmitted signal was dependent on the orientation of the tail. The receiving aerials, both the omnidirectional and the Yagi, were mounted vertically. Thus, when the goshawk was perched (tail vertically oriented) the signal was received with maximum strength, whereas when the hawk was flying (tail horizontally oriented, and often close to the ground) the signal strength was weaker. This, plus the general signal pattern, made it possible to determine whether the hawk was sitting still or flying. Activity separation of this type was first tested by continuous, close-range radio tracking, where every change of position could be timed and mapped simultaneously with activity recording, in order to recognize the signal patterns typical for locomotion and stillness, respectively. The material was then classified into activity, which only includes locomotive activity, i.e. flying, and non-activity, including what might be called activity-in-place, e.g. eating and preening, as well as resting. No attempt was made to distinguish further between different types of activity, since that would require more detailed tests to enable interpretation of the different signal patterns. This would necessarily involve visual observation of free-flying, hunting goshawks, which is extremely difficult in this habitat.

To obtain good registration, the signal strength must be adjusted carefully. The hawks could sometimes move long distances very quickly, resulting in drastic changes of the signal strength. This was a major problem, and since the equipment was generally inspected only once a day, the recording was frequently interrupted by periods of either too weak or too strong signal reception. These periods have been excluded from the material. During 10 January—3 March the activity of three goshawks, one male and two females, was recorded. In total, 853 hours (35.5 days) of activity recording of usable quality were collected.

With a chart speed of 5 mm s^{-1} activity periods could normally be observed when as short as 12 s, while registrations of shorter duration could not be distinguished. For inactivity to be observed, a longer uninterrupted period was needed. Thus, periods shorter than 48 s, not classified as activity on the basis given above, could not be treated as separate units of inactivity either. Such periods were therefore classified according to preceding and succeeding activity. Although very little material had to be treated in this way, some inactivity of short duration may have been treated as activity.

In order to record the winter predation of goshawks, radio tagged hawks were monitored using a car with a roof-top Yagi-aerial (Cederlund et al., 1979), and approached when they had made a kill (Widén, in press). A description of this technique was given by Kenward (1980). Of the kills recorded in this way, 14 were obtained simultaneously with continuous activity recording, and thus included in this study.

RESULTS AND DISCUSSION

Goshawk activity was confined to the daylight hours. The hawks left the roost on average 22 min before sunrise (± 4.8 95% confidence limits, $n = 54$), and settled for the night on average 8.5 min before sunset (± 13.2 95% confidence limits, $n = 60$). In total, the goshawks were active 3.9% of the recorded time. Estimated over the daylight hours only, total activity was 7.1%. It was highest in the morning, with a maximum of 15.2% between 8 and 9 a.m. (Fig. 1).

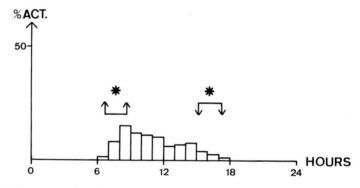

FIG. 1. Percentage of active time per hour during the period 10 January–3 March 1980.

The activity took place in short intervals, separated by longer periods of inactivity. The activity periods averaged 60 s (± 4 95% confidence limits, $n = 1963$), whereas the intervening periods of inactivity averaged 9 min (± 49 s 95% confidence limits, $n = 1975$). The long periods of inactivity during the nights are not included in this calculation.

A hunting goshawk is typically described (e.g. Brown & Amadon, 1968; Glutz, Bauer & Bezzel, 1971; Brown, 1976; Kenward, 1976; Brüll, 1977) as attacking from a short distance, taking the prey by surprise and with great dexterity, either from a perch or approaching flying low over the ground, usually in thick cover. The hunting method may be dependent on habitat though, and in open country goshawks have been observed to use other techniques, sometimes in a falcon-like manner, circling at high altitudes and attacking flying birds in the open after a long dive (Erzepky, 1977).

Kenward (1976) showed that goshawks in deciduous woodland areas of lowland Britain flew most frequently at 4—6 min intervals, for an average 100 m in woodland and 200 m in open country. His observations were probably biased against recording the longest flights. My data indicate a longer interval between flights, and longer flight lines than in Kenward's observations. This may reflect a bias in activity monitoring against recording short flights, since a goshawk moves 100 m in less than 12 s, but it may also reflect a difference between the hunting techniques used in deciduous woodland and coniferous forest.

The activity registration gave only information about when the hawks were active and inactive. Therefore, it could not be determined when the hawks were in fact hunting, and when their behaviour, whether they were sitting still or flying, was not aimed at prey capture. Thus, the material does not allow very far-reaching conclusions about the hunting method of the goshawk. However, with the hawks spending so little of their time flying, it is likely that their hunting consisted to a large extent of waiting on a perch.

It can be assumed that hunger increases the hunting efforts of the goshawk. To test this, total activity was estimated in relation to the time passed since last food intake (Fig. 2). This was used as a measure of hunger, since, with the prey species occurring in this study, every kill was sufficient to give a goshawk a full crop. Very roughly estimated, daily food consumption of the goshawks was 166 g, and every meal gave 200—250 g (Widén, in preparation). The material was pooled day by day, starting on the day a kill was made and after it had been consumed, and continuing with the following days, until

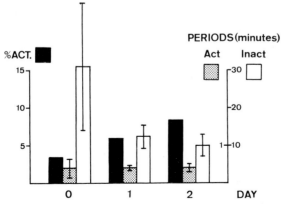

FIG. 2. Percentage of activity, and duration of periods of activity and inactivity, per day after last food intake. 95% confidence limits are indicated for the periods.

a new kill was recorded. In only one of the observed cases ($n = 14$ kills) had the goshawk not taken a new prey when two days had passed.

The goshawks increased their total activity as their hunger increased ($x^2 = 104$, $p < 0.001$). On the same day as a kill total activity was 3.2%, on the next day 5.8%, and on the second day after a kill 8.3%. Such an increase in activity was not reported by Kenward (1976).

In Fig. 2 it can also be seen that the duration of the activity periods was not altered as the goshawks increased their activity. Thus, the increased activity was due to increased frequency of the activity periods, i.e. shorter periods of inactivity. Thus, when the hawks became more hungry, they moved more often from one hunting site to another, but maintained their general pattern of activity, and probably hunting technique. The constant length of the flight periods may be taken as an indication that it was the more frequent changes of hunting site that were important when hunting effort was increased, not the total increase of flying time. A similar relationship, where lack of hunting success influenced predatory behaviour in raptors, was documented by Warner & Rudd (1975).

It is striking that even on the second day after taking a prey, the total activity was as low as 8.3%. At this time, their main interest must have been to find food, yet they spent more than 90% of their time sitting still on a perch. This would be expected if the most important hunting tactic was to hunt from a perch.

So far the material has been used to see how observed kills influenced the activity. However, it is also possible to put it the other way around, using activity registration on chart rolls to determine when a prey has been taken. After registration of activity and recording of predation had been carried out simultaneously for some time, it became evident that the signal pattern on the chart rolls was very typical when the hawks made a kill. A period of intense activity was abruptly replaced by a long period of inactivity with poor signal strength (the hawk on the ground, feeding). Although not used in this study, this may become a valuable way of simultaneously monitoring the predation of several goshawks from a central, well positioned receiving station.

ACKNOWLEDGEMENTS

I am grateful to Kjell Larsson for assistance with the field work, and Karin Åström for assistance with the compilation of the material. I also thank my other colleagues at Grimsö for valuable discussions, and Erik Lindström, Gunnar Markgren and Robert Kenward for comments and criticism on the manuscript. Financial support was provided by the National Swedish Environment Protection Board and Stiftelsen Olle Engkvist.

REFERENCES

Brown, L. (1976). *Birds of prey, their biology and ecology.* London: Hamlyn.
Brown, L. & Amadon, D. (1968). *Eagles, hawks and falcons of the world.* Feltham: Hamlyn.
Brüll, H. (1977). *Das Leben europäischer Greifvögel.* Stuttgart, New York: Gustav Fischer.
Cederlund, G., Dreyfert, T. & Lemnell, P.-A. (1979). *Radio tracking techniques.* National Swedish Environment Protection Board SNV PM 1136.
Cederlund, G. & Lemnell, P.-A. (In press). Activity recording of radio-tagged animals. *Biotelemetry.*
Dunstan, T. (1973). A tail feather package for radio-tagging raptorial birds. *Inl. Bird Band. News* **45**: 3—6.
Erzepky, R. (1977). Zur Art des Nahrungserwerbs beim Habicht (*Accipiter gentilis*). *Ornith. Mitt.* **29**: 229—231.
Glutz v. Blotzheim, U., Bauer, K. & Bezzel, E. (1971). *Handbuch der Vögel Mitteleuropas*: **4**. *Falconiformes.* Frankfurt am Main: Akademische Verlagsgesellschaft.
Kenward, R. (1976). *The effect of predation by Goshawks,* Accipiter gentilis, *on Woodpigeon,* Columba palumbus, *populations.* D.Phil. thesis: University of Oxford.
Kenward, R. (1978). Radio-transmitters tail-mounted on hawks. *Ornis scand.* **9**: 220—223.
Kenward, R. (1980). Radio-monitoring birds of prey. In *A handbook on biotelemetry and radio tracking*: 97—104. Amlaner, C. J. & Macdonald, D. W. (Eds). Oxford and New York: Pergamon Press.
Lindquist, B. (1966). Vegetation belts and floral elements. In *Atlas of Sweden.* Stockholm: GLA.
Tarboton, W. (1978). Hunting and the energy budget of the black-shouldered kite. *Condor* **80**: 88—91.
Wakeley, J. (1978a). Activity budgets, energy expenditures and energy intakes of nesting Ferruginous Hawks. *Auk* **95**: 667—676.
Wakeley, J. (1978b). Factors affecting the use of hunting sites by Ferruginous Hawks. *Condor* **80**: 316—326.
Wakeley, J. (1978c). Hunting methods and factors affecting their use by Ferruginous Hawks. *Condor* **80**: 327—333.
Warner, J. S. & Rudd, R. L. (1975). Hunting by the White-tailed Kite *Elanus leucurus. Condor* **77**: 226—230.
Widén. P. (In press). Goshawk predation in a central Swedish coniferous forest area. *Int. Wildl. Congr.* **14**.

DISCUSSION

V. B. Kuechle (Minnesota) — Although your technique sounds useful certain caution is necessary when receiving signals at a distance in woodland as the polarization may be affected. There would be no problem in open country but in woodland the signal may change from a vertical to a horizontal polarization.

P. Widén — I think this will depend to some extent on the height of the receiving antennae. My receiving antennae were about 20 m above the forest canopy.

Kuechle — If the bird is on the ground and the aerial above the canopy the signal will tend to pass through the trees and over the top of the canopy rather than in a straight line and such conditions are exactly when polarization changes occur.

Widén — I am convinced that my method works but would agree that this point should be tested.

R. Kenward (Institute of Terrestrial Ecology) — I wonder how much of your signal change is due to polarization change and how much is due to a bird dropping down from a perch and flying along at ground level. The latter would also cause a fall in signal strength. By the same argument an increase in height would raise the signal strength but in my experience goshawks rarely fly high in winter.

Widén — Goshawks were not observed flying high in January and February. I think the fall in signal strength is due to a combination of change in polarization and the transmitter being closer to the ground.

Symp. zool. Soc. Lond. (1982) No. 49, 161–173

Radio Tracking Greater Horseshoe Bats with Preliminary Observations on Flight Patterns

R. E. STEBBINGS

Institute of Terrestrial Ecology, Monks Wood Experimental Station, Abbots Ripton, Huntingdon, England

SYNOPSIS

The greater horseshoe bat (*Rhinolophus ferrumequinum*) in Britain has undergone a substantial decline in range and numbers in the past century and is now considered to be endangered. An autecological study in Wales has been in progress for four years, but little information has been gathered on flight and feeding behaviour. A variety of techniques were used to learn more about flight patterns, but none gave more than a glimpse of the bats' total activity.

Twenty-seven greater horseshoe bats weighing from 17 to 27 g have been marked with radio transmitters. Packaged transmitters with 15 cm whip aerials weighed about 2.0 g (up to 12% of body weight) when glued mid-dorsally to the fur. These radio marked bats have provided much new information on roosts, flight patterns and feeding areas, and roosting associations, that could not be gathered in any other way. The transmitters do not appear to affect the bats' flight, behaviour or survival.

INTRODUCTION

An autecological study has been in progress for four years on the greater horseshoe bat, *Rhinolophus ferrumequinum*, in south-west Wales. Results have been gathered on population dynamics and distribution, but little on the flight paths, foraging behaviour and roosting associations. This chapter describes the first radio tagging trials on this small insectivorous bat: the problems, some solutions and requirements for improved equipment.

CONSERVATION AND RESEARCH PROBLEMS

R. ferrumequinum is considered to be endangered in Europe because of the very large decline in numbers (e.g. in Britain *c*. 99% loss in the past century). The study colony in Wales was estimated at many thousands in the 1950s, but now consists of about 350 bats occupying an area of 2500 km². Because of the decline and precarious status of the populations, research must be conducted very cautiously. However, there is also an urgent need to measure present trends and find the causes of declines. *R. ferrumequinum* are very sensitive to disturbance in roosts, although they do not appear to be under stress when caught and handled.

One of the most useful tools in ecological studies is to be able to identify individual animals. Recognition of individuals from natural markings is perhaps the ideal, but is limited to a few species (Pennycuick, 1978). Other methods often involve capture/recapture, but this may affect behaviour and survival. Ringing bats has been the most used and reliable system, but is limited in usefulness for studying the ecology of bats, because the individuals cannot be identified at a distance.

All bats are individually marked using magnesium alloy rings around the forearm, but the animals need to be recaught each time in order to read the number. Ringing was the cause of much mortality before better designed rings were made (Stebbings, 1978). Although ringing and recapturing gives information on individual survival and on population dynamics and movements between capture sites, it cannot give more than a glimpse of the bats' activity. Longer glimpses can be obtained using ultrasonic detectors (range up to 10 m) and image intensifiers (effective range up to 50 m), but neither of these devices copes adequately with bats that fly rapidly close to the ground and through dense woodland. Discrete flight paths have been plotted up to 7 km from a major roost, and by capture/recapture we have found some bats moving a straight line distance of up to 41 km between roosts in a few hours. Food is caught and eaten both in the air and on the ground, and bats may spend all night away from the day roost.

Until recently we had no information on the bats' foraging behaviour and whether they had temporary night roosts, and if so, what roosting associations were formed. Although about 40 roosts were known throughout the area, simultaneous visits to all would reveal at best 60% of the colony in summer, but only 10% in winter, and it was obvious that there were other unknown roosts. It was important to find these other sites, because loss of roosting places is thought to be one of the major causes of population decline.

Tagging bats with radio transmitters was thought to be a possible means of answering questions on where the bats foraged and roosted. Few bat studies using radio transmitters have been published. Williams & Williams (1970) in Trinidad successfully marked

Phyllostomus hastatus weighing 60—110 g with 7—8 g transmitters (6—13% of body weight). Morrison (1978) in Panama fitted *Artibeus jamaicensis* weighing 45—50 g with radios of 3.5—4.5 g (7—10%), and these were carried for 8—22 days before dropping off. Thomas & Fenton (1978) in Rhodesia tagged another fruit bat, *Epomophorus gambianus*, with 8 g transmitters (7—8%) and these had a maximum range of 11 km from a high point.

Small insectivorous bats had not previously been marked in this way, and it was vital to ensure that the radios did not substantially alter normal behaviour or survival. Encouragement to try the technique was provided by publication of a successful radio tracking study in Costa Rica on the small fruit-eating bat *Carollia perspicillata*, which weighed 17—22 g (Heithaus & Fleming, 1978). In Ecuador another fruit bat, *Artibeus hirsutus*, weighing 37 g, was caught carrying *Ficus* sp. fruits weighing 18 g in its teeth (Stebbings, personal observations). Also, pregnant *R. ferrumequinum* carry up to 7—8 g, and hence some bats are capable of carrying considerable loads, often displaced from the normal centre of gravity.

RADIO EQUIPMENT

R. ferrumequinum are small (15—30 g), highly mobile animals, and the requirement was for a compact, lightweight transmitter with long range. Initially we used 12 AVM-SM1 transmitters* (173.3 MHz) attached to small mercury cells with welded solder tags and 150 × 0.28 mm steel guitar string whip aerials. These were encapsulated in epoxy resin, and weighed 1.4—1.5 g. Their performance was very variable, with some having three times the range of others, as measured on direct line of sight between transmitter and receiver. Range varied from 500 to 1500 m using a hand-held Yagi 3-element aerial and an AVM-LA12 receiver. Battery life was not measured prior to field trials, but three transmitters were recovered and monitored until the signals ceased. Those lasted 14, 19 and 20 days.

Improved transmitters have recently been supplied by Biotrack[†] with increased power output at the expense of battery life. Fifteen transmitters were static tested in the same site as the AVM-SM1s, and found to have maximum ground ranges of 1300—1740 m. Six transmitters were clearly audible at 1740 m and would probably have been heard up to 2000 m if direct sight could have been maintained.

---

[*] AVM Instrument Company, 3101 West Clark Road, Champaign, Illinois, USA.
[†] Biotrack, Manor Farm House, Church Causeway, Sawtry, Huntingdon, England.

The Biotrack transmitters weighed 0.87 ± 0.03 g ($n = 10$) and were encapsulated with a mercury 312 cell of 0.84 g, weighing 1.78 ± 0.07 g. When glued to bats, the total weight was 1.9–2.0 g, representing up to 12% of the bats' weight. This is more than the normal recommended maxima on other animals of 10% (Brander & Cochran, 1969) or 5% (Macdonald, 1978). Four of these had transmitter lives varying from five to 22 days (Table I).

TABLE I

Transmitter life in field use and in laboratory trials. Variations were due to battery capacity and transmitter current drain. Transmitters 1 and 3 were recovered in the field after transmissions ceased. They were then tested in the laboratory with new batteries. Transmitter 4 had two separate batteries attached in succession from the same batch

Transmitter no.	Pulse rate/min.	Life in days	Field or laboratory
1	62	5+ 6	F L
2	70	22	L
3	78	10+ 9	F L
4	144	5 8	L L

RADIO TAGGING

Transmitters were usually connected to the cells one to two days before use, because frequency shifts up to 12 KHz were recorded during encapsulation. Also, a further slight shift occurred due to change in capacitance and temperature when the packages were glued to the bats. It was important to measure and record frequency and pulse rate before releasing the bat. Subsequent shifts in frequency were small, but if a bat was lost for a few days the shift was sometimes sufficient for the observer to miss the bat unless the whole band was swept. In practice, frequency and pulse rate were checked daily. A pulse rate of about 80 per minute was found to be best for tracking bats. Slower rates meant that bats could travel a long distance between pulses before a fix was obtained, and faster rates further reduced tag life.

Bats were mostly caught by hand in roosts during the day, radio tagged and released back in the roost in good time to allow them to settle down before evening emergence. Others were caught in nets as they emerged, and were marked and released immediately near the roost. Bats were lightly held in one hand and the transmitter, smeared with rapid-setting epoxy, was placed mid-dorsally over the distal half of the scapulae. Fur was matted into the resin (Epoxy-patch 608*), which set in about 3 min, after which the bat was released. The aerial trailed about 5 cm beyond the bat's feet.

A total of 27 bats was tagged in four sessions: 27 June—10 July (8 bats), 16—30 September (6 bats), 1979; 30 June—7 July (6 bats) and 22 September—2 October (7 bats), 1980. Ten were males (3 juvenile, 5 immature, 2 adult) and 17 were females (3 juvenile, 4 immature, 10 adult). A summary of the duration of observations is shown in Table II. The total number of day-records obtained was 92. There was no significant difference between the number of records obtained for males and for females (males total 32 days, mean 3.2; females 60 days, mean 3.5).

TABLE II

Summary of study periods obtained from 27 radio tagged bats

	Number of days (n)														
	1	2	3	4	5	6	7	8	9	10	11	12	13	14	15
No. of bats observed on n days	6	6	2	6	2	3	1	–	1	–	–	–	–	–	–
No. of bats observed for a time span of n days	6	2	3	4	4	4	–	2	2	–	–	–	–	–	–
No. of bats with maximum possible period of observation of n days	1	1	–	1	3	7	–	–	–	3	8	–	–	2	1
No. of bats observed for maximum possible period	1	–	–	1[a]	2	4	–	–	–	–	–	–	–	–	–

[a] This bat was caught and the transmitter cut off.

Radio tagged bats weighed 17.1—27.0 g, males mean weight 18.8 g ($n = 10$) and females mean weight 19.7 g ($n = 17$). Bats that were low in weight when tagged were later recaught with a similar frequency to those that were heavy when tagged. The initial weight of seven recaught female bats averaged 20.8 g, while five which were not recaught had initially averaged 21.0 g.

In order to assess whether survival or behaviour of radio tagged bats were being affected substantially, the recapture rates for ringed

---

* Epoxy patch from W. J. Furse & Co. Ltd, Wilford Road, Nottingham, England.

bats, and tagged and ringed bats, were examined. The assessment was for bats of both classes caught and released at the same time. There was no significant difference in the percentage recaught between the two groups (Table III). The average number recovered by the end of the first year was about 50%.

TABLE III

Recapture of ringed and ringed and radio tagged bats in the first year from marking

	Ringed		Ringed and radio tagged	
	Number	%	Number	%
Recaught within 6 months	32	36	7	35
Recaught within 12 months	42	47	10	50
Bats not recaught	47	53	10	50
Total	89		20	

On several occasions radio tagged bats were seen roosting either solitarily or in tight clusters, and their behaviour appeared normal.

R. ferrumequinum moult in June and July, and tags glued to fur at that time were likely to drop off sooner than at other times. One dropped off after four days, and another lasted 11 days. Another radio was loose after six days and was cut off. On 19 October 1980, a torpid bat was found still carrying a transmitter which was attached 26 days earlier. This period is probably exceptionally long, because the weather had been cool and stormy, and it was likely that the bat had been hibernating for much of the time. The aerial was rusty, suggesting that the bat had spent some time in wet tunnels where hibernation occurs, rather than in dry roofs which form the main summer roosts.

When the radio drops off, a hairless patch about 1 cm^2 remains. Bats marked in June/July and recaught in September were found to have regrown their hair totally. Autumn-marked bats were not recaught until the following autumn, but because of hibernation, a bare patch probably remains throughout the winter. Since the body temperature of these bats in hibernation normally approximates the ambient (normally 6—11°C), the lack of fur should not cause inconvenience.

Few weights were obtained of bats immediately after radio loss.

One gravid four-year-old female gained 2.6 g in 11 days, and of two one-year-old immature males, one increased 0.8 g in 26 days and the other 0.1 g in six days.

RADIO TRACKING

Before radio tagging bats, we placed radio transmitters in different parts of the study area so as to become familiar with the varying effects which terrain and objects such as trees and buildings have on signal strength and direction. It was vital to be completely familiar with local topography, roads, tracks, buildings and roosts. In the previous two years, we had already plotted some of the major flight paths, and knew that the bats preferred sheltered to exposed sites. This field tracking experience proved its worth when radio tracking, because the bats fly at 6—7 m s^{-1}, and can quickly fly out of receiver range, so it was helpful to be able to predict where they were likely to be going. Most tracking was done using a single receiver, but a little was done with two, combined with walkie-talkies, and this helped to confirm our predictive abilities.

Initially two people did the tracking, but later one person with a continuously-running tape recorder was found to be sufficient.

An observer stood on a high point a few hundred metres from the roost in the expected flight path, and listened to the transmitter signal. Because the roosts were often in buildings with walls up to 1 m thick, the signal was considerably attenuated while the bat remained inside. Bats often flew around inside for some minutes before emerging, and these internal movements could be detected by slight variations in the signal strength. Immediately the bat emerged, the signal strength suddenly increased. The observer constantly swung the aerial from side to side to the null points, with the gain control set as low as possible. A constant commentary was given on where the bat was, and what it seemed to be doing, and an assistant with a stopwatch and maps wrote down a summary. Before we started radio tracking, we had been advised that our position should be changed before the bat went out of range. This we did, and in many cases it resulted in us losing the bat for the rest of the night. Later we found it was better to wait until up to one minute after losing contact, because foraging bats fly back and forth over a feeding area, and would often come back into range. When tracking, we were always with a vehicle, so we could change our position quickly. Most movements to new listening positions took 3—4 min, and if the bat was not heard within another minute, we found it best to return to the initial site. In the first 30 min of activity, our predictions were usually correct, but later on we often failed to relocate the bat. We believe this was caused by the bat changing from a slow twisting and turning foraging flight to a rapid direct flight to a roost.

R. ferrumequinum were known to catch and eat some food on the ground, and often the signal would disappear from being clearly audible to nil within one pulse. The aerial was kept pointing at the same spot where the signal disappeared, and eventually the signal would reappear, suggesting that the bat had been on the ground. It was already known that transmitter range was very short with radios placed on the ground in wet grass. At other times the signal faded and disappeared within two or three pulses, and on these ocassions we presumed that the bat had flown behind a hill or into a steep-sided valley. This was usually confirmed when we moved to the predicted area.

Initially, predictive mistakes were usually made after the first move to a new listening site, because of the difficulty of deciding on the signal direction. We were often wrong by 180°, and this resulted in our failing to locate the bat after the next move. As knowledge of the bats' likely behaviour improved, such errors became far fewer.

At the end of a night's tracking, the results were plotted on maps, with an interpretive written summary. Only large-scale movements in excess of about 5° laterally and range ±10% could be detected, but this still allowed adequate assessment of the bats' behaviour and location. In this way, the time each bat spent foraging over various habitats and at rest could be measured.

In some areas it was impossible to radio track bats because of interference on 173.3 MHz due to electricity power lines. Also two nights' observations were lost because of powerful radio transmissions on 173.35 MHz.

BAT MOVEMENTS

In the two years prior to radio tagging, a large number of caves, mines, tunnels and buildings were visited to look for, ring and recover bats. In this way a picture of movements was gradually built up, but it was clear that we were obtaining only a glimpse of the total (Fig. 1).

As soon as we started radio tagging, *Rhinolophus ferrumequinum* were discovered using roost sites totally different from those previously known. We also lost many of the radio tagged bats, and despite frequently searching 300 km² by road, stopping about every

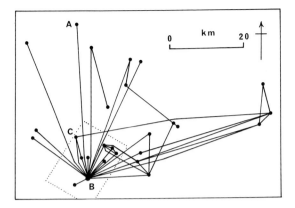

FIG. 1. Pattern of movements recorded by mark/recapture for *R. ferrumequinum* in S.W. Wales. The large dots are roosts. Most of these movements were recorded with time span between recaptures of days, weeks or months. One bat moved from the roost at A to the breeding roost B and then to C in 24 h. The rectangular area enclosed by the dotted line is shown in Fig. 3.

1 km, we failed to relocate most of them. In an effort to find some of these bats, the receiver was attached to a light aircraft in September 1979, and 850 km² was searched on two successive days. Three radio tagged bats in known roosts were heard from 8 km at a height of about 300 m, but none of the missing three bats was found. From this we presumed that either the lost transmitters had failed, or the bats had flown beyond the search area, or that bats were in underground roosts which shielded the signals. In fact, the three missing bats were relocated in known roosts a few days later in the centre of the searched area. While it was possible that they had travelled beyond this area (See Fig. 1), it seemed more likely that they were deep in caves. There are a large number of caves in cliffs close to the areas where the bats were radio tagged, and the aircraft was flown directly towards these cave entrances, but nothing was heard.

In one cave, a test was carried out to find whether radio signals could be detected outside. The limestone cave runs horizontally into a cliff face for about 36 m (Fig. 2). The passages are about 2 × 0.5 m in section. Transmitters hung throughout from the walls and roof were just heard from the entrance, but when placed on the ground at the end (A) nothing was heard. A transmitter hung halfway along could be heard from an angle of up to 20°, 50 m from the entrance. It seems that the bats must have been in longer or more complex caves.

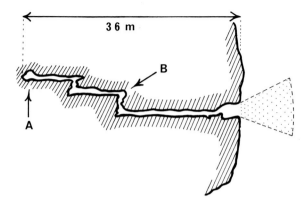

FIG. 2. Plan of limestone cave in which radio transmitters were placed. With the aerial at the entrance and directed into the cave signals could be heard from transmitters hung in all areas. When transmitters were placed on the ground at A (as might happen if a transmitter fell off a bat, or if the bat died still carrying it) no signal was heard, but faint signals were detected when the transmitter was similarly placed at B. Transmitters hung at B could be heard at least 50 m from the entrance up to an angle of 20° either side of the entrance passage (within the stippled area).

Further evidence suggesting that some alternative roosts were in nearby caves was obtained by maintaining a watch for missing radio tagged bats at dusk near the caves. For several nights some of these bats were detected foraging in the area, but the day roosts remained unknown. These and other foraging areas discovered with the help of the transmitters are shown in Fig. 3, together with the longer movements between roosts. Figure 4 shows examples of the movements of one bat on three successive nights.

CONCLUSIONS

Identifying individuals in a population is a useful step in studying the ecology of bats, as of any animal. Radio tagging is one such means to that end, which is particularly useful for a fast-moving nocturnal animal. *R. ferrumequinum* appear to carry these tags without obvious harm, and much new information is being obtained that could not be obtained in any other way. However, as stressed by Lance & Watson (1980) and others, it is important to ensure that best use is made of this expensive technique.

In the 42 days spent radio tracking *R. ferrumequinum*, many new roosts were found, often in sites previously considered unsuitable. Also, clues have been obtained as to where other important under-

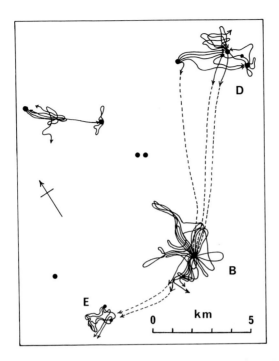

FIG. 3. Some of the large-scale movements recorded in 23 nights by radio tracking *R. ferrumequinum* in 1979–1980. The lines show the range and area where the bats fed and movements between roosts. Large dots indicate known day roosts which were checked daily. Two bats that were tagged in breeding roost B left and did not return. They were later found foraging at dusk at E for six nights. Several temporary night roosts were used (small dots), but the day roost was not found. Almost certainly the day roost was very close to E and underground. Similarly in area D, some bats were missing during the day but foraged there at night. Dashed lines indicate rapid movements between roosts. The routes taken were not known.

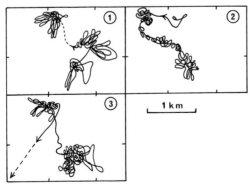

FIG. 4. Recorded foraging movements of an adult female *R. ferrumequinum* for three nights; 26, 27 and 28 June 1979 (Area D on Fig. 3). Just before dawn on 29 June the bat moved 10 km to the breeding roost. The foraging flights lasted 1–2 h per night interspersed with varying lengths of rest in roosts.

ground roosts will be found. We are beginning to discover where bats feed and which habitats are preferred, and we can now study the foraging behaviour in some detail.

The main shortcomings of radio tracking for small insectivorous bats is the short range and relatively great weight of the package. Battery life is also very variable and difficult to predict (Macdonald & Amlaner, 1980). To be most effective, we need a transmitter with a range about 50% better than at present, and a packaged weight of about 1 g. A reliable life of ten days would be sufficient, but longer life would increase the chance of recovering the package after it had fallen off the bat.

ACKNOWLEDGEMENTS

Many people in the West Wales Naturalists' Trust and the Nature Conservancy Council have generously helped with field tracking these elusive bats, organized by Tom McOwat who also provided considerable help with the radio tracking. Robert Kenward has kindly provided much help and advice on radio tracking, which saved a great deal of time and expense.

The work was commissioned by the Nature Conservancy Council as part of its programme of research into nature conservation.

REFERENCES

Brander, R. B. & Cochran, W. W. (1969). Radio-location telemetry. In *Wildlife management techniques*; 95—103. Giles, R. H. (Ed.). Washington: Wildlife Society.

Heithaus, E. R. & Fleming, T. H. (1978). Foraging movements of a frugivorous bat, *Carollia perspicillata* (Phyllostomatidae). *Ecol. Monogr.* **48**: 127—143.

Lance, A. N. & Watson, A. (1980). A comment on the use of radio tracking in ecological research. In *A handbook on biotelemetry and radio tracking*: 355—359. Amlaner, C. J. & Macdonald, D. W. (Eds). Oxford: Pergamon.

Macdonald, D. W. (1978). Radio tracking, some applications and limitations. In *Animal marking: Recognition marking of animals in research*: 192—204. Stonehouse, B. (Ed.). London: Macmillan.

Macdonald, D. W. & Amlaner, C. J. (1980). A practical guide to radio tracking. In *A handbook on biotelemetry and radio tracking*: 143—159. Amlaner, C. J. & Macdonald, D. W. (Eds). Oxford: Pergamon.

Morrison, D. W. (1978). Foraging ecology and energetics of the frugivorous bat *Artibeus jamaicensis. Ecology* **59**: 716—723.

Pennycuick, C. J. (1978). Identification using natural markings. In *Animal marking: Recognition marking of animals in research*: 147—157. Stonehouse, B. (Ed.). London: Macmillan.

Stebbings, R. E. (1978). Marking bats. In *Animal marking: Recognition marking of animals in research*: 81—94. Stonehouse, B. (Ed.). London: Macmillan.
Thomas, D. W. & Fenton, M. B. (1978). Notes on the dry season roosting and foraging behaviour of *Epomophorus gambianus* and *Rousettus aegyptiacus* (Chiroptera: Pteropodidae). *J. Zool., Lond.* **186**: 403—406.
Williams, T. C. & Williams, J. M. (1970). Radio tracking of homing and feeding flights of a Neotropical bat *Phyllostomus hastatus. Anim. Behav.* **18**: 302—309.

DISCUSSION

J. White (Loughborough University) — Did the transmitters affect the bats' behaviour?

R. E. Stebbings — I think not. The transmitters weigh only 12% of the body weight of a bat and I think they could carry 20—30% quite easily. A pregnant bat does this. Some species of bat can carry up to 180% of their body weight over short distances. A stomach full of food can weigh up to 30% of body weight.

R. Ransome (Dursley) — As the wing loading of a heavily pregnant bat is critical transmitters should not be used on these animals.

Stebbings — We avoided putting transmitters on highly pregnant animals. One transmitter-carrying bat was in the early stages of pregnancy. Normally the transmitter would fall off before the latter stages are reached.

D. C. Seel (ITE Bangor) — Using only one receiver how could you be sure where a mobile creature such as the bat was at any given moment?

Stebbings — Experience and knowledge of the area are most important. When it was possible to check locations we were invariably spot on. Experience enabled us to judge the distance of the bat from us from the strength of the signal. While in flight we were concerned particularly with knowing in which field, rather than where in that field the bat was.

A. Watson — Can your equipment be adapted for species smaller than the Greater Horseshoe Bat — e.g. Natterer's Bat?

Stebbings — The transmitter used is too heavy for species under 15 g.

V. B. Kuechle (Minnesota) — Did battery life in the field correspond to the predicted life?

Stebbings — No, it was usually considerably less. I have no idea why.

Symp. zool. Soc. Lond. (1982) No. 49, 175–196

Techniques for Monitoring the Behaviour of Grey Squirrels by Radio

R. E. KENWARD

Institute of Terrestrial Ecology, Monks Wood Experimental Station, Abbots Ripton, Huntingdon, England.

SYNOPSIS

Methods for detecting adverse effects of transmitter packages are reviewed, and used in comparing radio tagged grey squirrels with toe-clipped controls. After an initial radio collar design change, inter-season recapture rates and weight changes did not differ between radio tagged squirrels and controls at any time of year, but squirrels newly equipped with collars had less tendency than others to lose weight within a six-day trapping session. The collars did not reduce male sexual status in spring, or summer litter frequency in females, but may have reduced spring litter frequency. The recovery rate for collars was 72%, which enabled causes of failure to be identified and design improvements made. Techniques are described for systematic radio location to determine range size, for radio-assisted surveillance of foraging, and for automatic biotelemetry of activity using collars sensitive to environmental temperature.

INTRODUCTION

If a study using radio tagged animals is to succeed, the following three conditions must be met. (i) The radio equipment must be capable of meeting minimum range, length of life and reliability criteria. (ii) The transmitters must not have an adverse effect on the animals. (iii) Adequate data collection techniques must be developed. This chapter is an example of a study on the grey squirrel (*Sciurus carolinensis*), in which equipment performance, its effect on the animal, and its use in data collection have been continually assessed and improved. Details of the study area and of the methods of trapping and handling used in the study are given in Appendix 1.

RADIO EQUIPMENT

General Considerations for Receivers and Transmitters

With the variety of reliable radio monitoring receivers now available commercially (addresses in Appendix), the cost in time of building and correcting one's own is likely to be greater than the cost of purchase. When buying equipment, however, certain criteria should be kept in mind. Receiver sensitivity is obviously important: minimum perceived signal should be at least -140 dBm for most radio tracking. Selectivity, which indicates ability to distinguish between signals on adjacent channels, should preferably permit a transmitter frequency spacing of 5–10 kHz, but should not be too great. Depending on crystal specifications, simple transmitters may drift about 2 kHz over a 25°C temperature range, so with high receiver selectivity they may drift off frequency and the signal be lost. This problem is accentuated if there is much drift in receiver frequency due to variation in temperature or battery voltage. The receiving frequency of portable equipment can be adjusted if it is known that signals are received, say, 3 kHz lower at -5°C than at 20°C, but with a fixed receiver frequency for automatic recording, the signal may be lost at temperature extremes.

Choice of receiver is influenced by its mode of use. The shape and weight of field-portable receivers should enable comfortable carriage over long distances. These receivers should be rugged enough to withstand the occasional knock against a vehicle floor or a tree. Commercial receivers are not completely waterproof at present; some let water in so easily that they must be carried in polythene bags during rain.

Some receivers now contain programmable memories for storing a number of transmitter frequencies. It is easier to remember and set frequencies as a one- or two-digit memory address than as a three-digit kilohertz definition; however, for routine tracking of up to 15 squirrels at a time, this advantage is over-ridden by increased receiver bulk and weight. A receiver which steps (scans) automatically through a series of programmed frequencies at selected time intervals has proved useful in rapid-moving searches (e.g. from a car) for several dispersing animals, and in an automatic recording system.

Of the 104 MHz and 173 MHz bands available for wildlife research in Britain (Skiffins, this volume p. 20), 173 MHz is most suitable for monitoring grey squirrels. Although this band is shared with some industrial users, no British researcher seems to have been inconvenienced by interference. Aerial dimensions are smaller at the higher frequency. This enables the use of a three-element hand-held Yagi, which gives a -3 dB beam width of 60°, and a 6 db gain over the $\lambda/2$ dipole. An equivalent 104 MHz Yagi is too large for portable use. Whip or collar transmitting aerials are more efficient, at the same dimensions, on 173 MHz than on 104 MHz, which makes the higher frequency desirable for most small mammal and bird studies. Receiving aerials are easy to build, but require sophisticated equipment to ensure optimum tuning. The extra cost of commercial units will often be less than the cost in extra distance travelled and time taken to detect weak signals when using inefficient aerials.

Transmitters can be obtained commercially, as ready-to-mount packages including batteries and aerials if required. Building one's own transmitters saves money and allows freedom to modify signal characteristics as well as package design, but is practical only for those able to solder miniature components and prepared to accept some transmitter failures as they gain experience. Moreover, building radio equipment requires a Home Office testing and development licence in the UK, and is generally only worthwhile for those using large numbers of transmitters. The performance of the completed units must satisfy the Home Office specifications (see Skiffins, this volume, p. 22; Beach, this volume, p. 38). Another option, which allows flexibility in package design and some financial saving, is to buy commercial transmitters, then add batteries and the chosen attachment. It is worth noting that poor package design can lead to failure (Sargeant, 1980), whose source can often only be identified if the package is recovered after use. For the squirrels, trapping two or three times a year has led to a 72% package recovery rate, and enabled several design improvements to the radio collar developed by Wood (1976).

The Squirrel Collar

The collar is formed from an 8–10 mm wide tuned loop aerial of 21 gauge brass strip, which gives rigidity and some protection to the transmitter and battery (Fig. 1a). The transmitter circuit (Fig. 1b) follows Cochran (1967), partly revised as in Amlaner (1978). For maximum range with minimum current drain, the tank circuit is selected for each transmitter as follows. Rather than using an expensive miniature variable capacitor (at C_4) in this circuit, combinations of 15–22 pF capacitors with 12 or 14 turn coils can be chosen by trial and error. It is usually 15 pF with a 14 turn coil for the squirrel collar transmitter. Too low a capacitor value gives a slow, lengthy "bleep" when heard on a beat frequency oscillator (BFO) equipped receiver and causes excessive current drain. Too high a value gives a fast "blip" with reduced signal strength. Tank circuit selection is done before encapsulation (potting), bearing in mind that potting with Rapid Araldite detunes the circuit by an amount equivalent to a 1–2 pF capacitance increase. The loop tuning capacitor (C_3) can also be selected for each collar, but is usually 5.6 pF

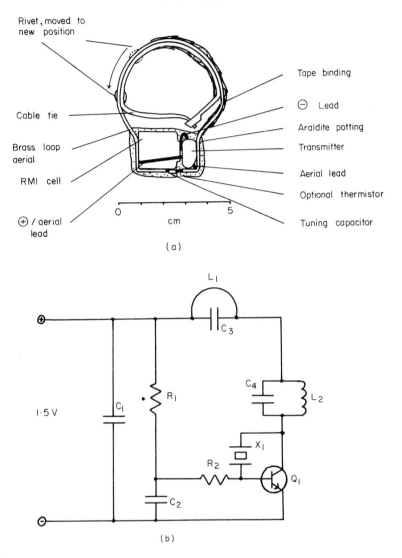

FIG. 1.(a) Transverse section through squirrel radio collar, showing position of components. Binding tape is omitted from one side for clarity. (b) Transmitter circuit. L_1 = loop aerial; L_2 = 12–14 turns 40 SWG on 3 mm diameter; R_1 = 330 k or thermistor; R_2 = 8.2 k; C_1 = 1n; C_2 = 2.2 μ; C_3 = selected 3.3–5.6 p; C_4 = selected 15–22 p; Q_1 = MMT 74; X_1 = 3rd overtone crystal.

for this collar size. Collars with a larger diameter or wider brass strip for the aerial require lower value tuning capacitors, and *vice versa*.

The transmitter is potted, with a Mallory RM1 mercury cell*
giving 1.35 V, in Rapid Araldite. This epoxy resin is viscous enough
to ensure a thick protective layer in one application, and is hard
enough to withstand nibbling (but not persistent gnawing) by other
squirrels, but can be chipped away fairly easily with a strong knife
to replace a discharged cell. The circuit is completed prior to potting,
a few days before a trapping session. When in use it should run from
the 1000 mAh cell for 8—11 months; as a rule of thumb these transmitters have a drain of 3—4 mAh per day. The negative supply lead
passes close to the resin surface inside the collar. Squirrels cannot
chew the lead there, but the wire can be exposed and cut, to stop
transmission if the collar is not used or is recovered with several
months' cell life remaining. Completed transmitter collars weigh
about 30 g (4—7% of squirrel weight).

Some collars include a miniature bead thermistor ($470\,k\Omega$ at
$25°C$), which replaces the resistor (R_1) and is set with its tip just
protruding from the resin at the bottom of the collar. This circuit
increases the signal pulse rate as temperature increases at the bottom
surface of the "thermistor collar".

Factors Affecting Length of Tag Life

Collars were put on squirrels 120 times in the study, up to four times
per squirrel and three times per collar. Sixty-seven collars, including
18 thermistor collars, have stopped transmitting and 48 have been
recovered.

Most (73%) of the 48 recovered radio collars ceased functioning
because their batteries were exhausted, and functioned again when
these were replaced. Component or connection failure was the next
most frequent (17%) cause of failure: one crystal failed, five packages
contained poor connections and three transmitters failed for unknown
reasons. Damage by squirrels was rare (8%), and none has been
recorded after leads were soldered to the inside of the aerial loop,
rather than to the outside where they could be more easily severed
by gnawing at the corners of the package. Three of the 19 unre-
covered collars were lost by the squirrels.

Operating life of recovered standard collars ended at a fairly constant rate (Fig. 2a). Collars which were not recovered are excluded
because two of 13 were known to have been on squirrels which
emigrated, and the recorded mean life of the remainder was only
67% of that for recovered collars, which suggests that additional

* See discussion

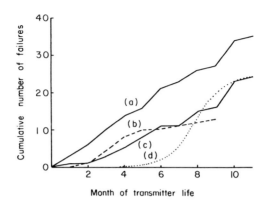

FIG. 2. Rate of failure, with time, of recovered radio collars. Curves are for (a) all standard collars, (b) standard collars with exhausted cells, (c) thermistor collars, and (d) the expected standard collar cell exhaustion rate if all cells performed to manufacturers' specifications.

emigration may have gone undetected. Sixty percent of recovered collars reached 50% of the eight month minimum expected cell life, 40% operated for at least six months and 26% functioned for eight months. The longest recorded life was 325 days.

The operating life of thermistor collars (Fig. 2b) averaged 70% of that of standard collars. The relationship for standard collars with exhausted cells Fig. 2(c) may be contrasted with the sigmoid function Fig. 2d, expected if signal life depended entirely on cell exhaustion and if these cells functioned to specification. Although (c) has some sigmoid tendency, it is clear that cells were not always functioning satisfactorily. This was a major cause of premature failure, since 15 (58%) of the 25 collars which failed before eight months had exhausted the cells. It is possible that some cells were overheated when leads were soldered* to them; a few were noticeably warm despite heat shunting, perhaps because local heating had caused some internal shorting (Harding, Chute & Doell, 1976). This result indicates a need for special care when connecting to cells, and those with solder tabs should be used where possible. The manufacturers also reported some unsatisfactory performance and changed the cell design during the life of these collars.

The cells of all recovered thermistor collars were exhausted, except for one which failed in its fourth month and another in its fifth. Therefore the sigmoid curve (b) mainly reflects cell exhaustion. Pulse rates indicated that these collars were drawing 1.5–3.0 times the current of standard collars when squirrels were in dreys. Moreover,

* See discussion.

thermistor collar tags had a life in mid-winter only 48% of that in autumn, when squirrels spend much more time out foraging. Thermistor collars are currently being modified to reduce current drain when in dreys, without risking too high a load resistance which might upset the circuit function on frosty mornings. Using one resistor in series and one in parallel to the thermistor reduces sensitivity to temperature change. A better solution, which would also conserve battery life in the inactive (drey-bound) animal, would be to use positive temperature coefficient thermistors (in which resistance increases with rising temperature), if these were available in a suitable size and temperature range.

THE EFFECT OF RADIO PACKAGES ON STUDY ANIMALS

Effects Recorded in Mammal Studies

There have been few quantitative comparisons of radio tagged and non-tagged mammals. Habituation of rodents to transmitters has been shown in the laboratory by Bohus (1974). In a more extensive study, Smith (1980) showed that implants* which averaged 10% of body weight had no long-term effect on growth rates of juvenile *Peromyscus leucopus* in the laboratory (despite a slight weight decrease just after implantation) and that the transmitters did not affect pregnancy rates, litter sizes or survival in wild adults. With such a traumatic attachment technique, one would want to be sure that any short-term effects were caused by the radio package rather than by the handling; trapping itself may sometimes cause detectable effects compared with untrapped controls (Amlaner, Sibly & McCleery, 1978). Leuze (1980) found that putting radio collars on wild water voles (*Arvicola terrestris*) had no effect on sighting-based range sizes, on litter size or on social status, although activity was reduced for a day. In contrast, Hamley & Falls (1975) showed a long-term reduction of exercise wheel activity by radio collared *Microtus pennsylvanicus*. The adverse effects in this latter study may have resulted from poor equipment design, since the *Microtus* collars had 10 cm whip aerials and weighed 6–8% of body weight. Tester (1971) recommended that mammal collars should not exceed 6% of body weight. Highly conspicuous or strong smelling collars may have influenced mule deer (*Odocoileus hemionus*) and pronghorn antelope

---

* This procedure, when carried out in the United Kingdom, requires Home Office authority under the *Cruelty to Animals Act 1876*. Persons wishing to carry out such procedures should consult the Home Office Inspectorate at 50 Queen Anne's Gate, London SW1H 9AT.

(*Antilocapra americana*) which refused to accept their radio tagged fawns in some studies (Beale & Smith, 1973; Goldberg & Haas, 1978). On the other hand, Coah, White, Trainer & Glazener (1971) found the same mortality in 81 collared white-tailed deer (*Odocoileus virginianus*) fawns as in untagged fawns. In one large carnivore study, leopards (*Panthera pardus*) captured while wearing radio collars were the same weight as when marked (Hamilton, 1976).

Since it is important to be sure that a data-gathering technique is not itself biasing the observations, it is surprising that only a small proportion of mammal studies have investigated effects of the attached transmitters. Animals which can be captured easily enough for many to be marked can usually be recaptured in sufficient numbers for analysis of weight changes, recapture rates (which indicate survival if there is little emigration), and signs of tissue damage by the radio package. Even if weights and survival are not being studied, a preliminary investigation may suggest package design improvements likely to minimize adverse effects in general.

Effects of Squirrel Collars

Short-term results

When released, about one squirrel in ten rolls up into a ball trying to remove the collar with its hind leg. These squirrels move off "normally" after a few minutes if left alone, and no further difference in behaviour from other radio tagged squirrels has been detected. Tagged squirrels usually run off to a drey, in which they remain after release for 1–5 h.

The majority of squirrels are caught more than once, and some up to ten times during a six-day trapping session. Weight changes within trap sessions in which new squirrels were radio tagged are shown in Table I. Apart from May 1979, untagged and previously tagged squirrels tended to lose weight during a trap session, whereas newly tagged squirrels tended to gain weight, or at least to lose less than the other groups. There was a tendency to tag squirrels caught early in trapping sessions, but the weight change differences remained if the analysis was confined to tagged and untagged squirrels caught in the first two days of a session. Weight changes of previously tagged squirrels were closer to those of untagged squirrels than to those of newly tagged individuals. A possible explanation of this short-term phenomenon is that newly tagged squirrels had a greater food deficit than others because they were feeding less outside the traps, and therefore ate more maize within the traps. Thirty-one killed squirrels from similar traps contained up to 42 g (mean 14.7 ± 0.6 g) of maize in their stomachs.

TABLE I

Weight changes within a trap session of squirrels first tagged with radio collars that session, squirrels wearing collars from previous sessions, and untagged squirrels. Mean weight change (g) is given with 95% confidence limits

Session	Newly tagged				Previously tagged				Untagged			
	1st capture to next capture		1st capture to last capture		1st capture to next capture		1st capture to last capture		1st capture to next capture		1st capture to last capture	
	N	mean	N	mean	N	mean	N	mean	N	mean	N	mean
March 1979	11	1±22	7	−22±31					38	−4±8	22	−19±14
May 1979	4	6±49	3	7±56	11	10±26	10	16±26	23	3±11	20	9±18
August 1979	6	6.7±15[a]	2	12	6	−5±16	3	−7±40	11	−11±8	6	−9±12
May 1980	6	−1.7±21	5	−3±13[b]	7	−10±15	7	−15±17.	22	−8±10	16	−19±11
August 1980	6	10.8±13[c]	2	12	4	−5±19	2	−5	19	−12±11	10	−16±19

[a] Newly tagged differ from untagged (t-test, $P < 0.02$). [b] Newly tagged differ from untagged (t-test, $P < 0.05$). [c] Newly tagged differ from untagged (t-test, $P < 0.01$). Other differences not significant ($P > 0.05$)

TABLE II

Weight change and recapture rate between trap sessions of radio tagged and untagged squirrels. Mean weight change is given with 95% confidence limits

| Period | Weight change (g) | | | | | | Recapture frequency (%) | | | |
| | Radio tagged | | | Untagged | | | Radio tagged | | Untagged | |
	N	mean	minimum	N	mean	minimum	N	% retrapped	N	% retrapped
August 1978– January 1979	3	63±157	15	8	79±29	40	9	44	11	91[a]
March 1979– May 1979	12	−17±18	−40	28	−5±17	−105	16	87	42	83
May 1979– August 1979	8	11±39	−30	17	12±24	−55	17	76	31	90
August 1979– December 1979	8	126±32	10	7	80±43	15	22	55	26	58
December 1979– March 1980	10	−92±33	−125	26	−99±13	−170	17	94	30	97
March 1980– May 1980	11	18±29	−60	23	21±14	−60	12	100	35	94
May 1980– August 1980	8	41±34	−20	15	45±15	5	57	57	29	52

[a] Radio tagged squirrels recaptured less frequently than untagged squirrels (Fisher exact test, $P = 0.038$). Other differences not significant ($P > 0.05$).

Long-term results

Change in first capture weight between trap sessions did not differ significantly between radio tagged and untagged squirrels at any time (Table II). Juveniles were excluded from the analysis, because they were rarely radio tagged and usually increased markedly in weight between sessions.

Recapture rates are based on squirrels recaptured in the next and subsequent sessions. The recapture rate for radio tagged squirrels was below that of untagged animals in the first period (P < 0.05). Two of the radio tagged squirrels were known to have died, but the fate of four which disappeared is unknown, because they had been marked with dummy transmitters to test the collar design. One dead squirrel was recovered with lesions on its neck where the cable tie had rubbed. Lesions from the cable tie may have contributed to the death of this squirrel and others which were not recovered. No neck lesions have occurred since the rivets anchoring the cable tie have been moved down the collar (Fig. 1) from their position in Wood's (1976) design, and the recovery rate for tagged and untagged squirrels has been similar.

As Smith (1980) pointed out, increased mortality in radio tagged animals could be concealed if compensated by reduced emigration. However, preliminary evidence from trapping in neighbouring woods indicates that the emigration rate is as low for untagged squirrels as for those with radio collars. It therefore seems likely that the present collar design, put on squirrels in the months shown in Table II, has not markedly influenced survival. Nevertheless, it cannot be assumed that a radio package without detectable adverse effect in one situation will always lack adverse effect. If tagged when subject to severe food stress, a short-term feeding reduction might be enough to bring an animal below the starvation threshold.

The mean weight change between trap sessions was similar between radio tagged and untagged squirrels, and there was no evidence of radio tagged individuals showing extreme weight loss. Although the radio tagged squirrels had maintained body weight when trapped, they may not always have done so during intervening periods. This may be why no radio tagged adult females bred in the spring (Table III), although the spring pregnancy rate did not differ significantly between groups and there was no effect of radio collars on summer pregnancy rates. Spring pregnancies occur during a period in which resources are probably more scarce than in summer (compare December—March weight changes with May—August in Table II).

Although neck lesions no longer occur in squirrels with radio collars, 12% have a 4—8 mm diameter callus on the chin, where the

TABLE III

Breeding status of adult squirrels with and without radio collars

		Females		Males	
Months	Radio collars	Pregnant or lactating	Not breeding	Scrotal testes	Regressed testes
March–	Present	0	7[a]	3	1
May	Absent	7	15	4	4
August	Present	6	2	0	4
	Absent	27	12	1	4

[a] Difference not significant (Fisher exact test, $P = 0.11$)

package has rubbed, and a further 25% have smaller but detectable calluses there. These calluses have no effect on weight changes or recapture rates; the largest was observed on a female which had become pregnant since collar attachment.

It is difficult to investigate possible long-term effects of radio tags on behaviour in animals as shy as these squirrels. Trap range sizes were the only data on which this could be tested. There was no consistent range size difference between radio tagged and untagged squirrels (Table IV). The significant difference for sub-adult females $(0.05 > P > 0.02)$ was based on only five tagged squirrels and may be a spurious result.

TABLE IV

Trap range diameter of squirrels captured in two or more traps during a session. Squirrels already wearing collars at the start of a session are compared with untagged squirrels. Mean range in metres is given with 95% confidence limits

	Sub-adults				Adults			
Radio collars	N	Males mean	N	Females mean	N	Males mean	N	Females mean
Present	4	282 ± 229	5	102 ± 42[a]	13	261 ± 94	13	147 ± 32
Absent	13	227 ± 77	21	155 ± 38	13	220 ± 68	23	178 ± 31

[a] Ranges of radio tagged and untagged sub-adult females differ (t-test, $P < 0.05$). Other differences not significant $(P > 0.05)$

DATA COLLECTION

Terminology

Radio transmitters attached to wildlife enable three types of data collection. One is "biotelemetry" of variables such as heart rate, temperature, movement and vocalization. "Radio location" gives observations of an animal's position. "Radio-assisted surveillance" (called predictive tracking in Macdonald, 1978) is using hand-held equipment to find an animal so that its behaviour can be observed without direct reference to radio equipment. Radio surveillance and radio location are often grouped as "radio tracking". This section considers briefly some techniques for radio biotelemetry, radio location and radio surveillance of grey squirrels.

Radio Biotelemetry Techniques

The first radio tracking observations showed that although they are a diurnal species, squirrels spend much time in their dreys during daylight. It would be pointless to make range and feeding observations at such times, so drey use was monitored automatically to indicate when foraging was likely to occur. Squirrels quite frequently changed the dreys they used overnight, and often used a different drey for resting during the day, so initial recording was of lactating females. Having young, these might be expected to use the same drey all the time. A three-element Yagi aerial was fixed pointing at the drey, so that the signal would peak when the squirrel was there. The aerial was coupled to an AVM Instrument Co. LA 12 receiver, with its output taken to a Rustrak 2.5 cm/h paper chart recorder fitted in a plastic dustbin with a 12 V car battery. One battery charge ran the equipment throughout the chart period of one month, but the recorder timing was inaccurate and had to be reset every few days. Moreover, receiver temperature drift was sufficient to warrant adjustment of the tuning to anticipate temperature change. The tape record for the first female, summed over 12-day periods, showed that she foraged between 05:00 and 09:00, and for a longer afternoon period from 12:00 to 20:00 (Fig. 3a,b,c). Checks on other squirrels indicated that they too were in dreys between 09:00 and 12:00, and the activity pattern resembled that recorded visually in Canada by Thompson (1977). Systematic checks of location and feeding activity were therefore scheduled for 07:00—08:00, 13:00—14:00 and 17:00—18:00.

Recording of a female with a summer litter revealed a similar activity pattern at first, but as the days became shorter her foraging periods tended to merge (Fig. 3d,e,f). She was foraging for longer than the spring female, and the general impression from radio tracking was that squirrels are seldom in dreys during daylight in the autumn. It would be interesting to know what factors influence their foraging at different times of year, without being dependent on females with young. This problem is being tackled with the thermistor

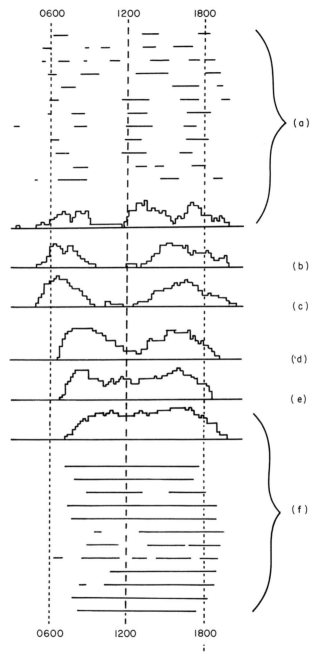

FIG. 3. Diurnal pattern of absence from a litter drey during 12-day periods for two adult female squirrels. Squirrel I in late April (a), early May (b) and late May (c); squirrel II in early September (d), late September (e) and early October (f).

collars. Two prototype collars were used to investigate mid-winter conditions in dreys during 1978. Each collar contained two transmitters: one to monitor squirrel skin temperature and the other to monitor the ambient drey temperature. These collars showed that drey temperature near the squirrel was close to body temperature and not strongly correlated with air temperature, and that there was no drop in body temperature (torpor) during a sub-zero air temperature period which was the coldest for several winters (Proby, 1979). More important, it became clear that an externally-orientated thermistor would indicate when a squirrel was in a drey, no matter which drey, at any fixed receiving station which could register the pulse rate. Use of thermistors in this way has been independently developed by Osgood (1980). For any animal with a warm nest, this technique can give a less ambiguous measure of foraging than detection of collar motion by mercury switching (e.g. Wood, 1976), because it does not matter if the animal moves in its nest. A number of squirrels can be monitored at once, by either plotting by hand at regular intervals (Fig. 4), preferably with the aid

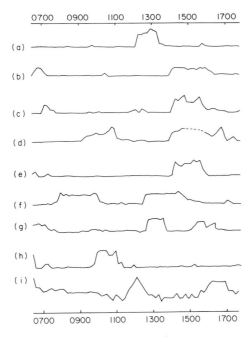

FIG. 4. Activity of nine squirrels on 4 March 1980, recorded as thermistor collar pulse rates at 10-min intervals. Elevated pulse rates indicate absence from a drey or den (hole in tree). Note that (h) and (i) were foraging when recording started, about 20 min before sunrise; (b), (c), (d) and (e), which started afternoon foraging simultaneously, were sharing the same den; the pulse rate of (i) was slow and falling prior to cell exhaustion.

of a pulse interval timer, or recording with a programmable scanning receiver linked to an automatic data logging system for computer plotting and analysis. Initial results indicate that whereas squirrels forage for long periods each day during the warm weather and when the food of autumn is abundant, in late winter (Fig. 4) they spend most time in their dreys, presumably to conserve energy.

Radio Location and Surveillance Techniques

Although signals from squirrel collars can normally be detected in flat woodland at 1 km tree-to-ground and 400 m ground-to-ground, long distance triangulation is impractical for accurate radio location because there is so much signal reflection, or diffraction from regular planting patterns. Squirrels are therefore approached until the signal is heard at minimum receiver gain. They are then normally within about 60 m in trees, or about 30 m on the ground. A low rate of signal strength increase during the approach usually indicates that a squirrel is in a tree, in which case the precise tree is found by moving until the strongest signal occurs directly overhead. The aerial elements are held horizontal during the approach, since this usually gives the most accurate indication of transmitter direction in woodland. Continually swinging the aerial slightly from side to side helps to determine the average peak signal direction despite the effects of reflection and scatter amongst the trees. Overhead signal directions are determined with the elements in one plane, and then checked for spurious effects by checking that the direction is the same with the elements at right angles to their original plane.

A high rate of signal strength increase during the approach, often coupled with failure to detect any signal more than 400 m from the squirrel's home range, indicates that the animal is on the ground. Squirrels on the ground frequently run up into a tree when approached, in which case there is a characteristic marked increase in signal strength. Visual observations and pulse rates of thermistor collars indicate that if squirrels are approached directly they usually hide in the nearest suitable tree, and do not move from tree to tree back to a drey, although they may move away when the observer does. It is therefore important, when tracking a number of animals, that a sequence is followed for approaching the nearest first with minimum disturbance to the remainder. An experienced operator can locate 15 squirrels within an hour, whereas an inexperienced person may take half an hour for each and disturb many others in the meantime.

Squirrels approached stealthily in trees can very occasionally be observed feeding, but more frequently their food must be deduced from chewed remains on the ground. Determining what food is eaten on the ground has proved almost impossible if there is a choice of many alternatives, although remains of dug-up mast or roots can sometimes be detected. Radio tracking makes it easy to locate litter dreys of lactating females, but examining the drey contents by hand had to be abandoned after three dreys, because two females then moved 2 to 4-week-old young to other dreys, and one deserted them. Radio tagged squirrels were presumably more exposed than untagged animals to disturbance during radio tracking. There was no tracking during trapping sessions, and any effect of the short-duration disturbances at other times was not severe enough to influence weight changes or recovery rates.

Although most of the radio monitoring is of squirrels within the wood, dispersal from the study area is also investigated. For long-distance tracking, the gently undulating topography provides excellent high points for detecting signals. After two squirrels had moved at least 8 km from the study wood, one was detected from a distance of 7 km, and one from 5 km, with the receiving system on hills 50 m high.

Detection distances of radio tagged squirrels or other animals can be substantally reduced in flat country, especially if there is much woodland, and searches are therefore impractical without aircraft or collapsible masts to increase aerial height. In general, rugged terrain also makes long-distance tracking difficult, partly because access to high points may be limited and partly because of misleading signal reflections from steep faces. If the expected "minimum detection distance" in one terrain is $1/x$ that in another $(x > 1)$, the number of listening points to be visited as the search pattern expands is increased by x^2.

During observation periods, locations and behaviour are recorded three times a day for each squirrel. This provides data which are comparable for every animal and for every period, without the bias likely if animals are not located systematically and some are more easily found than others. To determine the minimum duration of observation period for home range recording, range diameter (which correlates very strongly with convex polygon range area) was plotted against number of locations (time) for two weeks of the first two observation periods (Fig. 5). Average range size might be expected to continue to increase slightly with time *ad infinitum*, owing to "real" small increases in the area each animal covers. However, the initial "observational" increase, as the animal is recorded throughout

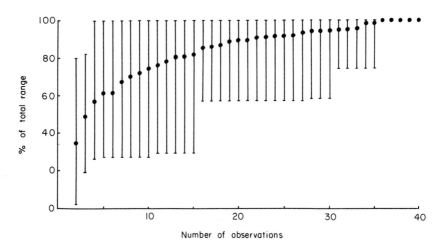

FIG. 5. Percentage of total 40-observation range diameters for 24 squirrels, as a function of consecutive observation number, shown as mean and range of observations. There were three observations per day.

its current range, is mostly complete by ten days (30 locations). The observational increase was very variable, with the maximum range diameter being recorded for two squirrels the fourth time they were located, but 92% of the squirrels had reached 80% of the 40-location range diameter by ten days and all observation periods are now this length.

CONCLUDING REMARKS

Results presented in this chapter show how equipment performance, its effects on the study animal and its use in making observations, can be optimized by collecting data on each aspect. It can be argued that quantitative assessment of the effects of transmitter attachment should be made in every study, yet such assessments are rare, possibly because of lack of time or for fear of the outcome. The results also show that, even with suitable radio equipment and an attachment design available, there can still be "teething problems". Beyond these problems with equipment there is a need to develop field skills in using it. This is perhaps the least emphasized among the many requirements for successful radio tracking (reviewed by Sargeant, 1980); yet the lack of such skills can easily result in few useful data being collected during the first year of a project. Radio tracking, like many other techniques, is seldom as easy as it looks.

ACKNOWLEDGEMENTS

I am most grateful to J. P. Dempster, D. Jenkins, D. W. Macdonald, and A. Village for comments, and to M. B. Haas for typing the manuscript.

REFERENCES

Amlaner, C. J. (1978). Biotelemetry from free-ranging animals. In *Aminal marking: Recognition marking of animals in research*: 205–228. Stonehouse, B. (Ed.). London: Macmillan.

Amlaner, C. J., Sibly, R. & McCleery, R. (1978). Effects of telemetry transmitter weight on breeding success in herring gulls. *Biotelemetry* 5: 154–163.

Baumgartener, L. L. (1940). Trapping, handling and marking fox squirrels. *J. Wildl. Mgmt* 4: 444–450.

Beale, D. M. & Smith, A. D. (1973). Mortality of pronghorn antelope fawns in western Utah. *J. Wildl. Mgmt* 37: 343–352.

Bohus, B. (1974). Telemetered heart rate responses of the rat during free and learned behaviour. *Biotelemetry* 1: 193–201.

Coah, R. S., White, M., Trainer, D. O. & Glazener, W. C. (1971). Mortality of young white-tailed deer fawns in south Texas. *J. Wildl. Mgmt* 35: 47–56.

Cochran, W. W. (1967). *154–160 MHz beacon (tag) transmitter for small animals.* American Institute of Biological Sciences Bioinstrumentation Advisory Council Information Module 15.

Goldberg, J. S. & Haas, W. (1978). Interactions between mule deer dams and their radio-collared and unmarked fawns. *J. Wildl. Mgmt* 42: 422–425.

Hamilton, P. H. (1976). *Movements of leopards in Tsavo National Park, Kenya, as determined by radio-tracking.* M.Sc. Thesis: Nairobi University.

Hamley, J. M. & Falls, J. B. (1975). Reduced activity in transmitter-carrying voles. *Can J. Zool.* 53: 1476–1478.

Harding, P. J. R., Chute, F. S. & Doell, A. C. (1976). Increasing battery reliability for radio transmitters. *J. Wildl. Mgmt* 40: 357–358.

Kenward, R. E. (1980). Grey squirrel foraging. *Rep. Inst. terr. Ecol.* 1979: 65.

Leuze, C. C. K. (1980). The application of radio tracking and its effect on the behavioral ecology of the water vole, *Arvicola terrestris* (Lacépède). In *A handbook on biotelemetry and radio tracking*: 361–366. Amlaner, C. J. & Macdonald, D. W. (Eds). Oxford: Pergamon.

Maconald, D. W. (1978). Radio-tracking: some applications and limitations. In *Animal marking: Recognition marking of animals in research*: 192–204. Stonehouse, B. (Ed.). London: Macmillan.

Osgood, D. W. (1980). Temperature sensitive telemetry applied to studies of small mammal activity patterns. In *A handbook on biotelemetry and radio tracking*: 525–528. Amlaner, C. J. & Macdonald, D. W. (Eds). Oxford: Pergamon.

Proby, C. E. (1979). *The influence of temperature and wind on the activity of the grey squirrel*, Sciurus carolinensis. Unpublished Undergraduate Thesis: Cambridge University.

Rowe, J. J. (1973). *Grey squirrel control.* London: HMSO (Forestry Commission Leaflet no. 56).

Sargeant, A. B. (1980). Approaches, field considerations and problems associated with radio tracking carnivores. In *A handbook on biotelemetry and radio tracking*: 57—63. Amlaner, C. J. & Macdonald, D. W. (Eds). Oxford: Pergamon.

Sharp, W. M. (1958). The art and technique of live-trapping gray squirrels. *Spec. Rep. Penn. Coop. Wildl. Res. Unit* No. 3: 29—34.

Smith, H. R. (1980). Growth, reproduction and survival in *Peromyscus leucopus* carrying intraperitoneally implanted transmitters. In *A handbook on biotelemetry and radio tracking*: 367—374. Amlaner, C. J. & Macdonald, D. W. (Eds). Oxford: Pergamon.

Tester, J. R. (1971). Interpretation of ecological and behavioural data on wild animals obtained by telemetry with special reference to errors and uncertainties. *Proc. Symposium organised by the South African Council for Biotelemetry, Pretoria*: 385—408. Pretoria: Council for Scientific and Educational Research.

Thompson, D. C. (1977). Diurnal and seasonal activity of the grey squirrel (*Sciurus carolinensis*). *Can. J. Zool.* 55: 1185—1189.

Wood, D. A. (1976). Squirrel collars. *J. Zool., Lond.* 180: 513—518.

APPENDIX

Study Area

This is a 36 ha isolated area of mixed woodland on the Cambridgeshire—Northamptonshire border. One side of the wood is 10 ha of 25-year-old beech (*Fagus sylvatica*) or oak (*Quercus* sp.), alternating with rows of spruce (*Picea* sp.) or thujas (*Thuja* sp.). The rest of the wood is mainly mature oak, ash (*Fraxinus excelsior*) and sycamore (*Acer pseudoplatanus*). The wood is on a ridge rising 80 m above the surrounding valleys, and the nearest neighbouring woodland is 500 m away across open fields. Squirrels are studied in the area because the young plantation has had severe bark-stripping damage in recent years (Kenward, 1980).

Trapping

Legg multiple capture traps are used (Rowe, 1973), at a density of two per hectare and as regularly spaced (70—100 m apart) as is consistent with placement close to likely squirrel pathways. Bait is whole maize. Traps are pre-baited for five days, then set three times for two days at one or two day intervals. When set, the traps are checked mid-morning and evening.

Handling

Squirrels are coaxed from traps into 220 cm long handling "cones" of 1.25 cm^2 grid weld-mesh. A 5 × 6.25 cm cross-section fits the fattest adult yet restricts all but the youngest juveniles from turning round to try and bite the fingers closing the open end of the cone. A 30 × 150 cm "U" of 4 mm diameter wire closes this open end for weighing. Squirrels are aged as juvenile until the winter after their birth (when they reach full body size), as sub-adult until they are a year old (after which they can breed), and then as adult. Squirrels caught after their first winter can only be classed definitely as adult if they have achieved breeding condition (large teats in females, enlarged scrotum in males). I have recorded tail and fur markings supposedly typical of sub-adults (Sharp, 1958) in squirrels otherwise known to be adult. Squirrels are marked by toe-clipping (Baum-gartener, 1940).

For radio tagging, the cable tie of the collars is first threaded but not tightened. The squirrel is then allowed to back slowly out of the handling cone under the arched fingers and thumb of the left hand. This hand, wearing a thick yet supple leather glove (a falconer's glove) closes to grasp the animal just behind the front legs. If the legs are forced forward slightly the squirrel cannot move its head much, and the collar can be slipped over the head with the transmitter uppermost. The cable tie's free end points backwards and, with the right hand steadying the collar, can be safely grasped in the teeth to tighten it. The free end of the tie is then cut off with sharp scissors and the collar rotated so that the transmitter is under the chin. Occasionally a squirrel will "faint" and go completely limp in the hand. Such an animal must be rested on the ground and held loosely until it recovers. If this is not done immediately, the breathing ceases and the heart stops beating not long afterwards.

Equipment Suppliers

AVM Instrument Co., 3101 West Clark Road, Champaign, Illinois 61820, USA

Mariner Radar Ltd, Bridle Way, Camps Heath, Lowestoft, England

Biotrack, Manor Farmhouse, Church Causeway, Sawtry, Huntingdon, England

Telemetry Systems, Inc., 5830 N. Shore Dr., Milwaukee, Wisconsin 53217, USA

Advanced Telemetry Systems, Inc., 23859 Ne. Hwy. 65, Bethel, Minnesota 55005, USA

Wildlife Materials, Inc., R.R. 2, Reed Station, Dillinger Roads, Carnondale, Illinois 62901, USA

DISCUSSION

R. Mitson (MAFF) — One should never solder directly to the cases of either mercury or lithium cells. This is dangerous and will reduce the life of the cell. We keep batteries refrigerated and before use carry out an instantaneous discharge rate test on each cell. If they pass this test connections are made by welding.

R. E. Kenward — This is an important point, but it is possible to use heat shunting to prevent the cell becoming heated and theoretical battery lives can be achieved using this method. It is obviously preferable to use batteries with connecting tags.

V. B. Kuechle (Minnesota) — It is possible to buy batteries with connecting tags. One possible reason for not achieving theoretical battery life is using batteries designed for high drain under low drain conditions.

Kenward — The mercury cells used were designed for low drain (i.e. the Mallory RM range). I suspect that the reason is probably the variable quality of batteries supplied.*

A. R. Hardy (MAFF) — We use methoxyfluorane as an anaesthetic for rats. Would the use of an anaesthetic overcome the shock problems you experienced when handling squirrels?

Kenward — I no longer experience this problem which was probably due to holding the squirrels too tightly. I do not favour the use of anaesthetic because of the risks of injury as a result of release before complete recovery.

C. J. Feare (MAFF) — In my experience with birds, those that are caught in baited traps tend to represent a biased sample of the population, being the lower weight individuals. By providing these birds with bait over a long period in traps, they may increase in weight owing to the presence of food that would otherwise not be available. Could this explain the increase in weight of squirrels that were caught on several occasions?

Kenward — I agree that trapping may provide a biased sample of the population. However, this would not explain why squirrels which were trapped and radio tagged tended to gain weight relative to squirrels which were similarly trapped but not radio tagged.

* The RM1 battery has now been replaced by type RM1/PX1. The new type gives consistently low collar life, possibly because it is designed for higher current drain (PX cells) than the original RM cells. I now use only Saft LC 02 lithium/copper oxide cells at 1.5 V.

Symp. zool. Soc. Lond. (1982) No. 49, 197–205

A Semi-automated System for Collecting Data on the Movements of Radio Tagged Voles

I. J. LINN and P. WILCOX

Department of Biological Sciences, University of Exeter, Hatherly Laboratories, Exeter, England

SYNOPSIS

This chapter describes a transportable radio tracking system which can be operated by one person and which allows the collection of a large volume of range and movement data on up to ten animals simultaneously. In the first instance the system will be used with field voles (*Microtus agrestis*). Radio tagged animals are located by triangulation using static, remotely controlled, rotating aerials, thus minimizing observer disturbance in the study area. Data on position and time of location are recorded directly on to magnetic tape by means of a data logger and can later be read directly into a computer file. The operation of the equipment is simple and rapid, permitting a large number of fixes per unit time. A computer program is being constructed which will allow a variety of analyses of the data in terms of home range characteristics, including spatial and temporal factors, and inter-individual interactions. The system has potential for use with many species, particularly small, secretive mammals which up to now have mainly been studied in the field by trapping methods. It could be adapted to include information on activity state.

INTRODUCTION

Of the variety of techniques used to study home range and movements of small mammal species, all except isotope tracking and radio tracking depend on the siting of traps or boards for data collection and so impose a limited, artificial pattern of sites from which a picture of home range is derived. These methods also yield data points at rather long time intervals, and the repeated confinement of animals in traps is an obvious drawback. Although isotope tracking can avoid these problems, it involves the repeated disturbance of the habitat since the observer has to approach within a few metres of the tagged animal in order to locate it. Moreover, the isotope method permits only one animal at a time to be followed in a given area. The system described here aims to maximize the benefits of small radio transmitters in the study of field voles,

Microtus agrestis, by removing observer disturbance in the study area, allowing the simultaneous tracking of up to ten voles, and having the capacity to deal with the large volume of data collected by intensive monitoring.

Small rodent radio tracking so far reported has been almost exclusively in North America (Chute, Fuller, Harding & Herman, 1974; Banks, Brooks & Schnell, 1975; Herman, 1977; Madison, 1977, 1978, 1980; Mineau & Madison, 1976) and involves microtine species somewhat larger than *Microtus agrestis*, under different ecological conditions. The only such work so far reported in Britain was a comprehensive radio tracking study of water voles by Leuze (1976). Information on home range, movements and social interaction of *Microtus agrestis* is scattered and live trapping has been the predominant study technique. A small number of workers have used isotope tracking on this species, for example, Godfrey (1954) using leg rings and Myllymäki, Paasikallio & Häkkinen (1971) using radioactive bait. D. Bell at Exeter University is currently studying the movements of *Microtus agrestis* using isotope tagging.

DESCRIPTION AND OPERATION OF THE SYSTEM

Receiving Equipment

The receiving system is based with minor modifications on that of Smith & Trevor-Deutsch (1980). A major difference overall is in the automation of data recording. The following is a brief description of the system: two telescopic aluminium masts, about 100 m apart, each support a receiving aerial consisting of paired Yagi aerials, and are positioned so that the line joining them runs about 10 m beyond the study area (these distances can be modified; they depend on the dimensions and topography of the site). The motor-driven aerials are rotated by remote control (see Fig. 1). Signals received by the aerials

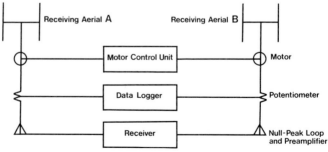

FIG. 1. Diagram of a radio tracking system.

are pre-amplified at the masts, then relayed by cable to a receiver (AVM LA12) at the control point. The aerials are aligned as quickly as possible one after the other, in order to minimize inaccuracy of the radio fix due to movement of the tagged animal. Triangulation follows conventional procedure, the operator locating the direction of the peak signal from each aerial by listening through headphones while rotating the receiving aerial by remote control. The peak is rapidly identified with the paired aerials in phase; they are then switched out of phase, again by remote control, so that a very narrow silent null occurs when the aerial is orientated directly towards the transmitter (the phase difference is obtained by means of a loop being switched into the co-axial cable of one aerial).

Data Recording

The bearing of each receiving aerial is monitored as the change in voltage across a variable linear potentiometer attached to the rotating aerial mast. For each aerial reference the voltage (1 V) applied to the potentiometer, and the output voltage from the latter, are recorded directly on to magnetic tape using an electronic data logger (Cristie CD6). The logger is modified to permit activation by an external scan button; when this is pressed the logger records the two voltages from each aerial together with the time of the scan and a one-digit animal reference number. All this information is recorded in digital form on magnetic tape. The logger has six input channels, each allowing the recording of three digits. The first four channels are used for the four voltage inputs. On the remaining two channels, five digits carry the time in terms of hours, minutes and tens of seconds (measured on an electronic clock which is started when the experiment starts) and the last digit carries a selected animal identifier (any digit from zero to nine) which is set manually. Before the start of an experiment, a ten-digit display is set manually. This carries the following information: experiment identifier (up to two digits), and the absolute time at the start of the experiment — month, day, hour and minute. This information is recorded on the tape as soon as it starts.

Calibration of Aerials and Calculation of Position

The operations described above comprise all that is carried out in the field. The information recorded on the magnetic tape is not translated into data on the actual position of radio tagged animals until entered into the computer. However, to clarify the use of the system, the reasoning behind position calculation will be described here.

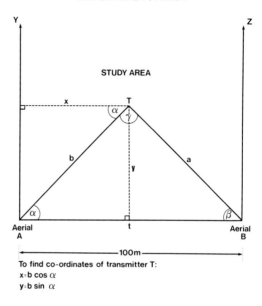

FIG. 2. Calculation of x–y co-ordinates of transmitter location.

The two aerial masts are positioned at opposite ends of the study area baseline, the area being partitioned by an accurate rectangular grid of canes. This grid forms the basis of the computer representation of the site (see Fig. 2). The length of the baseline AB(t) is known, so in order to calculate the position of a transmitter T, the angles α and β must be derived from the voltage data recorded for each fix. To do this, the voltage change at each aerial is related to the degree of rotation of the aerial. The calibration for each aerial is carried out by placing transmitters at known bearings with respect to the aerial, the baseline being designated as 0°. Each bearing is determined a number of times by radio location, applied voltage and potentiometer voltage being noted on each occasion by activating a visual LED display on the logger. After calculating the ratio of applied and output voltages for each reading, the mean ratio is calculated and this figure corresponds to the known bearing. A calibration curve of voltage ratio against bearing (in degrees) can then be drawn, which should be linear. This calibration is incorporated into the computer program so that bearings can be calculated from voltage data recorded during radio tracking.

It should be noted that the *ratio* of output voltage/applied voltage is used to determine the bearing of the transmitter, rather than the output voltage alone. This is to ensure that the calibration remains valid even if the applied voltage should fluctuate.

Once the bearings of a fix, α and β, have been calculated, the rectangular co-ordinates of the position can be determined by simple trigonometry. For transmitters within the rectangular grid AYZB, co-ordinate $x = b \cos \alpha$ and co-ordinate $y = b \sin \alpha$ (length $b = t \sin \beta / \sin \alpha$). The co-ordinates of points falling outside the rectangle AYZB can be determined by similar calculations. All the calculations described will be carried out by the computer program. The fixes resulting from these calculations are points. However, the triangulation can only pinpoint a transmitter within limits of accuracy determined by the degree of precision of bearings that the equipment can produce. The result is an error polygon within which the transmitter is situated. Since it is not practical to record a specific error polygon for each fix, standard error polygons will be determined experimentally, using a transmitter placed in known positions. This information will be included in any published reports. Triangulation becomes increasingly inaccurate with very obtuse or acute bearings (Heezen & Tester, 1967) and ideally voles being tracked should reside largely within the study grid rectangle, at least 10 m from the baseline.

A second source of error is that arising from the time lag between bearings when locating a moving animal. This error will also be investigated experimentally.

Transmitters

Two transmitter designs are being tested for use on this project. The first type is attached around the neck of the vole by means of a collar made from thin plastic-coated wire, which also forms the external aerial. A small piece of heat-shrink tubing secures the collar, with the transmitter body positioned ventrally. The transmitter components — crystal, battery and circuitry — are embedded in a hard, waterproof encapsulating medium. The entire assembly weighs 2 g. The second model has a trailing whip aerial 5 cm in length and is secured to the animal by gluing the transmitter package on to the back of the vole using a non-toxic cyanoacrylate glue. As before, components are embedded in a suitable medium. This transmitter assembly is lighter (1.7 g) and the whip aerial gives it a superior range (150 m compared to about 100 m for the collar transmitter). Both models emit a pulsed signal. The frequency is crystal-controlled and lies within the 173.2—173.5 MHz range. Maximum life span of the transmitters is three weeks.

Preliminary laboratory trials with collar-type dummy transmitters indicate that the weight and bulk make them unsuitable for animals

under 25 g in weight; but larger voles do not appear to be impeded
by the collar transmitter. Initially only large individuals (about 30 g)
will be radio tracked. If attachment of the dorsally-mounted transmitters proves successful, their lower weight may allow the technique
to be extended to include smaller voles.

While fitting the collars, the voles have been anaesthetized using
Penthrane (methoxyfluorane). Recovery has been swift and no
adverse effects have been noted.

ANALYSIS OF DATA

With the advent of intensive monitoring using radio tracking, the
concern with simple line boundaries of home range and the modifications necessary when interpreting trap-revealed home ranges have
become less relevant. Sanderson (1966) stressed the need to relate
home range and movement studies to ecological factors. Recently
Macdonald, Ball & Hough (1980) and Voigt & Tinline (1980) have
reiterated this point with reference to radio tracking studies. The
FORTRAN program being developed for use with this system will
include a number of strategies for analysing the data, which it is
hoped will give real insight into the factors influencing patterns of
space use in voles.

The first step in the program is to convert the recorded voltages
for each radio fix into grid co-ordinates and thus obtain the position
of the tagged animal, as explained on pp. 200–201. This information
will be stored in a computer file where, for each animal under study,
fixes are listed along with the time at which they were made. Using
these stored data a number of different representations of home
range can be derived, yielding information on temporal and spatial
patterns of use intensity. One approach to be used is a plot of
chronologically linked fixes (Voigt & Tinline, 1980); this is particularly useful in showing use intensity shifts over the radio tracking
period. Another approach is to allocate each fix to a small square, so
that various kinds of use patterns can be computer-printed, based on
numbers of fixes per square (Voigt & Tinline, 1980). Use patterns, of
whatever kind, could be related to habitat characteristics such as
vegetation, cover, food supplies, etc.

The system has the capacity to record data on up to ten voles con-
currently. Once it has been adequately tested for tracking individuals,
data on interactions between voles living in close proximity can be
obtained by using consecutive fixes on two or more animals. The
program could then output information on degree of overlap in

home ranges of voles, on whether mutual use of a particular area occurred at different times and on how much actual interaction took place. Interaction data can be related to sex, age, size and reproductive state of the animals involved, as well as to particular habitat features.

In the future it should be possible to modify the logger to record the activity state as well as the position of an animal, if suitable biotelemetry transmitters are developed. If the system proves successful in studying *Microtus agrestis*, it could be used to investigate other small mammal species such as *Apodemus sylvaticus* and *Clethrionomys glareolus*.

SUMMARY OF IMPORTANT FEATURES OF THE SYSTEM

(i) The system is transportable. It can readily be moved to any required site, although the aerial assemblies are intended to remain in position for the duration of a particular study.

(ii) It permits the monitoring of several radio tagged voles simultaneously with no interference in the study area.

(iii) Such a system generates a large volume of data and the inclusion of the data logger is intended to reduce the problems of dealing with data collection and subsequent analysis manually.

(iv) Note that the system is *semi*-automated. Full automation (automatic frequency location and scanning) in a small-scale transportable system is a possible future development, but is beyond the scope of this project. The basic radio location still involves the experimenter fully.

LIMITATIONS

The accuracy of triangulation is a crucial factor in determining the usefulness of this system, since the animals under study are to be tracked from a distance. This point has been dealt with earlier (pp. 200–201). Performance of transmitters and the responses of animals to wearing them provide the other important limitations of the study. First, the transmitters have a lifespan of three weeks at most. The picture of home range and movements obtained thus applies to conditions during this time only. Care should be taken about generalizing from such short-term data. This is one aspect where isotope tracking is superior to radio tracking. However, particular voles may be fitted with a succession of transmitters in order to extend the period of monitoring; whether this is done depends on the specific objectives of the tracking.

The second point is that the weight and bulk of the transmitters may affect movement and social behaviour of the tagged voles so that results are not completely representative of free-ranging animals. Hamley & Falls (1975) and Webster & Brooks (1980) investigated possible effects of fitting transmitter collars on the activity of meadow voles, *Microtus pennsylvanicus*. Neither study established serious, permanent disruption to activity. However, the results from one species have limited relevance to the response of another. It is thus important to look for possible adverse reactions to the carrying of transmitters in the species under study. Even so, especially with smaller mammals, one cannot be certain that the tagged animals are not showing modified behaviour. This problem will only be minimized by further reduction in size and weight of transmitters. Laboratory studies are in progress aimed at assessing the adverse effects, if any, on vole behaviour caused by the presence of the transmitter.

ACKNOWLEDGEMENTS

This project is funded by a grant from the Natural Environment Research Council. The valuable assistance of J. Whitworth and A. Haley in constructing the system is gratefully acknowledged. Illustrations were prepared by the Graphics Unit of Exeter University Teaching Services Centre.

REFERENCES

Banks, E. M., Brooks, R. J. & Schnell, J. (1975). A radio tracking study of home range and activity of the brown lemming (*Lemmus trimucronatus*). *J. Mammal.* 56: 888–901.
Chute, F. S., Fuller, W. A., Harding, P. R. J. & Herman, T. B. (1974). Radio-tracking of small mammals using a grid of overhead wires. *Can. J. Zool.* 52: 1481–1488.
Godfrey, G. K. (1954). Tracing field voles (*Microtus agrestis*) with a Geiger-Müller counter. *Ecology* 35: 5–10.
Hamley, J. M. & Falls, J. B. (1975). Reduced activity in transmitter-carrying voles. *Can. J. Zool.* 53: 1476–1478.
Heezen, K. L. & Tester, J. R. (1967). Evaluation of radio-tracking by triangulation with special reference to deer movements. *J. Wildl. Mgmt* 31: 124–141.

Herman, T. B. (1977). Activity patterns and movements of subarctic voles. *Oikos* 29: 434–444.

Leuze, C. (1976). *Social behaviour and dispersion in the water vole*, Arvicola terrestris (*Lacépède*). Ph.D. thesis: University of Aberdeen.

Macdonald, D. W., Ball, F. G. & Hough, N. C. (1980). The evaluation of home range size and configuration using radio tracking data. In *A handbook on biotelemetry and radio tracking*: 405–424. Amlaner, C. J. & Macdonald, D. W. (Eds). Oxford: Pergamon Press.

Madison, D. M. (1977). Movements and habitat use among interacting *Peromyscus leucopus* as revealed by radio telemetry. *Can. Fld Nat.* 91: 237–281.

Madison, D. M. (1978). Movement indicators of reproductive events among female meadow voles as revealed by radio telemetry. *J. Mamm.* 59: 835–843.

Madison, D. M. (1980). Space use and social structure in meadow voles, *Microtus pennsylvanicus. Behav. Ecol. Sociobiol.* 7: 65–71.

Mineau, P. & Madison, D. M. (1976). Radio-tracking of *Peromyscus leucopus. Can. J. Zool.* 55: 465–468.

Myllymäki, A., Paasikallio, A. & Häkkinen, U. (1971). Analysis of a 'standard trapping' of *Microtus agrestis* (L.) with triple isotope marking outside the quadrat. *Annls Zool. Fenn.* 8: 22–34.

Sanderson, G. C. (1966). The study of mammal movements – a review. *J. Wildl. Mgmt* 30: 215–235.

Smith, R. M. & Trevor-Deutsch, B. (1980). A practical, remotely-controlled, portable radio telemetry receiving apparatus. In *A handbook on biotelemetry and radio tracking*: 269–274. Amlaner, C. J. & Macdonald, D. W. (Eds). Oxford: Pergamon Press.

Voigt, D. R. & Tinline, R. R. (1980). Strategies for analysing radio tracking data. In *A handbook on biotelemetry and radio tracking*: 387–404. Amlaner, C. J. & Macdonald, D. W. (Eds). Oxford: Pergamon Press.

Webster, A. B. & Brooks, R. J. (1980). Effects of radiotransmitters on the meadow vole, *Microtus pennsylvanicus. Can. J. Zool.* 58: 997–1001.

DISCUSSION

V. B. Kuechle (Minnesota) — How far apart did you place the Yagi antennae?

P. Wilcox — A half of one wavelength.

M. J. Delany (Bradford University) — How many animals can you monitor simultaneously?

Wilcox — Up to ten. This is limited by the number of digits available on the data logger.

Symp. zool. Soc. Lond. (1982) No. 49, 207–230

The Home Range of the Hedgehog as Revealed by a Radio Tracking Study

N. J. REEVE

Department of Zoology, Royal Holloway College, Alderhurst, Englefield Green, Surrey, England

SYNOPSIS

The hedgehog (*Erinaceus europaeus*), despite its familiarity and abundance in Britain, remains relatively poorly studied with many facets of its ecology and behaviour inadequately understood. One of the aims of this study was to investigate the home range and movements of wild hedgehogs. The study site was a golf course of about 40 ha in the suburbs of West London.

The study combined a conventional capture–mark–recapture method with radio tracking and a new, cheap and simple radio tracking system was used (Reeve, 1980). A method of computer analysis has been usefully applied to a substantial part of the home range data.

There was a considerable sexual difference in seasonal home range size, the males having substantially larger ranges ($\bar{x}$ 32 ha) than the females ($\bar{x}$ 10 ha) and the sub-adults ($\bar{x}$ 12 ha). This greater range was achieved by more rapid and extensive movement during the activity periods. The nightly distance travelled was found to be a more useful expression of home range in the short term than a conventional area representation. Overnight males were found to travel a mean distance of 1690 m, females 1006 m, and sub-adults 1188 m. Ranges were found to overlap extensively irrespective of sex; territoriality and associated behaviour (e.g. scent marking) were apparently absent.

INTRODUCTION

The hedgehog, despite its familiarity and abundance in Britain, remains relatively poorly studied with much of its ecology and behaviour inadequately understood. Morris (1969), recognizing this, carried out a multi-faceted study which included a small capture–mark–recapture (CMR) program and an experimental radio tracking program. This work showed the feasibility of a field study of hedgehogs, using radio tracking to augment conventional CMR, and thus provide a greater insight into aspects such as home range, territory and the general behaviour of wild hedgehogs.

Few workers have attempted to investigate these aspects, presumably hampered by the nocturnal and secretive habits of the hedgehog. The acute senses of hedgehogs prevent close observation without detection, and their spines make handling unpleasant without gloves. Their habit of rolling up when handled can frustrate the collection of physical data. Day nesting sites are difficult to detect, even with practice, and trapping or searching for active individuals is very laborious.

Because the hedgehog has been very successful in colonizing the suburban environment, a study site was chosen in a built-up area so that the project might also add to our understanding about the ways in which hedgehogs are apparently so well suited to life in the suburbs.

THE STUDY SITE

Many problems arise when attempting to study a nocturnal and secretive mammal in a built-up area, the greatest of these being restrictions of access to private land such as gardens. In order to minimize such problems an open area was sought in suburban West London which would be easily accessible, where unimpeded field-work could be carried out, and yet which would be entirely contained within a built-up area. Ashford Manor Golf Course, Ashford, Middlesex (grid ref. TQ 0771) was found to be ideal in many ways for such a study. The area of 40 ha was relatively easy to cover on foot. It was surrounded by a complex of roads and housing estates to the north, east, and west, and bounded by a major dual carriageway (A308) to the south (see Fig. 1). Hedgehogs could pass freely through most of the perimeter of the area hence their movements would not be artificially restricted. The surrounding gardens provided a typical suburban environment to which the hedgehogs had free access. The golf course itself provided a mosaic of diverse vegetation, copses, woods, areas of undergrowth and rough grass intersected by fairways of short sward with scattered trees and bushes. An abundance of landmarks aided the precise plotting of locations even at night and in poor light.

METHODS

Capture

The usual way of capturing small and medium sized mammals is by trapping; however, this method has many drawbacks, including:

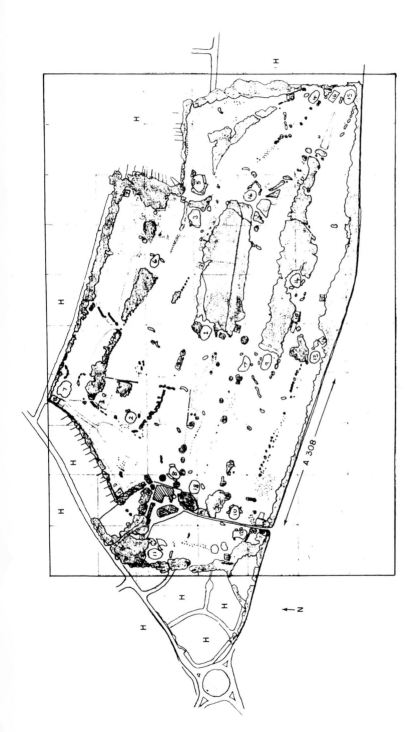

FIG. 1. A map of the study site, Ashford Manor Golf Course. The grid shown superimposed on the study area measures 1000 × 700 m and each grid square is one hectare (100 × 100 m). The areas labelled H are houses and gardens. The A308 runs along the length of the southern perimeter. Shaded areas represent woods, copses, thickets, trees and other tall vegetation; the white areas are the fairways (mown grass). The numbered areas are greens and tees, which provided useful points of reference.

(i) Animals are confined for at least part of their activity period.

(ii) The use of baited traps can influence results

    (a) by altering the animals' dependence on natural food resources;

    (b) by causing "trap addiction" where an animal will enter traps at every possible opportunity (Delany, 1974); conversely the "trap shy" animal will avoid traps.

(iii) Trap deaths and injuries can add to the mortality of the study population.

(iv) Traps require regular checking and maintenance.

For these main reasons, and also because the dissemination of traps large enough to catch hedgehogs over a 40 ha area would be very laborious, trapping was not used. Baited "Havahart" traps (placed outside known nest sites) were used on a few occasions in 1978 but were unsuccessful.

Searching directly for active animals by night was found to be a successful strategy, usually producing on average one capture per man-hour. The method used was to walk quietly and methodically around the search area, listening carefully for noises in the under-growth and scanning the open areas with a handlamp. Hedgehogs do not react to red light so when it was wished to observe their behaviour without disturbance a dark red filter was attached to the lamp.

Marking

Several possible methods of marking were considered; see Twigg (1975) for a review. Methods that have been used to mark hedgehogs include: leg rings (Herter, 1938), ear tags (Kristiansson & Erlinge, 1977), paint applied to the spines (Brockie, 1958; Campbell, 1973), and spine clipping (Morris, 1969).

Paint marking was tried but found to be relatively impermanent. Morris (1969) found spine clipping an effective method and this was the method adopted in the present study. The animal was anaesthetized with ether, and between one and three patches of spines clipped from the dorsal surface, following a clearly defined system of dividing the spiny part of the coat into eight areas (Fig. 2).

The spines are shed and replaced slowly and marks lasted for several months. Advantages of this method include easy recognition with no conspicuous mark to attract predators, and minimal confusion with marks acquired in any other way. However, there were some disadvantages. First, the skin of the hedgehog's back is so mobile that the animals had to be anaesthetized on the first marking

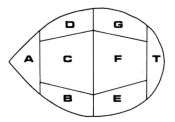

FIG. 2. The marking scheme. The hedgehog's dorsal area was divided into eight zones from which spines could be clipped: A = the crown of the head; B = the left shoulder; C = centre back; D = the right shoulder; E = the left hindquarter; F = rear back; G = the right hindquarter; T = the tail area. Marks in zones B, D, G, E and T extended to the extremity of the spiny area.

occasion to ensure that the mark was in the right area of the back. Secondly, the mark, though not attracting predators, would be a disadvantage in the event of a confrontation with one. For this reason the clipped patches were kept as small and as far apart as possible. Observation of marked animals showed that this risk, though real, was less than might be imagined. The considerable mobility of the skin allowed the animals to bunch up the skin in the marked area when touched (a natural reaction to increase the density of spines in any threatened area), thus reducing the area of skin exposed by the clip. Each clipped area was about 2.5×2.5 cm when the animal was in a relaxed state.

The tail clip was not used after 1977 (except in one case) because it led to some confusion with the two rear hindquarter marks. In addition the spines in the tail area are rather sparse so the mark was never very clear after some time had elapsed. The spines were clipped using a pair of sprung side cutters, and were cut quite close to the skin so that the marks were easily visible. In time new spines gradually grew in the clipped area, but many stubs of the clipped spines (usually easily distinguishable from broken spines by their regularity) remained as a characteristic indication of a mark.

In some cases (especially in juveniles) the marks became obscured by new growth in a few weeks. Whenever any animal was recaptured the mark was retrimmed if any significant growth had occurred; this was easily carried out in the field without anaesthetic so that the mark was always maintained despite growth.

One of the advantages of this system of marking is that of easy identification with the minimum of interference. Often it was possible to identify individuals without handling them at all. The marking system described above provided a series of 68 distinctive marks which was a sufficient number for the present study.

The Radio Tracking Equipment

A receiver was developed which can be built without detailed knowledge of electronics, using commercially available parts. A "frequency converter" (obtained from Microwave Modules Ltd*) was used to convert the high frequency pulse signal from the transmitter (104.7 MHz) to a much lower frequency (7.3 MHz) which could then be received by a conventional shortwave transistor radio (a Sony ICF 5900W). The Ashwell design of transmitter was used (Taylor & Lloyd, 1978) with minor modifications (see Reeve, 1980). An Adcock "H" aerial as described by Taylor & Lloyd (1978) was used, with telescopic elements. This aerial has good directional properties and can be collapsed when not in use.

The total cost of the receiver in 1977 was approximately £120 and the transmitters cost under £5 each, excluding labour[†].

Attachment of the Transmitter

The hedgehog has no well defined neck and must be free to roll up in defence, thus a collar or other rigid harness, and therefore a tuned loop aerial such as used by Anderson & DeMoor (1971) could not be used. The transmitter weighed about 20—25 g and its maximum external dimensions were 30 × 30 × 15 mm, thus it was too large to be implanted and in any case the batteries could not then be changed. An elastic harness of the type used by Morris (1969) was tried, but it was found that this prevented the hedgehog from rolling up completely and the animals frequently escaped from the harness after only a few hours. When the harness was tight enough to prevent escape there was a restriction of the legs. Even apparently well-fitting harnesses caused chafing of the skin of the axilla. The harness was abandoned and gluing was adopted as the most satisfactory of the few methods available (Kristiansson & Erlinge, 1977).

An epoxy resin (Rapid Araldite) was applied liberally to the base of the transmitter. The hedgehog was then anaesthetized with ether until completely relaxed. The transmitter was pressed on to the spines between the animal's shoulders, with great care to place it centrally, and was held lightly in place while the animal recovered from the ether, by which time (about 10 min) the resin had set and the transmitter was firmly attached.

---

* Microwave Modules Ltd, Brookfield Drive, Aintree, Liverpool L9 7AN.
† It is worth noting that the crystal now recommended for this circuit costs £13 for quantities of 5 in 1981.

Once attached the transmitter tended to flop onto one side or the other because of the very loose skin on the back. However, the hedgehogs showed no overt signs of discomfort or restriction.

The transmitters usually remained firmly attached for several weeks, but after a month or so enough spines had often moulted from underneath the transmitter to loosen it somewhat. In most cases it was easy to reglue the transmitter to the newly grown spines. Sometimes, however, the load of the transmitter on only a few spines caused a slight inflammation of the skin. In these cases, the transmitter was removed and an antibiotic powder applied to the affected area. Occasionally when very few spines were left to support the transmitter the animal was able to pull it off. This was usually achieved when the transmitter caught on grass or a fence. Very few injuries resulted from this method of attachment, but a slight wound was sometimes caused when the last few spines holding the transmitter on were pulled out.

Removal of the transmitter was easily accomplished. Two methods were used:

(i) A knife blade inserted between the glue and the base of the transmitter allowed it to be prised off, leaving a plate of glue still attached to the animal's spines. This method was very easy and quick.

(ii) When method (i) did not work, the spines were neatly trimmed away thus freeing the transmitter. This method was used as little as possible because the clipped spines might confuse the marking system.

The Radio Location of Hedgehogs in the Field

Using the radio receiver and Adcock "H" directional aerial in the field, the range obtained in good conditions could be as much as 400 m, and sometimes more. A good range could be ensured when: (i) there were no obstructions between the transmitter and the receiver (line of sight); (ii) the aerial was elevated using a pole or by standing on higher ground.

However, it was found that the searching pattern adopted by the investigator was governed not by the best figures for range and performance, but the worst. No more than 200 m could be relied upon, and in some cases of severe screening of the signal the range could be cut to as little as 100 m. This limitation affected the searching strategy (described below). When searching for a radio tagged animal in the field the following methods were adopted.

During the day

During the day the object was to find the animal in its nest. The signal strength was constant because the animal was stationary. Firstly the area was transected methodically until the signal was detected. (Experience of the animal's habits later helped to reduce this period of searching.) The directional aerial was then used to indicate the bearing of the signal. This was done by rotating the aerial until a null in the signal was heard. When the transmitter was about 10 m away the signal was at a maximum and it was no longer possible to distinguish a directionality from the signal. At this stage a smaller aerial could have been substituted, but this would have meant carrying extra equipment, so instead the receiver was retuned to 104.7 MHz on the FM waveband to detect the signal as weak breakthrough. This sounded like short silent periods through the general background hiss. The telescopic FM aerial built into the receiver was used for this purpose with the Adcock removed. As the transmitter was approached, the FM aerial could be progressively collapsed thus keeping the signal at below maximum on the signal strength meter. When the aerial was fully collapsed and the signal was at a maximum, the receiver was directly over the nest. This was such a sensitive method that it could be used to find detached transmitters buried deep in the leaf litter, and even underground as in one case where a transmitter became detached below ground and it was possible to dig down (0.5 m) exactly to the transmitter and recover it. Hence it was possible to locate nest sites which to the unaided observer would have been undetectable. The method though lengthy to explain was simple in practice and gave good results in the field.

At night

At night the animals were usually active and the signal strength was not constant. Hence it was necessary to take slightly longer over taking a bearing, to avoid being deceived by a false null in the signal. All the initial stages of the location are the same as above, but the final location of the animal was usually much easier. When the signal was so strong as to make the Adcock aerial unusable, the animal was often already visible on the fairway or audible in the undergrowth. Once it was thus located an animal could be recaptured or observed as required. Observation was much aided by the use of Beta Lights which were temporarily taped to the transmitter for the duration of the observation period, and were then removed so that they would not attract the attention of predators or people.

RESULTS AND DISCUSSION

Population Size

Many methods have been used to estimate population size, but usually the Petersen method (= Lincoln Index) or its derivatives are employed (Begon, 1979; Southwood, 1978; Strandgaard, 1967). Such methods each have their own constraints but they all require that consistent sampling efforts be made at regular intervals during the study. This requirement could not be met by the present study. Trapping was considered impractical and the efficiency of the searching method was variable because it depended on many factors, e.g. the weather conditions, the nature of the vegetation in the search area, the number of animals that were found, and whether or not they were observed (perhaps in a courtship encounter) before being intercepted. Additionally in 1978 and 1979 searching was progressively abandoned in favour of radio tracking a small number of selected individuals, thus any captures or recaptures obtained were considered supplementary.

Inconsistency was thus difficult to avoid and no method of population estimation is really suited to such data; however Krebs (1966) described a method using the minimum number of animals known to be alive (MNA) during the sampling period to estimate population size. This method was used by Hilborn, Redfield & Krebs (1976) who found it to be a reliable technique but giving an underestimate of about 10—20% for a simulated population of *Microtus*. However when the probability of capture fell below 0.5 (50%) the method became unreliable.

Using the data from the present study no reliable way of calculating the probability of capture was found, therefore although the MNA method was used to provide an estimate of the population size this was supplemented by an estimate of the residential population obtained by counting the number of individuals caught more than once in the study area (per season). The results indicate a population size of about 33 individuals within the 40 ha study area. Allowing for the uncaught portion of the population, a density of one per hectare seems a reasonable estimate.

Home Range Area

Home range is a concept commonly employed to indicate the area used by an animal during its routine activities. Definitions by Burt (1943) and Jewell (1966) exclude the occasional forays (perhaps exploratory) outside the animal's normal range, whereas such forays were included in Allen's (1943) definition of "individual range". Time is also an important factor in the definition of home range: should an animal's range alter with time (e.g. seasonally) it becomes necessary to introduce a prescribed time scale into the definition.

During the present study it was not possible to distinguish between the hedgehog's normal activities and excursions, hence no attempt was made to separate the two. The range areas presented below represent a total of all the captures and radio fixes during one active season (i.e. a yearly range).

Home range is conventionally expressed as an area which can be calculated in a variety of ways; most methods pertain to grid trapping field methods. The literature relating to this subject is extensive and no attempt is made to review it here. Sanderson (1966) and Jennrich & Turner (1969) are among the numerous review papers available. More recently some authors have reviewed and assessed several methods of home range estimation with relevance to radio tracking, e.g. Trevor-Deutsch & Hackett (1980), Macdonald, Ball & Hough (1980), and Voigt & Tinline (1980).

In the present study one of the simplest methods available, the Minimum Area Method (MAM) (using a convex polygon), was used despite some of its inherent problems — see Macdonald *et al.* (1980). Many of the alternative methods are somewhat complex and usually involve complex statistics and often the services of computer experts. It was felt that area measurements of home range are often unrealistic (see below) and frequently only serve to provide an estimate of the area required to meet an animal's general needs. For animals that apparently do not have territorial boundaries and whose excursions are not distinguishable from their rather erratic normal activities (as with the hedgehogs in this study), it seemed inappropriate to devote a great deal of effort to the precise statistical definition of their ranges.

The method used to provide area measurements of home range made use of a grid system. The grid used had a cell size of 25 x 25 m (0.0625 ha) and measured 40 cells (1000 m) on the x axis, and 28 cells (700 m) on the y axis. Captures and fixes were each assigned a co-ordinate corresponding to one cell: e.g. 40/28 (x/y). All the used cells for an individual (over one season) were marked and a convex polygon drawn to enclose all the used cells. The resultant area was measured with a planimeter.

It is important to note that the measured range area is heavily dependent on the number of captures obtained (see Haugen, 1942). Stickel (1954) discussed this in depth, comparing simulated grid trapping in an artificial population with field data. Because of the great variation in home range size between individuals it was difficult to assess how many captures and fixes were required to fully reveal the animals' ranges. Subjective interpretation of the results obtained indicated that when more than 20 locations had been obtained (provided that these had not been closely spaced in time) the range area would be adequately revealed. In fact although more than 20 locations was set as a lower limit the majority of areas measured were based on many more locations (up to 532), and none was based on fewer than 30 locations.

The results obtained from the present study contrast markedly with those found in the literature. Most previous estimates indicate the home range of hedgehogs to be no larger than 4 ha (e.g. Berthoud, 1978; Kristiansson & Erlinge, 1977); however Parkes (1975) calculated theoretical ranges, using a probability ellipse method, of 7.0 and 12.9 ha for males and females respectively.

The present study showed the average range ($>$ 20 fixes over one active season) for male hedgehogs to be 32 ha ($S = 8.9$, $n = 6$, range 15.5–41.5 ha), for females 10 ha ($S = 2.2$, $n = 7$, range 5.5–12.0 ha), and sub-adults 12 ha ($S = 2.5$, $n = 3$, range 10.0–15.0 ha). Thus males are shown to have approximately three times the range area of females, and the males especially have a much larger range than previous work leads one to expect. Both the males and females show a considerable variation in range area between individuals.

A Mann-Whitney "U" test (Siegel, 1956) was used to check the significance in the difference between the male and female range areas. There was no overlap in the two groups, therefore $U = 0$ ($P <$ 0.002) indicating a clear statistically significant difference between male and female range sizes. The sub-adult sample was too small to test but inspection of the data suggests that sub-adult range size was similar to female range size.

Territoriality

In the present study, and as also noted by Campbell (1973) and Parkes (1975), the ranges all overlapped considerably and often completely in both sexes. This was true for both adults (Figs 3 and 4) and sub-adults. This overlapping of ranges implies a lack of territoriality, at least in the conventional sense of an exclusively defended area (Burt, 1943).

Leyhausen (1963) described a territorial system in cats where mutual avoidance allowed the non-simultaneous use of the same paths and areas by several individuals and Morris (1969) suggested

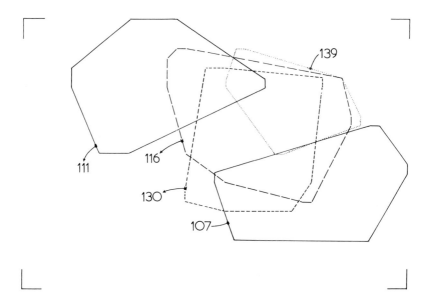

FIG. 3. The ranges of five female hedgehogs radio tracked in 1978. The four corner markings indicate the perimeter of the 1000 m × 700 m grid superimposed on the study area (See Fig. 1 and text). Representations of the home ranges of five females (nos 107, 111, 116, 130 and 139) tracked in 1978 are shown. The ranges overlap considerably indicating a lack of territoriality.

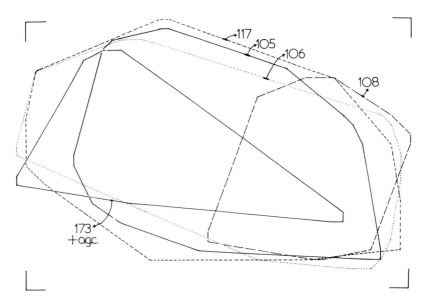

FIG. 4. The ranges of five male hedgehogs radio tracked in 1979. The four corner markings indicate the perimeter of the grid (see Fig. 3). Representation of the home ranges of five males (Nos 105, 106, 108, 117 and 173) tracked in 1979 are shown. The ranges overlap considerably indicating a lack of territoriality. Note that the male ranges are considerably larger than the female ranges shown in Fig. 3. Male 173 also extended his activities off the golf course (OGC) therefore the range shown is incomplete.

that hedgehogs may have a similar system. Hedgehogs forage for diverse and diffusely distributed food and are sexually promiscuous, therefore the defence of an area to avoid competition between conspecifics may be a waste of energy and impractical if the diversity and distribution of the food (or other resource) means that they rarely compete in any case. Direct competition where animals meet in the same area may be avoided by the maintenance of a "personal space" by mutual avoidance. The evidence from the present study indicates that this may be the pattern found in hedgehogs.

Hierarchies are commonly found to be formed in captive groups of hedgehogs (Dimelow, 1963; Lindemann, 1951) but have not been demonstrated in the wild and therefore may be artificial, induced by a localization of the animals' food source and a degree of confinement which would not occur in the wild. However, the facility to form hierarchies argues that, despite the lack of corroborative evidence in the wild, there may indeed be a social order of some kind in the wild which has so far defied observation. A similar situation is found in the kinkajou (Ewer, 1973); in the wild they were found to be non-territorial with overlapping ranges. In captivity they were found to be mutually tolerant but formed a hierarchy in relation to access to food.

Scent marking of some kind may possibly occur in the hedgehog but has not been demonstrated positively either in captivity or in the wild. During the present study no scent marking behaviour was ever knowingly observed, and there was a very low incidence of aggressive encounters.

Although the hedgehogs were not territorial they did show a marked tendency to remain in the same general locality from year to year. Figure 5 illustrates this and shows the range of a female (No. 111) in each of the three years of the study. Several other individuals of both sexes also showed this tendency to remain in approximately the same locality throughout the study.

Distance Travelled as a Representation of Home Range

As radio tracking has become more readily available to those investigating animal movements, conventional expressions of home range area can now be supplemented by more detailed information. The use of radio permits the continuous tracking of individuals and thus can rapidly provide detailed information about the movements of the animal. The distance travelled by an animal during one activity period is of great biological significance, particularly for animals such as hedgehogs who forage for diffusely distributed food and therefore

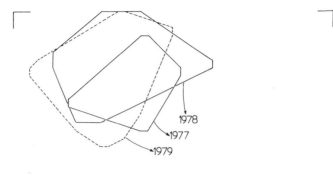

yearly ranges of female 111

FIG. 5. The yearly ranges of female hedgehog No. 111. The four corner markings indicate the perimeter of the grid (see Fig. 3). Representations of the range of a female (111) in each of the 3 years of the study are shown. Note the strong tendency to remain in the same locality in each year.

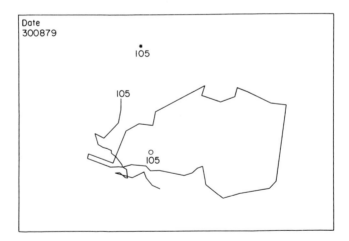

FIG. 6. A computer plotted route-map. The rectangle in which the route is plotted represents the perimeter of the 1000 × 700 m grid used previously. The starting nest is marked with an asterisk and the animal number. The first fix is marked with the animal number. The final fix is shown simply by the termination of the line representing the route travelled by the animal. A circle (also marked with the animal number) represents the geo-metrical centre of activity (the means of the x and y co-ordinates of all the fixes). The route shown is for a male, 65 fixes being made in 7 h 25 min of activity.

must travel a certain distance to get enough to eat. Distance travelled, combined with accessory data such as speeds, time spent in various activities (e.g. foraging), and differential use of habitat types within a range, can be used to provide the basis for further investigations into such fields as foraging strategies and ecological energetics. Expressions of home range area lack this versatility. The distance travelled by a hedgehog within its range may be quite independent from the area derived using the animal's locations and the MAM.

Maps of the routes taken by animals within their ranges have been termed "trackograms" by Mauget (1979, 1980) and expression of the animals' daily range in terms of distance travelled is becoming more common (Deat et al., 1980; Kristiansson & Erlinge, 1977; Mauget, 1979, 1980; Voigt & Tinline, 1980).

During the present study, 12 hedgehogs were tracked over a total of 27 whole nights in 1979, and substantial data were obtained on eight individuals (three males, three females, one male sub-adult, one female sub-adult). This resulted in the collection of data on 83 animal-nights with an average of approximately three individuals tracked on each night.

A computer was used to analyse the data obtained in the following way. The location of each fix was drawn on maps in the field which were subsequently digitised on a 4000 × 3000 unit grid using a TEKTRONIX 4662 Plotter/Digitizer which was connected via an IMLAC PDSI-G Interactive Graphics Terminal to the Imperial College Computer Centre. Fixed points on the boundary were also digitized to ensure the precise matching of each animal track.

The co-ordinates could then be used to plot maps of the nightly routes taken by each animal with straight lines linking the fixes in chronological order. An interactive program was developed so that further relevant data could be added to each fix; these were: time, habitat and behaviour. It was then possible for the computer to calculate:

(i) the distance between each fix;
(ii) the time between each fix;
(iii) the total distance travelled;
(iv) the total time elapsed;
(v) the speed between each fix;
(vi) the average speed.

These parameters were calculated for each animal on each night, and also additional information about habitat use and behaviour. An example of a computer plotted route map is shown in Fig. 6.

The greatest distance travelled by an animal in one night (male 105) was 2828 m. No female was found to travel more than 2116 m (sub-adult No. 180). The total distance travelled per night measured with straight lines between fixes (D(S)) is dependent to a degree upon the number of fixes taken over the time of the activity period. For this reason only the figures for nights in which 20 fixes or more were taken are shown below. In the group > 20 fixes per night the mean number of fixes per night is 44.4, the numbers ranging from 20 to 70; however, relatively few (less than a quarter) had fewer than 30 fixes.

The mean values for D(S) are as follows:

Males, 1690 m (S = 627) (three individuals over 14 nights)
Females, 1006 m (S = 451) (three individuals over eight nights)
Sub-adults, 1188 m (S = 468) (two individuals over nine nights)

A Mann-Whitney "U" test showed that there was a sexual difference in the values of D(S), $P < 0.02$, but that subadults were indistinguishable from the females $(P > 0.10)$.

Thus it seems that males travel further per night (on average) than females and sub-adults. This conclusion, though apparently based on rather few data, is strongly supported by further data obtained from nights when fewer than 20 fixes were taken, and also supported by the subjective impression gained from a large number of part-night observations on many additional individuals.

Kruuk (cited by Morris, 1969) following the tracks of a hedgehog in the sand at low tide at Ravenglass found that it continued for 4 km. Kristiansson & Erlinge (1977) found that the movements of hedgehogs totalled between 700 and 1000 m during one night, and Berthoud (1978) noted that individuals may travel several kilometres per night and instanced a male which travelled a route of 3.2 km on five occasions. The figures for distance travelled per night obtained during the present study are thus broadly in line with those few reported in the literature.

Average Speed

The males may achieve their somewhat longer nightly travels by either staying active for longer, or by travelling faster, than the females and sub-adults.

There was no apparent sexual difference in the time at which the hedgehogs retired to their nests in the morning, and although on most occasions the animals were not observed leaving their nests the observations that were made gave no grounds for suggesting that emergence time differs between the sexes. Campbell (1975) and Kristoffersson (1964) both studied the length of the hedgehog's activity period, but neither made any suggestion that there might be a sexual difference in its duration.

The idea that males travel faster than females and sub-adults is supported by an analysis of the average speeds of each group. As with distance measurements, the revealed average speed varies with the number of fixes per night, hence only groups with >20 fixes per night are discussed here.

The mean values of average speed are as follows:

Males, 3.73 m/min (S = 1.5) (three individuals over 14 nights)
Females, 2.19 m/min (S = 1.0) (three individuals over eight nights)
Sub-adults, 2.17 m/min (S = 0.8) (two individuals over nine nights)

A Mann-Whitney "U" test confirmed that there was a significant sexual difference in average speed, $P < 0.02$, and that subadults were indistinguishable from females, $P > 0.10$.

Again the conclusion that males travel faster than females is here based on relatively few data; however, it is strongly supported by data collected during a large number of part-night observations and subjective impression.

Although on average hedgehogs do not move particularly fast, they are capable of short bursts of relatively high speed. There were several cases of hedgehogs travelling at speeds in excess of 30 m/min for between 3 and 10 min. All these hedgehogs were males. One male was observed to cross a fairway spanning at least 60 m in one minute. This was the fastest that any hedgehog was seen to run; the action appeared to be spontaneous and not in response to any evident fright stimulus.

Differential Use of Home Range Area

Data obtained from radio tracking studies can help to elucidate the internal structure of the home range. Two methods were used to represent the differential use of parts of the home range.

The grid method

The grid method used to measure home range area was again used but instead of simply marking the used cells, a score was kept of the number of locations (captures and radio fixes) in each cell. These data were then presented using different symbols denoting different intensities of use. The resulting diagram provides a clear indication of which areas within the range were heavily used and which were not (Fig. 7). Similar methods were described by Voigt & Tinline (1980).

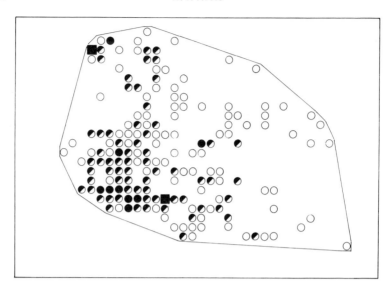

FIG. 7. A representation of the internal structure of the home range of a male (No. 105) in 1979. The rectangle in which the range is shown represents the perimeter of the 1000 × 700 m grid used previously. The perimeter of the home range is shown as a convex polygon. The differential use of the area is indicated using the grid method (see text). The symbols show the number of locations in each grid cell as follows: ○ = location, ◖ = 2–5 locations, ● = 6–10 locations, ■ = 11 + locations.

The superimposition of route maps

This method was also used by Mauget (1979, 1980) and involves simply superimposing the "trackograms" for each individual. This builds up a picture of the intensity of use of the different areas within the range; where the lines of the trackograms are densely packed a heavier use is indicated than where the lines are sparse.

Both these methods were applied to the data from the present study and the two methods compared visually. They were found to closely resemble each other in appearance.

From the diagrams produced by either of these methods it is clear that the "home range" represented by the area within the convex polygon is unevenly used. Animals may concentrate their activities around certain areas while other areas remain unvisited or sparsely used. The fact that the MAM is thus deficient in its representation of home range is well known and several authors (e.g. Voigt & Tinline, 1980; Macdonald *et al.*, 1980) make this point. However many of the deficiencies of the MAM are offset by its simplicity and its usefulness is enhanced when combined with an indication of internal structure.

Seasonal Changes in Home Range

The suggestion that male hedgehogs expand their range in the breeding season, presumably in search of females, has been put forward by several authors, e.g. Campbell (1973), and Berthoud (1978), and has often been held to account for the large numbers of males killed on the roads during the Spring. During the present study it was found that the breeding season (as defined by the period in which courtships were observed) lasted from late May until September. Thus little of the hedgehog's active year lies outside this period and therefore it is difficult to determine whether males "normally" have small ranges which are expanded for breeding purposes or whether their ranges are always large.

A large number of fixes and captures were obtained for only two males before mid May (1979); unfortunately the results from one of these cannot be used to display a range expansion because only two captures follow the cessation of radio tracking in mid May. This leaves only the data for one male (No. 106) for this analysis. Between 10.4.1979, when this hedgehog was first captured, and 20.5.1979, 274 locations were obtained. During this time the hedgehog's range had remained relatively small at about half its total yearly range and was centred around the south-eastern corner of the study area. The 258 subsequent fixes (up to 26.8.1979) show a much enlarged range encompassing almost the whole of the earlier range plus an extension over to the north-western corner of the study area. The date 20.5.1979 was chosen as the dividing point in the data because it was at this time that the first excursion over to the north-western part of the golf course was made. If there is any truth in the idea that this expansion is due to the onset of sexual activity then it may be more than coincidence that only six days earlier No. 106 and female No. 107 had been observed participating in the first sexual encounter of the season.

Thus from the data supplied by this one animal there does seem to be some support for the idea that males expand their range when sexually active. If this issue is to be resolved further work needs to be carried out involving a greater amount of radio tracking early in the hedgehog's active season.

DISCUSSION OF SOME PRACTICAL ASPECTS OF MEASURING DISTANCE TRAVELLED

The merits of using distance travelled as a substitute for area expressions of short-term home range are briefly discussed above. In the same way that the revealed home range area increases with the number of locations used in its calculations, the measured distance travelled also increases with the number of locations per night. Theoretically it should be possible to determine the optimum number of position fixes per night, so that minimum effort will result in the best approximation to distance actually travelled. This could be done by plotting measured distance travelled against the number of fixes per night. However, under field conditions an animal's activity is likely to alter considerably during the night, and when several animals were being tracked at once the intervals between the fixes were often irregular. Thus a simple graphical approach based upon standard fix intervals was inappropriate.

Calculations showed that the mean D(S) values for animals with 10–19 fixes per night were an average of 85% of the values obtained for the group with 20 or more fixes per night; where there were fewer than 10 fixes per night the D(S) values were reduced to 58% of the 20 or more fixes group. It seems desirable to obtain as many fixes as possible per night, within practical constraints. If fixes were taken every minute, the small movements would be difficult to plot on field maps. Moreover, very close contact would have to be maintained, with consequent risk of disturbance to the animal. One fix per hour ($<$ 10 fixes per average eight hour night) seems to result in a marked reduction of D(S), whereas the 10–19 fix per night group (about one fix per 30 min) seems better.

The mean number of fixes per night in the 20 or more fixes group is 44.4, so for an average eight hour night the interval between fixes is then about 11 min. Using the mean average speed values given earlier, the straight line displacement of a hedgehog in this time would be 24–41 m, convenient to plot on field maps with a scale of about 2 cm to 100 m. On this basis the average interval between fixes is appropriate. In practice, fixes were often made at as little as five minute intervals, near to the practical minimum interval because it took almost that amount of time to note down the relevant field data for each fix. A minimum of two fixes per hour is recommended for tracking hedgehogs.

The total distance travelled per night obtained by the computer measuring straight line distance between the fixes results in an underestimate and represents the minimum possible distance travelled. In an attempt to quantify by how much D(S) was an underestimate of the *actual* distance travelled an equal sex sample of eight routes was measured manually to include all the observed deviations in the route taken by the animal. On average the distance including deviations was 10.4% (S = 3.96) greater than D(S). Thus is would be reasonable to increase the calculated values of D(S) by this figure. This would mean that the maximum distance walked by a male hedgehog (No. 105) was 3123 m. For a mammal the size of a hedgehog this is a considerable feat for one night's walking. Only major deviations in route can be treated in this way; during foraging a hedgehog will wander about making small movements of a metre or so from side to side, and will often turn back on itself. Attempts to quantify these very small movements met with failure.

Some attempts were made to follow dew trails but these were impossible to see in torch light and by morning were often mingled with the tracks of other hedgehogs and my own footprints. Dew trails were also unreliable when they crossed several vegetation types and where intensive foraging had taken place they had become generalized smudges in the grass. Thus the measurements of distance travelled and average speed both suffer from the same inherent problem, namely that females and sub-adults may not actually move less far or more slowly than the males, but only less linearly, making up the apparent difference by adopting a meandering gait resulting in a highly convoluted route. Subjective impressions formed during the study do not resolve this problem. Males seemed to travel more linearly and forage more rapidly than the females and sub-adults which apparently spent more time meticulously foraging, but it could not be ascertained whether the total minor meanders of the latter were equivalent to the males' more extensive and direct movements. Further work may resolve this issue, perhaps by the use of some method of trail laying, or a suitable time lapse photographic technique with a small light source attached to the animals.

ACKNOWLEDGEMENTS

This work was carried out as part of a Ph.D. project at Royal Holloway College (University of London), Egham, Surrey.

I would like to thank for their financial support: Mr and Mrs J. A. Reeve; the Principal of Royal Holloway College (for discretionary awards in the second year); and the Clothworkers Company who generously provided (through the University of London) a Postgraduate Scholarship in the third year.

I would also like to thank Dr P. A. Morris for his invaluable support, advice and general assistance, and Mr G. Rentmore whose technical abilities were of great value in the development of the radio tracking system and without whom the project could not have succeeded.

Mr John Anderson wrote the programs used in the analysis of the radio tracking data and instructed me in the use of the computer.

The staff and members of Ashford Manor Golf Club allowed the use of their grounds as a study site and were very co-operative throughout the study.

I would like to particularly thank my wife, Kati, for all her patient support and assistance during the project.

REFERENCES

Allen, D. L. (1943). Michigan fox squirrel management. *Publs Mich. Dept. Conserv., Game Div.* 100: 1—404.

Anderson, F. & DeMoor, P. P. (1971). A system for radio tracking monkeys in dense bush and forest. *J. Wildl. Mgmt* 35: 636—643.

Begon, M. (1979). *Investigating animal abundance.* London: Arnold.

Berthoud, G. (1978). Note préliminaire sur les déplacements du hérisson européen (*Erinaceus europaeus* L.) *Terre Vie* 32: 73—81.

Brockie, R. E. (1958). *Ecology of the hedgehog* (Erinaceus europaeus *L.*) *in Wellington, New Zealand.* M.Sc. Thesis: University of Wellington (N.Z.).

Burt, W. H. (1943). Territoriality and home range concepts as applied to mammals. *J. Mammal.* 24: 346—352.

Campbell, P. A. (1973). *Feeding behaviour of* Erinaceus europaeus *in New Zealand pasture.* Ph.D. Thesis: University of Canterbury (N.Z.).

Campbell, P. A. (1975). Feeding rhythms of caged hedgehogs. *Proc. N. Z. ecol. Soc.* 22: 14—18.

Deat, A., Mauget, C., Mauget, R., Maurel, D. & Sempéré, A. (1980). The automatic continuous and fixed radio tracking system of the Chizé forest: Theoretical and practical analysis. In *A handbook on biotelemetry and radio tracking*: 439—451. Amlaner, C. J. & Macdonald, D. W. (Eds). Oxford: Pergamon Press.

Delany, M. J. (1974). *The ecology of small mammals.* London: Arnold.

Dimelow, E. J. (1963). The behaviour of the hedgehog (*Erinaceus europaeus*) in the routine of life in captivity. *Proc. zool. Soc. Lond.* 141: 281—289.

Ewer, R. F. (1973). *The carnivores.* London: Weidenfeld & Nicholson.

Haugen, A. O. (1942). Home range of the cottontail rabbit. *Ecology* 23: 354—367.

Herter, K. (1938). Die Biologie der europäischen Igel. *Z. Kleintierk. Pelztierk.* 14: 1—222. [Monographien der Wildsäugetiere 5].

Hilborn, R., Redfield, J. A. & Krebs, C. J. (1976). On the reliability of enumeration for mark and recapture census of voles. *Can. J. Zool.* 54: 1019—1024.

Jennrich, R. I. & Turner, F. B. (1969). Measurement of non-circular home range. *J. theoret. Biol.* 22: 227—237.

Jewell, P. A. (1966). The concept of home range in mammals. *Symp. zool. Soc. Lond.* No. 18: 85—109.

Krebs, C. J. (1966). Demographic changes in fluctuating populations of *Microtus californicus. Ecol. Monogr.* 36: 239—273.

Kristiansson, H. & Erlinge, S. (1977). Rörelser och activitetsområde hos igelkotten. *Fauna Flora, Upps.* 72: 149—155.

Kristoffersson, R. (1964). An apparatus for recording activity of hedgehogs. *Ann. Acad. Sci. fenn.* (4) No. 79: 1–8.

Leyhausen, P. (1963). The communal organisation of solitary mammals. *Symp. zool. Soc. Lond.* No. 14: 249–263.

Lindemann, W. (1951). Zur psychologie des Igels. *Z. Tierpsychol.* 8: 224–251.

Macdonald, D. W., Ball, F. G. & Hough, N. G. (1980). The evaluation of home range size and configuration using radio tracking data. In: *A handbook on biotelemetry and radio tracking*: 405–424. Amlaner, C. J. & Macdonald, D. W. (Eds). Oxford: Pergamon Press.

Mauget, R. (1979). Mise en évidence, par capture-recaptures et radio-tracking, du domaine vital chez le sanglier (*Sus scrofa*) en forêt de Chizé. *Biol. Behav.* 1: 25–41.

Mauget, R. (1980). Home range concept and activity patterns of the European wild boar (*Sus scrofa*) as determined by radio tracking. In: *A handbook on biotelemetry and radio tracking*: 725–728. Amlaner, C. J. & Macdonald, D. W. (Eds). Oxford: Pergamon Press.

Morris, P. A. (1969). *Some aspects of the ecology of the hedgehog* (Erinaceus europaeus). Ph.D. Thesis: University of London.

Parkes, J. (1975). Some aspects of the biology of the hedgehog (*Erinaceus europaeus*) in the Manawatu, New Zealand. *N.Z. Jl Zool.* 2: 463–472.

Reeve, N. J. (1980). A simple and cheap radio tracking system for use on hedgehogs. In *A handbook on biotelemetry and radio tracking*: 169–173. Amlaner, C. J. & Macdonald, D. W. (Eds). Oxford: Pergamon Press.

Sanderson, G. C. (1966). The study of mammal movements. *J. Wildl. Mgmt* 30: 215–235.

Siegel, S. (1956). *Nonparametric statistics for the behavioural sciences.* Maidenhead: McGraw Hill.

Southwood, T. R. E. (1978). *Ecological methods* (2nd edn.) London: Chapman & Hall.

Stickel, L. F. (1954). Comparison of methods of measuring home range. *J. Mammal.* 35: 1–15.

Strandgaard, H. (1967). Reliability of the Petersen method, tested on roe deer population. *J. Wildl. Mgmt* 31: 643–651.

Taylor, K. D. & Lloyd, H. G. (1978). The design, construction and use of a radio tracking system for some British mammals. *Mammal Rev.* 8: 117–141.

Trevor-Deutsch, B. & Hackett, D. F. (1980). An evaluation of several grid trapping methods by comparison with radio telemetry in a home range study of the eastern chipmunk (*Tamias striatus* L.). In *A handbook on biotelemetry and radio tracking*: 375–386. Amlaner, C. J. & Macdonald, D. W. (Eds). Oxford: Pergamon Press.

Twigg, G. I. (1975). Techniques in mammalogy, Section 1. *Mammal Rev.* 5: 69–116.

Voigt, D. R. & Tinline, R. R. (1980). Strategies for analyzing radio tracking data. In *A handbook on biotelemetry and radio tracking*: 387–404. Amlaner, C. J. & Macdonald, D. W. (Eds). Oxford: Pergamon Press.

DISCUSSION

Gregory (MAFF) — Did the hedgehogs show any tendency to follow set routes?

N. J. Reeve — No. They will, however, make use of paths through hedges, for example, which are used by other species of animals.

B. Don (Oxford) — You posed the question why aren't males territorial. Your explanation was that they do not compete and your evidence for lack of competition is that they co-exist.

Reeve — This does seem to be a rather circular argument but the possibility that they do not compete is only offered as an explanation for the fact that they are not territorial.

M. J. Delany (Bradford University) — Was the population stable from one year to the next?

Reeve — Yes, it was fairly stable.

Delany — Do you know what happens to the young animals?

Reeve — Studies by Morris show that most juveniles die in their first winter.

McBride (UCL) — Do you have any information on self-anointing behaviour?

Reeve — I have seen this on several occasions but have no idea of the function of self-anointing.

S. Pickering (Durham) — (i) Do both sexes self-anoint? (ii) Have you any data to indicate that males that move longer distances per night get more matings than those which move shorter distances?

Reeve — (i) Both sexes self-anoint. (ii) I do not have enough data to relate distance travelled to reproductive success.

B. C. R. Bertram (Zoological Society of London) — The fact that males all seem to have the same home range boundaries suggests that either the habitat was limiting their movements or that there was a communal or group range.

Reeve — The difficulty of access to gardens surrounding the study area meant that tracking was confined to the golf course. On the other hand those animals which did venture into gardens did not seem to move very far out of the study area.

R. E. Kenward (ITE) — I have found that male grey squirrels expand their home ranges in the breeding season. Is there any evidence that male hedgehogs do this?

Reeve — There is a small amount of evidence that males expand their range when sexually active, but the hedgehogs' breeding season extends to almost their entire active year so it is very difficult to detect differences.

Symp. zool. Soc. Lond. (1982) No. 49, 231–257

Studies of Home Range of the Feral Mink, *Mustela vison*

J. D. S. BIRKS* and I. J. LINN

Department of Biological Sciences, University of Exeter, Hatherly Laboratories, Exeter, England

SYNOPSIS

Radio tracking techniques were employed in the collection of home range data from mink in three areas of contrasting habitat-type. Preliminary results from some individuals are presented in an earlier paper (Linn & Birks, in press); the present chapter contains a synthesis of the fully analysed results from all individuals radio tracked.

The stability of the social environment was found to influence home range sizes considerably; in socially stable areas in a linear riverine habitat, home ranges showed low size variation. A two-dimensional home range was exploited more efficiently than linear ones, and the constraints operating in the latter case are discussed. Home ranges were subject to considerable variations in use intensity, with mink spending, on average, about 50% of their time within a core area which occupied *c*. 10% of the total range. Altogether 104 dens were identified and categorized for each study area. Mink rarely excavated dens themselves, but utilized existing holes such as those associated with waterside trees or rabbit burrows. The mean number of dens used per home range was six. The extent of movements within the home range was related to den density, though the frequency of such movements showed more ambiguous individual variation.

There was evidence of seasonal variation in a number of parameters which suggests a reduction in the level of activity of mink during the winter months. Most mink were predominantly nocturnal although some, notably females, showed diurnal foraging activity. A sex difference was also apparent in the degree to which mink foraged away from water. Purely terrestrial foraging was exclusively a male activity (usually directed towards rabbit predation), and was most common on the oligotrophic study area where aquatic and riparian prey densities were low.

---

* Present Address: Department of Zoology, University of Durham, Science Laboratories, Durham, England

INTRODUCTION

The American mink (*Mustela vison*) is the only introduced carnivore to have successfully established itself as a widespread feral animal in Britain. Its arrival and spread have been documented by Thompson (1962, 1968) and Lever (1977). By 1969, feral mink were thought to be present in every county in Britain (Deane & O'Gorman, 1969), despite a widespread trapping campaign by the Ministry of Agriculture, Fisheries and Food throughout the 1960s. The impact of feral mink upon domestic stock and indigenous wildlife has been the subject of much debate in recent years (see, for example, Bourne, 1978; Lever, 1978; Linn & Chanin, 1978a,b), and a considerable amount of ecological research has been carried out on the animal, mainly in Britain and Sweden. The bulk of this work has involved dietary analysis (e.g. Gerell, 1967, 1968; Day & Linn, 1972; Akande, 1972; Wise, 1978; Cuthbert, 1979; Chanin & Linn, 1980) or live-trapping studies (e.g. Gerell, 1971; Chanin, 1976). Linn & Stevenson (1980) have described the establishment of an early mink colony in Devon, and Fairley (1980) has given some general data on Irish mink.

The study described here is the latest in a series of investigations into feral mink ecology at Exeter University. In a continuing attempt to elucidate the niche occupied by this controversial species, radio tracking techniques were employed to collect data on mink home range behaviour. A preliminary analysis of some results from this study is presented in Linn & Birks (in press). The present chapter explores in more detail certain aspects of home range behaviour, such as the sizes and shapes of home ranges, denning behaviour, movement patterns, use intensity, activity and seasonal variations, and attempts a synthesis of the results obtained from different individuals.

The term "home range" is used in this context because, although mink are known to be territorial (Gerell, 1970; Chanin, 1976; Birks, 1981), this part of the investigation is concerned less with the social behaviour of mink than with the details of habitat use shown by individual members of territorial systems. Much of this behaviour is, however, pertinent to considerations of territoriality and is interpreted accordingly.

STUDY AREAS

Radio tracking was carried out in three areas of contrasting habitat-type.

The River Teign

This oligotrophic river rises on the granite mass of Dartmoor and supports a salmonid-dominated fish fauna. The riparian habitat on the 12 km study area is largely dominated by steep-sloping mixed deciduous woodland. Waterside prey densities are generally low, though rabbits (*Oryctolagus cuniculus*) are common up on the valley-sides. Otters are present on the river, but are rarely resident on the study area, which extends from Dogmarsh Bridge (UK grid ref.: SX 712893) downstream to Steps Bridge (SX 804883). A detailed description of this stretch of river can be found in Chanin (1976) or Birks (1981). The dynamics of the territorial system were monitored by means of live-trapping mink over the study period, so that the social environment of each individual radio tracked was known.

Slapton Ley

This eutrophic lake system in South Devon (within the 10 km square SX 84) is managed as a nature reserve by the Field Studies Council. The reserve lies on the coast between the Dart and Kingsbridge estuaries. A shingle ridge *c.* 100 m wide separates an area of carr, marsh, reedbed and open fresh water from the sea in Start Bay. The area supports an abundance of mink prey, notably waterfowl, coarse fish and small mammals. Otters are also resident on the study area. As on the River Teign, the territorial system was monitored by live-trapping. A detailed description of this study area can be found in Chanin (1976) or Wise (1978).

The River Exe in Exeter

This study area is situated within the city limits of Exeter. The river supports moderate populations of coarse fish and some water-fowl; rabbits are also present in small numbers. The surrounding habitat consists of residential areas, allotments and recreation grounds (see Fig. 1), and footpaths or gardens border the river throughout much of the 3 km length of the study area. Although this represents a considerable degree of human disturbance, a fringe of emergent vegetation, alders (*Alnus glutinosa*) and willows (*Salix* spp.) provides some cover. A detailed description of this study area can be found in Birks (1981).

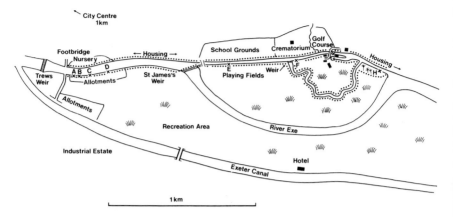

FIG. 1. Suburban study area on River Exe, including October 1979 home range of ♂543 (indicated by dot shading). Capital letters mark dens. Dens A to D are in rabbit burrows and den H is in an abandoned pigeon's nest 5 m up a willow tree.

METHODS

Mink were captured in standard cage traps and transported to the laboratory for the fitting of transmitters. All individuals were, with one exception, known to be resident territory-holders from recapture evidence. Mink were immobilized by intramuscular injections of ketamine hydrochloride (trade name "Vetalar"). A back-pack transmitter (supplied by Telemetry Systems Incorporated) with harness attachment was used initially and found to be unsuitable; all subsequent radio tracking involved the use of collar transmitters (supplied by DAV-TRON Inc. and AVM Instrument Co.). The mink's serpentine shape, with the neck a little smaller in diameter than the head, created some problems of collar retention until a foolproof procedure was developed. This involved clipping the guard hairs and underfur to a length of c. 5 mm at the site of collar attachment, and smearing the inside of the collar with a thin layer of adhesive to secure it to the fur remaining. Mink were kept under observation for several hours after transmitter attachment, then released at the point of capture.

Because of the shortage of equipment and labour, no more than one animal was tracked at a time. All radio tracking was carried out on foot by means of a hand-held directional H-Adcock aerial (supplied by Mariner Radar) and a portable 12-channel receiver on 102.3 MHZ (supplied by DAV-TRON Inc.). Radio tagged mink were located by the null-point determination method described by Taylor & Lloyd (1978). Most radio tracking was carried out by a single operator, working on a discontinuous shift basis. Since mink were located a minimum of twice per day (morning and evening), it was possible to construct a generalized continuous record of den-stays and movements from the discontinuous data. Where possible, faecal material was collected from outside dens used by radio tagged mink to provide dietary information on the tracking period.

In the case of one radio tagged mink (♂523), a feeding experiment was conducted to examine the effects of a food surplus on its pattern of movement. Five dead day-old chicks were planted daily outside one of its regularly used dens at the upstream end of the home range.

The Effects of Radio Tracking on Mink

Since the duration of tracking periods was never greater than four weeks, long-term parameters such as survival and reproductive success were of less concern than the possible short-term effects of radio tracking on mink behaviour. A consideration of these effects is especially pertinent to studies of species which indulge in sub-aquatic predation, since it is in this medium that a transmitter package is potentially most disruptive to speed and agility. The following effects, real and potential, of transmitter collars on mink behaviour were examined.

Post-release inactivity

On release at their point of capture, all radio tagged mink moved to a nearby den and rested for a variable period (3–50 h) before emerging to move about the home range. This recovery period, also observed by Gerell (1969) in radio tagged mink in Sweden, appeared to be a response to the stress of capture, handling and anaesthesia. Home range data were not recorded until the mink had vacated the "recovery den".

Collar irritation

Wild radio tagged mink were commonly observed to scratch at their collars on release, and tufts of fur were often found outside their dens, suggesting that vigorous grooming was occurring. A semi-tame captive mink was anaesthetized and fitted with a radio collar according to the standard procedure. She exhibited reduced activity for c. 48 h and showed signs of irritation (scratching and rubbing the collar) for c. 72 h. Her subsequent behaviour was apparently normal.

Weight change

Most mink were recaptured for removal of transmitters at the end of tracking periods. Any changes in the animals' condition were assessed by comparing their weights before and after tracking. Of the eight mink examined, five registered either weight increases or no change and three registered decreases. Only one of these decreases was in excess of 5% of initial body weight (10.25%), and was thought to have been caused by a bulky transmitter package. The animal's condition was still classified as "Good", despite its weight loss.

Diet

Although faeces ('scats') were collected from several individuals during tracking periods, in only one case was it possible to compare the diet during tracking with that before the collar was fitted. While tracking an adult female ($\female$522), 34 scats were collected from the dens which she frequented. These comprised 16 fresh scats (thought to have been produced during the tracking period) and 18 older scats (thought to have been produce prior to the tracking period). Fish remains occurred with a frequency of 60–70% in both samples, suggesting that the incidence of aquatic prey in the diet was not reduced by the presence of a transmitter collar in this case.

RESULTS

Transmitters were fitted to mink on 18 occasions, though useful quantities of data were obtained from only 11 individuals owing to premature transmitter failures. One-third of all known territory-holders were successfully radio tracked on the River Teign and Slapton study areas over a two-year period. All three individuals caught on the Exe study area were radio tracked. The results from these 11 animals are summarized in Table I (12 tracking periods are presented because one animal – $\male$543 – was tracked in two different seasons).

Optimum Tracking Duration

The tracking periods in this study were of variable length and, owing to premature transmitter failure, some were of only a few days' duration. It was, therefore, important to ascertain the minimum length of tracking period which was acceptable for the derivation of meaningful home range data. Of those mink tracked for 14 days or longer, all had revealed 80% or more of their total home range during that time within five days, and all had revealed their entire home range within ten days. On this evidence, it was felt that tracking periods of 5–10 days' duration yielded reasonable approximations to the entire home range in use at the time, while those of 10 days or

TABLE I

Summary of aspects of home range behaviour of 11 individual mink over 12 tracking periods

Mink	Study area	Home range length (km)	No. of dens used	Mean IDD (m)	Mean IDM distance (m)	Mean no. IDMs pr 24 h	RMR	Mean IDM distance per 24 h (m)	Estimated mean den-stay duration (h)	Predominantly nocturnal or diurnal	Date of tracking
♂520	Teign	2.79	9	314.4	563.8	0.93	1.79	523.5	23.6	Nocturnal	April/May
♂525[b]	Teign	5.94	8	885.7	1300	0.74	1.46	957.8	27.6	Nocturnal	June/July
♂523[b]	Teign	5.90	10	443.0	1585	1.4	3.57	2241.6	11.55	Nocturnal	May/June
♂536[b]	Teign	5.025	6	723.0	2973.6[c]	0.78[c]	6.71[c]	2336.4[c]	21.16[c]	Nocturnal	November
♀509[a]	Teign	0.55	2	—	1143	0.36	1.58	408.2	61.8	Diurnal	January
♀RA	Teign	2.87	9	385.5	681.8	1.7	1.76	1159	12.17	Nocturnal	July/August
♀522	Slapton	1.46	7	208.6	334	0.75	1.6	250	29.0	Both	March/April
♂505	Slapton	1.9	2	950	—	—	—	—	—	Nocturnal	November
♂543	Exe	2.9	8	467	590.7	1.4	1.26	826.9	15.2	Nocturnal	October
♂543	Exe	2.9	3	825	825	0.44	1.0	366.6	39.5	Nocturnal	February
♂548[a]	Exe	Transient	5	345	520	1.66	1.5	886.6	—	Both	February
♀547	Exe	0.50	3	200	260	—	—	—	—	Nocturnal	February

[a] Radio tracked for less than five days.
[b] Mink occupying unstable social environments.
[c] Data recorded during artificial feeding experiment (see text).
See text for explanation of abbreviated parameters presented here.

more were thought to be highly accurate. Two tracking periods of
only four days' duration have been included in Table I because the
animals concerned showed some interesting behaviour. Their data
have been excluded from calculations of the mean values of the
various home range parameters.

Home Range Size and Shape

Linn & Birks (in press) reported that riverine home ranges were
basically linear in shape with some extensions away from the main
river along feeder streams or cover lines to areas of high terrestrial
prey density. At Slapton, however, where the wetland habitat is
more two-dimensional, one animal (♀522) was found to occupy an
oblong-shaped area of marsh, reedbed and carr (see Fig. 2).

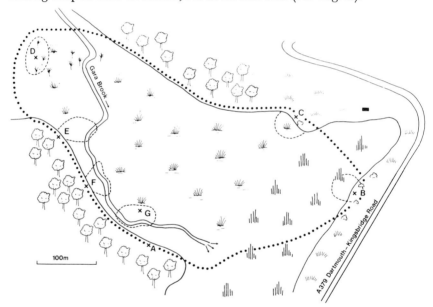

FIG. 2. Home range of ♀522 at Slapton. The capital letters mark the positions of dens
located around the periphery of an area of carr, marsh and reedbed. Dotted lines enclose
foraging areas associated with dens. The area occupied by this animal over a 12-day period
is indicated by dot-shading.

Home range sizes are expressed as the total length (in km) of river,
feeder stream, hedgerow or lake shore occupied by individuals (see
Table I). Since ♀522 mainly moved aound the periphery of her 9 ha
home range (Linn & Birks, in press), the perimeter length (c. 1.5 km)
is used in comparisons with linear home ranges. The mean values

TABLE II

Mean home range lengths of adult male and female mink, compared with those recorded by Gerell (1970) in Sweden and Chanin (1976) on the River Teign

Sex	n	Mean HR length (km)	Range	Author
Males	3	2.53[b]	1.9—2.9	This study
	5	2.48	1.6—4.4	Chanin (1976)
	4	2.64[a]	1.8—5.0	Gerell (1970)
Females	2	2.16	1.46—2.87	This study
	5	2.04	1.2—3.2	Chanin (1976)
	2	1.85	1.0—2.8	Gerell (1970)

Gerell's data were obtained by radio tracking, Chanin's by live-trapping.
[a] The juveniles tracked by Gerell have been excluded from consideration here.
[b] Males from unstable social environments excluded.

TABLE III

Mean home range lengths of adult male mink in stable and unstable social environments

Social environment	n	Mean HR length (km)	Range (km)
Unstable — HRs adjacent to mink control operations	3	5.62	5.02—5.94
Stable — no mink control	3	2.53	1.90—2.90

obtained for male and female mink home range lengths in this study compare favourably with those of Gerell (1970) and Chanin (1976) (see Table II). Male home ranges were, on average longer than those of females, though only by 17%. Home ranges of mink occupying unstable social environments (e.g. ♂536, see Fig. 3) have been excluded from Table II because they were found to be significantly longer $(t = 7.1; P < 0.002)$ than those of mink in stable social environments (see Table III). The causes of this instability, which only occurred on the Teign study area, were localized mink control, by trapping and shooting, and occasional meets of the Devon and Cornwall Minkhounds. ♂523 was observed to considerably extend his home range, to a new length of 5.9 km, after the removal of a neighbouring female by hounds (Linn & Birks, in press). Two other males (♂525 and ♂536 — see Fig. 3) occupied home ranges in excess of 5 km in length adjacent to the area of localized mink control (at different times). ♂520, a large male occupying a socially stable area

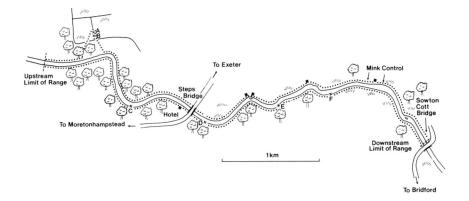

FIG. 3. Home range of ♂536 on the River Teign. The capital letters mark the positions of dens. Dens A and B are rabbit burrows some 225 m from the river, above the steep, wooded valleyside. The presence of localized mink control operations is thought to be responsible for the excessive length (5.0 km) of this home range.

on the River Teign, had a home range of approximately half the mean length of those occupied by the three males in unstable areas on the same study area.

Because of the incidence of other variables, such as sex, social environment and home range shape, the sample size is too small to allow a meaningful comparison of home range size in different study areas.

Denning Behaviour

Preliminary observations of mink denning behaviour (Linn & Birks, in press) have indicated that dens are usually located close to water and may be in temporary or permanent use. Each home range contained a number of different dens, and the duration of visits to dens was highly variable. Dens were found to be important foci of activity since mink apparently preferred to forage close to dens where possible. These findings are in accordance with those of Gerell (1970), who found that feral mink in southern Sweden foraged most intensively in the vicinity of dens.

The collection of additional data, together with further analysis of the original data, has brought the following new information to light.

Den types

During the course of radio tracking operations and extensive field searches, 104 mink dens were identified. The frequency of occurrence of five broad categories of den are compared for each study area in

TABLE IV

Frequency of occurrence of five den categories in the three study areas

Category	Teign	Slapton	Exe
In or beneath waterside trees	24 (42)	9 (28.9)	5 (31.2)
Rabbit burrows	10 (17.5)	1 (3.2)	3 (18.75)
Other holes or burrows	15 (26.3)	16 (51.6)	0 (0.0)
Human artefacts (walls, causeways, bridge parapets)	4 (7.0)	3 (9.6)	4 (25.0)
Above ground. In scrub, brambles, brushpiles, reedbeds, etc.	4 (7.0)	2 (6.4)	4 (25.0)

Figures in brackets are percentages of the total number of dens in each study area.

Table IV. Waterside trees are clearly important as potential den sites in all three study areas, with oak (*Quercus* spp.), willow (*Salix* spp.) and sycamore (*Acer pseudoplatanus*) the most important ones. Rabbit burrows are commonly used on the Teign and Exe, where occupation of this category of den is usually associated with predation upon the inhabitants (see Linn & Birks, in press). As expected, dens associated with human artefacts occurred most commonly on the suburban study area, where mink occupied holes in garden walls, under buildings or in garden hedges where these were close to the river. The anomalously high incidence of above-ground dens in this study area is entirely due to the behaviour of a transient male ($\male$548), radio tracked during the mating season. His lack of familiarity with the local geography apparently caused him to select less secure den-sites than those mink resident in the area. Almost all dens occupied by mink were considered not to have been excavated by them; instead, use was made of existing cavities or holes dug by other species.

There was considerable variation in the distance between dens and the water's edge. At Slapton and on the Exe, nearly all dens lay within 10 m of the water, and most were within 2 m. On the Teign, however, 17.5% of dens lay more than 10 m from the river (mean 131.5 m), and all were rabbit burrows.

Patterns of den use

The mean number of dens used by mink during tracking periods was six, ranging from two to ten (see Table I). A correlation was found between home range length and numbers of dens used ($t =$ 0.631; $P < 0.01$). In addition, a seasonal difference in the number of dens used was apparent; the mean number of dens used by mink tracked between November and April was four, as compared to 8.8 between May and October. Table I shows that ♂543 used eight dens when tracked in October, but only three in February. Linn & Birks (in press) showed that the distribution of dens within a home range can be highly variable. The mean inter-den distance (IDD) for all mink was 492.2 m, with females (mean IDD 264.7 m) showing a higher density of dens than males (mean IDD 589.7 m); this difference was statistically significant ($P < 0.02$).

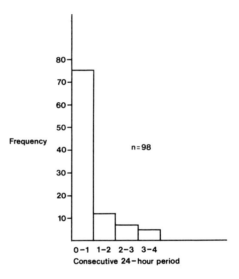

FIG. 4. Frequency of occurrence of den-stays of different duration. A den-stay is defined here as the period of time spent by a mink at or near a den, so it may include foraging activity in the vicinity of dens.

Linn & Birks (in press) observed that stays at, or in, the vicinity of dens were of highly variable duration. Figure 4 illustrates these results in more detail and shows that the majority of den-stays are of less than one day's duration. Those dens occupied for longer periods usually contained food surpluses (see Linn & Birks, in press). Further analysis has revealed that the mean duration of den-stays was significantly greater ($P < 0.001$) during the months November to April than during May to October.

TABLE V

Characteristics of eight core areas exploited by seven individuals in three study areas

Individual	Habitat type of core area	Principal prey taken	Core area length (m)	% of total HR length	% time in core area	Study area
♂543[a]	Allotments	Rabbits	360	8.9	42.1	Exe
♂543[a]	Marsh/willows	Birds?	200	6.9	45.2	Exe
♂536	Pasture	Rabbits	225	6.2	44.0	Teign
♀522	Marsh/stream	Fish	250	17.1	54.3	Slapton
♀RA	River pools	Fish	380	13.2	53.1	Teign
♂525	River pools	Ducks	600	10.1	57.3	Teign
♂523	Heathy hillside	Rabbits	180	3.0	52.3	Teign
♂520	Pasture	Rabbits	220	8.8	64.1	Teign

[a] Two core areas used by this animal within the same home range.

Home Range Use Intensity

Homes ranges were found to contain one or more core areas of high use intensity (Kaufmann, 1962), each associated with concentrations of suitable prey (Linn & Birks, in press). A finer analysis of the characteristics of eight core areas, exploited by seven individuals from all three study areas, is presented in Table V. Core areas generally occupied small percentages (3.0—17.1%, $\bar{x}$ 9.3%) of the total home range length, yet mink spent approximately half of their time within such areas. The fact that core areas are associated with particular habitat types and prey concentrations, components of which tended to predominate in the diets of the mink concerned (as indicated by dietary analysis) confirms their identity as that of primary foraging areas.

Four of the eight core areas in Table V were associated with rabbit predation, and three of these (all on the River Teign) were located at some distance from the river (see Fig. 3). The other was situated in an area of riverside allotments on the suburban study area (see Fig. 1); further details are presented in Linn & Birks (in press). A sex difference is apparent here in that all but one of the seven males radio tracked showed some activity away from water, yet all four females remained in close proximity to a river, lake or wetland. Similarly, five of the seven males spent some time preying on rabbits and occupying rabbit burrows, but this activity was never recorded in female mink. Those individuals concentrating on aquatic prey (see Table V) tended to exploit proportionately larger core areas than those preying on rabbits.

Movements Within the Home Range

Movements made by mink within home ranges were found to be of two types.

  (i) Small-scale movements associated with foraging in a restricted
      area.
  (ii) More extensive travels between dens or their associated forag-
      ing areas.

Since detailed observations were not made on foraging behaviour, the movements described here are all of type (ii) and are termed interden movements (IDMs; this term does not imply only those movements between two adjacent dens, but between any two dens in the home range). IDMs were generally made along linear habitat features such as riverbanks, lake margins, feeder streams or hedgerows. The extent, frequency and general pattern of IDMs showed considerable variation between individuals (see Table I). For example, the mean

IDM distances of ♂523 and ♀547 showed a six-fold difference. The mean IDM distance of males (1009 m) was significantly greater ($P <$ 0.01) than that of females (534 m). Much of the individual variation in mean IDM distances can be explained by variations in den density between home ranges, since a significant correlation was found between mean individual IDD and mean individual IDM distance ($t = 0.72; P < 0.01$).

The ratio of individual mean IDM distance to mean individual IDD has been termed the relative mobility ratio (RMR). Since the former two parameters are positively correlated, the RMR is relatively constant for most individuals (see Table I). All but one mink recorded RMRs of between 1.0 and 2.0, indicating that individual mean inter-den movement distances generally correspond to one to two times the respective mean inter-den distance. ♂523, however, made movements of relatively much greater extent, which were not apparently associated with obtaining food (see also Linn & Birks, in press).

Whereas the extent of movements was, in most cases, related to den density, the frequency of such movements (no. of IDMs per 24 h) showed more ambiguous individual variation (see Table I). The grand mean IDM frequency for all mink was 1.0/24 h, ranging from 0.357 for ♂536 to 1.7 for ♀RA. No differences were found between the sexes ($P > 0.05$); but a seasonal difference was found if only resident mink are considered (i.e. excluding the highly mobile transient ♂548). Resident mink made significantly more frequent IDMs between the months of May and October than between November and April ($P < 0.05$). If the parameters for the mean extent and frequency of movements are combined, an indication of the individual mean distance moved per day is obtained (mean IDM distance per 24 h, see Table I). This parameter varied between 0.25 km for ♀522 and 2.24 km for ♂523 (before artificial feeding). No significant difference was found between males and females for this parameter ($P > 0.1$), though there was evidence for a seasonal difference. The grand mean daily IDM distance between November and April was 0.34 km, as compared to 1.14 km between May and October ($P < 0.02$).

Against the background of any apparent sexual and seasonal variations, the sample size is too small to allow statistical comparisons of movement patterns within home ranges in different habitats. However, a subjective appraisal illustrates the nature of some basic differences. ♀522, for example, occupied an oblong-shaped home range in a two-dimensional area of marsh, willow carr and reedbed at Slapton (see Fig. 2; see also Linn & Birks, in press). Figure 5 illustrates the home range occupied by another female on the River

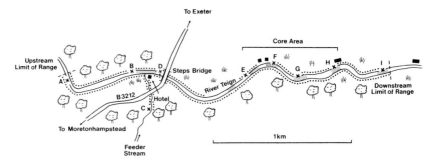

FIG. 5. Home range of ♀RA on the River Teign. The capital letters mark the positions of dens, and the locations of deep pools are indicated by heavy shading. The animal appeared to specialize on fish and no activity was recorded away from the river, except for an excursion up a feeder stream to den C.

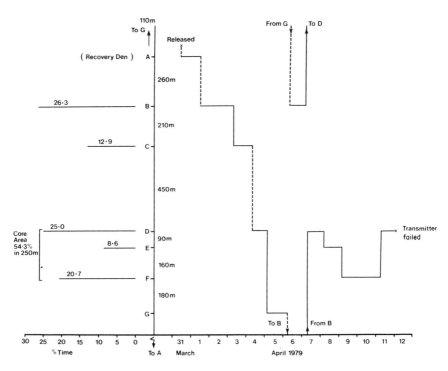

FIG. 6. Movement pattern of ♀522 at Slapton. The y axis represents the two-dimensional home range perimeter broken here between dens A and G for comparison with linear home ranges, along which movements occurred between dens indicated by capital letters (cf. Fig. 7). Movements over the 12-day tracking period are shown on the x axis to the right. Vertical lines represent movements between dens and horizontal lines represent den-stays. Broken lines indicate diurnal movements. To the left of the y axis is shown the percentage time spent in each den and associated foraging area. The stay at recovery den A is excluded from the data (see text).

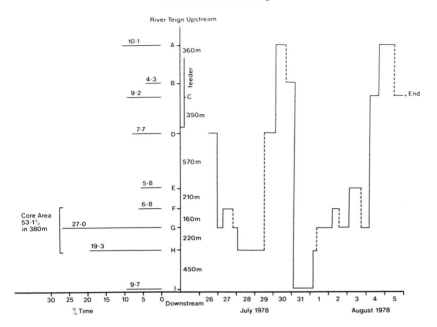

FIG. 7. Movement pattern of ♀RA on the River Teign. The same conventions are used here as in Fig. 6, except that the home range is linear in this case. Note the core area based around dens F to H (cf. Fig. 5) and the patrolling movements outside this core area to the extremities of the home range.

Teign (♀RA). Comparison of the home range parameters for these two mink in Table I shows that the Slapton female had more closely spaced dens, made less frequent and much shorter inter-den movements and spent longer at or near dens than the Teign female. The movement patterns of ♀522 and ♀RA are compared in Figs 6 and 7. Whereas ♀522 moved around the perimeter of her home range, visiting dens and foraging areas in rotation, ♀RA moved back and forth along the same axis as dictated by the linear home range shape. Any patrolling activity could be incorporated in the rotational home range use pattern shown by ♀522, whilst it appeared that special excursions were made by ♀RA to the extremities of her home range from the core area where most foraging occurred. Most mink occupying linear home ranges made these visits to the extremities with a greater or lesser frequency. ♂523 was exceptional in that he covered the bulk of his home range 12 times in 26 days; ♂543 was more representative, covering his home range twice in 10 days.

Activity

Since continuous tracking was not carried out, detailed information on activity patterns, of the kind recorded by Gerell (1969), was not collected in this study. However, sufficient information was obtained to label individuals as either primarily nocturnal or diurnal foragers (see Table I). Most mink were primarily nocturnal, though ♀509 showed a preponderance of diurnal activity when radio tracked in January. ♀522 was unusual in that she foraged regularly every 2—3 h, for 20—40 min, throughout both day and night. Other mink appeared to forage for a few hours at a time and then remained inactive for several hours. Those mink engaged in rabbit predation showed very low levels of activity for 2—3 days after kills were made (Linn & Birks, in press). It is estimated that mink spent, on average, 5—20% of each 24-h period engaged in foraging or travelling between dens. The remaining 80—95% was spent in dens, apparently inactive. The majority (grand mean 75%) of inter-den movements occurred by night, explaining why most captures of mink are made during the hours of darkness.

DISCUSSION OF RESULTS

Any assumption that the effect of carrying a transmitter package does not modify the behaviour of a free-ranging wild animal is a dangerous one. Such an assumption may be justifiable in the case of larger species but, where smaller animals are concerned, it is prudent to look for signs of adverse effects, and evaluate them where possible. In this study, the harness and back-pack transmitter assembly was rejected because it caused considerable discomfort to the wearer after some time, which led to a change in foraging behaviour. Transmitters carried on collars, however, represented only a mild and probably temporary irritant to the wearer, provided they were properly fitted and as slim as possible. No marked changes in foraging behaviour, comparable to that produced by the harness assembly, were observed when collars were used; nor did the limited dietary evidence suggest any reduction in sub-aquatic predatory success. Sanderson (1966) commented that "the transmitter attached to the animal being studied has an unknown effect on the animal's movements". Correctly fitted, transmitter collars were thought not to have seriously affected the behaviour of mink in this study, though subtler effects may have passed unnoticed. These possible unknown effects make it dangerous to assume that behaviour recorded from radio tagged individuals is entirely representative of general and normal behaviour. It is a safer policy to regard information from instrumented animals as a "best estimate" of the behaviour of the rest of the population.

Home Range Size and Shape

The two-dimensional home range occupied by ♀522 at Slapton was smaller, more compact and more efficiently exploited than linear home ranges on the riverine study areas. Prey distribution and density, as well as the nature and extent of the habitat, are thought to have been contributory factors to these differences. The effects of localized mink control on the Teign would suggest that the presence of neighbours is also an important determinant of home range size in this territorial species, since mink occupied much larger home ranges where neighbours were removed. The marked similarity of mean mink home range lengths found by Gerell (1970) in relatively eutrophic habitat in Sweden, Chanin (1976) and the present study suggest that the degree of variability in linear home range size, in stable social environments, is not great in this species. In contrast, Macdonald (1977) has revealed that the home range size of the red fox (*Vulpes vulpes*) may vary by a factor of well over 40, depending upon the density and distribution of available food. Similar data on mustelids occupying non-linear home ranges are sparse, though King (1975) was able to compare her weasel (*Mustela nivalis*) home ranges with those reported by Lockie (1966). Weasel home ranges were, on average, 65—85% smaller in the habitat with higher small mammal densities.

This apparent low plasticity of linear mink home range size in different areas is paralleled by a low variation in home range size between the sexes relative to other mustelids occupying two-dimensional home ranges. Weasels (*Mustela nivalis*), stoats (*M. erminea*) and mink all exhibit marked sexual dimorphism of body size, with males approaching or approximating to twice the weight of females. King (1975) found that female weasel home ranges were only 6.6—57% of the size of males', and Erlinge (1977) reported that female stoat home ranges were 20—33% of the size of males'. In contrast, the female mink in this study occupied home ranges which were, on average, 85.4% of the length of males' (excluding mink from socially unstable areas). The figures from Chanin (1976) and Gerell (1970) are 82.2% and 70.3% respectively.

Chanin (1976) pointed out that linear territories compare unfavourably with two-dimensional ones when the interplay between border patrol and foraging area is examined. To increase the foraging area within a strictly linear territory by four times would necessitate increasing the distance covered in border patrol by four times, because the regions of interaction with neighbours lie at the extremities. In a two-dimensional territory, however, a four-fold increase in area would only result in a doubling of the boundary length, if the same shape was maintained, since increase in area is proportional to the square of the increase in circumference.

These constraints might explain the low variability of mink home range sizes in linear habitat compared to other mustelids occupying two-dimensional home ranges. This effect is exemplified by the smaller difference between male and female mink home range sizes relative to the other species considered. Perhaps the constraints operating against lengthways expansion of linear ranges (outlined above) explain the fact that male mink do not occupy proportionately greater lengths of river (as predicted by differences in bodyweight) than females — unless neighbours are absent, when, however, patrolling effort may become excessive. In the light of this apparently more severe constraint facing male mink, it is not surprising that they showed a greater tendency to forage away from water than females, when conditions were suitable, thereby expanding their home ranges laterally instead of linearly to compensate for their greater size.

Clearly, if the time and energy costs of patrolling a linear home range are high (so placing a severe limit on the dimensions of the range) and this situation is coupled with a poor or patchy prey supply, the habitat will be sub-optimal. Such a situation appears to exist on the River Teign where the territorial system is relatively unstable, with mink occupying territories for a few months only before moving elsewhere (Chanin, 1976; Birks, 1981).

Denning Behaviour

The tendency for mink to utilize existing den-sites rather than excavating their own has been commented on by several authors (e.g. Errington, 1961; Sargeant, Swanson & Doty, 1973). This dependence on existing burrows or cavities, together with the mink's tendency to concentrate its foraging activity close to dens, would suggest that the presence of potential den-sites is an important requirement of suitable mink habitat. Where such sites are abundant, as on the rocky, tree-lined River Teign, the location of dens used by mink may be secondarily determined by other factors such as the distribution of prey (note, in Fig. 5, how the dens used by ♀RA are located close to pools). However, where the nature of the habitat results in limited den availability, but where prey is uniformly abundant, as in many marshy areas on the Slapton study area, the distribution of den sites may dictate the distribution of foraging activity and other home range characteristics. ♀522 exemplifies this effect because her dens were necessarily located on dry land around the edge of her marshy home range (see Fig. 2). The general abundance of prey enabled her to hunt only in the vicinity of these dens, with the result that the den-free interior of her home range was exploited very lightly.

The mink tracked in this study used, on average, twice as many dens as those tracked by Gerell (1970) in Sweden (six, as compared to three). There are three possible explanations for this difference:

(i) The flat, pasture-dominated study areas in which Gerell worked may have provided fewer potential den-sites than those used in this study.

(ii) The range sizes observed in this study are biased towards a large mean range by the inclusion of males whose social environment was unstable, and which therefore had exceptionally large home ranges. In Gerell's (1970) study, on the other hand, the bias was towards a small mean range, because of the inclusion of juveniles with small home ranges. If, as suggested by Linn & Birks (in press), the mean distance between dens is very similar in this study and in Gerell's, and if this distance between dens, though very variable, is determined by a biological requirement of the animal, then it would be expected that larger ranges would, other things being equal, have more dens in them.

(iii) A higher incidence of rabbit predation was observed in this study than by Gerell (1970) (45% of individuals tracked occupied rabbit burrows in this study as a consequence of rabbit predation; Gerell recorded only 18% of his mink preying on rabbits during tracking periods). Since each adult rabbit "kill" is usually followed by occupation of a different burrow, the number of burrow dens used by a mink specializing on rabbits will increase with time. This might explain the larger average number of dens used by mink in this study.

It seems possible that all three factors may have played some part in accounting for the observed differences.

Intensity of Use

Marked spatial variations in the intensity of home range use shown by mink have been reported by Gerell (1970) and Linn & Birks (in press), and there is evidence that these patterns may show temporal variation in response to seasonal changes in prey availability. The change in core area usage shown by ♂543 between October and February (Linn & Birks, in press), together with similar shifts in use intensity patterns recorded by Gerell (1970) for three mink tracked in different seasons, would appear to support this suggestion. The diet of mink has been shown to vary seasonally as the availability of different prey groups fluctuates (Wise, 1978). Movements outside core areas in use at any point in time may, therefore, not only serve the function of territory patrol, but may also be important opportunities for monitoring prey availability in the remainder of the home range. Such behaviour would allow a resident mink to respond rapidly to relative changes in prey availability by shifting its foraging pattern accordingly.

The tendency for core areas based on rabbit concentrations to be of more limited extent than those based on aquatic prey may be a function of the size and distribution of the prey species, and the necessarily different foraging strategies involved. Rabbit kills were made at intervals of a few days, in burrows which showed a clumped distribution. Disturbance to other members of the warren was small between successive kills because mink generally occupied only the burrow in which a carcase was being consumed. Fish predation, however, generally involved smaller prey items (see Wise, 1978) which would be taken with much greater frequency than in the case of rabbits. The regular movements between pools shown by ♀RA (see Fig. 7) in her core area suggests that the prey population in each pool became temporarily depleted, disturbed or otherwise less available as a result of her activities. This differential disturbance factor would explain why a chain of two or three pools may be a necessary requirement of a core area based on predation of small fish (to allow prey populations in each to recover between successive foraging bouts), while just one stretch of "rabbity" hedgerow may suffice in the case of a terrestrial core area.

The apparently more patchy distribution of prey on the marginal River Teign study area tended to produce more marked variations in intensity of home range use by mink than in the Slapton study area where prey densities were uniformly high.

Movements

The relationship between the extent of movements and den density within home ranges further points towards the importance of abundant den-sites as a prerequisite for suitable mink habitat. Areas with few dens would clearly involve more extensive travels by mink, and therefore greater risks and energetic costs, than an area well-supplied with dens.

The movements outside core areas to the extremities of home ranges have already been considered, in the discussion of use intensity, to be important for territory patrol. Gerell (1970) observed that the degree of patrolling activity was related to the status of the mink concerned. Linn & Birks (in press) have already reported on the unusually high degree of patrolling activity shown by the small ♂523 after he had expanded his home range in response to the removal of a neighbouring female. Comparison of this animal's activity (not suppressed by the provision of a food surplus) with that of other males of apparently higher status (e.g. ♂536, ♂525) would appear to support Gerell's observations, since these animals made much shorter daily movements, on average, even though both oc-cupied home ranges of similar length in unstable social environments.

With the exception of ♂523, the nature of the food supply tended to influence the pattern of movements shown by mink. Those engaged in rabbit predation (e.g. ♂536 and ♂520) were less mobile than others because they could afford to lie up with kills for two to three days at a time. ♀RA, on the other hand, was highly mobile between the different pools within her core area. The high mobility shown by the transient ♂548 was apparently associated with repro-ductively-orientated exploration, including visits to the home range of ♀547. The patterns of movement shown by different mink are clearly influenced by a diversity of factors, with the nature and distribution of the food supply, and the animal's status, being important ones.

Activity

In the light of the mink's rapid gut passage time of approximately two hours (Wise, 1978), one might expect the animal to show a pattern of foraging of similar periodicity. Only ♀522 conformed to this expected pattern, while the remaining animals tracked showed considerably longer periods of inactivity between bouts of foraging. Some individuals obviated the need to forage at regular intervals by selecting large prey items, such as rabbits, which enabled them to feed within dens for two days or more. The hoarding or caching of numbers of prey items by mink has been widely reported in America, and occasionally in Britain (Hill, 1964) and Sweden (Erlinge, 1969). Although the search for food caches was not possible in this study, since it would have involved destruction of dens, mink were ob-served to carry prey to dens on occasions. It is suggested that this behaviour enables mink to forage less frequently than might other-wise be dicated by the rapid emptying of the stomach in the absence of regular meals. This presumably also accounts for the large numbers of scats found commonly near the den openings.

Gerell (1969) hypothesized that mink activity patterns were primarily determined by the activity (and therefore availability) of prey species. Data from this study are not complete enough to confirm or refute this suggestion. However, the presence or absence of food surpluses and the status of the animal concerned are considered here to be important determinants of the level, if not also the pattern, of activity shown by mink, although it is probable that other factors are involved on which we have no firm evidence, such as the level of human disturbance, extent of cover and climatic extremes.

Seasonality

The results from this study suggest that a number of aspects of home range behaviour are subject to seasonal variation. The evidence points towards a general reduction in the level of activity in winter relative to summer. Fewer dens are used, stays at dens are longer, and daily distances travelled are shorter during the winter months. When tracked in February, ♂543 showed a reduction in the level of activity as well as a spatial concentration of activity relative to his October tracking period, though the total size of area occupied remained the same. Concentration of activity in winter has also been reported in mink by Gerell (1970) and in otters (Erlinge, 1967).

Three possible explanations for the apparent reduction and concentration of mink activity in winter are presented below.

(i) It has been shown that the serpentine body form of the smaller mustelids is energetically inefficient (Brown & Lasiewski, 1972). The importance of conserving energy when temperatures are low may, therefore, encourage the reduction of activity of mink in winter, so as to minimize loss of body heat. Marshall (1936) reported that mink were inactive during periods of very low temperatures and showed a high degree of diurnal activity. The incidence of diurnal activity in winter, also observed in ♀509 in this study, supports the energy conservation hypothesis, since it suggests that the colder parts of the 24-h period are being avoided by mink. Evidence presented by Gerell (1969) suggests that male mink show a lesser tendency towards reduced activity and diurnal foraging than do females in response to low temperatures. Perhaps this reflects the greater body size of males which allows greater energetic efficiency than in the case of females.

(ii) The presence of energy reserves, either in the form of fat stored during the autumn, or as food caches at dens, might permit a lower level of foraging activity during the winter months when conditions are unfavourable. Chanin (1976) found that the number of mink in excellent condition in Devon reached a peak in autumn and declined thereafter, suggesting some wastage of body reserves over the winter. Chanin also found that the weights of individual mink tended to rise in late summer and autumn, and decline from mid-winter.

(iii) The winter months coincide with a period of relative territorial stability (Birks, 1981) when the pressure from intruders is low in comparison with the late summer period of juvenile dispersal and the post-mating season period of male territory establishment. Extensive territory patrol is, therefore, less necessary at this time of year. In addition, male mink are at their least aggressive around November (MacLennan & Bailey, 1969). These two factors may be responsible in part for the reduction in movements within the home range shown by resident mink in winter in this study.

ACKNOWLEDGEMENTS

This research was supported by a grant awarded to one of us (J.D.S.B.) by the Natural Environment Research Council. The authors wish to thank the Field Studies Council, the Herbert Whitley Trust, the Devon Trust for Nature Conservation, the South West Water Authority and Fountain Forestry Limited for cooperation and permission to work on their land. Our thanks are also due to Messrs N. Barratt and P. Splatt and Mrs Anna Davey for technical assistance, and to Messrs P. Tree, S. Spelling, T. Birks and others for help in the field. We are grateful to the Graphics Unit of the Teaching Services Centre, Exeter University, for drawing the text figures.

REFERENCES

Akande, M. (1972). The food of feral mink (*Mustela vison*) in Scotland. *J. Zool., Lond.* 167: 475–479.
Birks, J. D. S. (1981). *Home range and territoriality of the feral mink* (Mustela vison *Schreber*) *in Devon*. Ph.D. thesis: University of Exeter.
Bourne, W. R. P. (1978). Mink and wildlife. *B.T.O. News* 91: 1–2.
Brown, J. M. & Lasiewski, R. C. (1972). Metabolism of weasels: the cost of being long and thin. *Ecology* 53: 939–943.
Chanin, P. R. F. (1976). *The ecology of feral mink* (Mustela vison *Schreber*) *in Devon*. Ph. D. thesis: University of Exeter.
Chanin, P. R. F. & Linn, I. (1980). The diet of the feral mink (*Mustela vison*) in southwest Britain. *J. Zool., Lond.* 192: 205–233.
Cuthbert, J. (1979). Food studies of feral mink (*Mustela vison*) in Scotland. *Fish Mgmt* 10: 17–25.
Day, M. G. & Linn, I. (1972). Notes on the food of feral mink (*Mustela vison*) in England and Wales. *J. Zool., Lond.* 167: 463–473.

Deane, C. D. & O'Gorman, F. (1969). The spread of feral mink in Ireland. *Ir. Nat. J.* **16**: 198—202.

Erlinge, S. (1967). Home range of the otter (*Lutra lutra*) in southern Sweden. *Oikos* **18**: 186—209.

Erlinge, S. (1969). Food habits of the otter (*Lutra lutra*) and the mink (*Mustela vison*) in a trout water in southern Sweden. *Oikos* **20**: 1—7.

Erlinge, S. (1977). Spacing strategy in stoat *Mustela erminea*. *Oikos* **28**: 32—42.

Errington, P. L. (1961). *Muskrats and marsh management*. Lincoln and London: University of Nebraska Press.

Fairley, J. S. (1980). Observations on a collection of feral Irish mink *Mustela vison* Schreber. *Proc. R. Ir. Acad.* **80B**: 79—90.

Gerell, R. (1967). Food selection in relation to habitat in mink (*Mustela vison* Schreber) in Sweden. *Oikos* **18**: 233—246.

Gerell, R. (1968). Food habits of the mink (*Mustela vison* Schreber) in Sweden. *Viltrevy* **5**: 119—211.

Gerell, R. (1969). Activity patterns of the mink (*Mustela vison* Schreber) in southern Sweden. *Oikos* **20**: 451—460.

Gerell, R. (1970). Home ranges and movements of the mink (*Mustela vison* Schreber) in southern Sweden. *Oikos* **21**: 160—173.

Gerell, R. (1971). Population studies on mink (*Mustela vison* Schreber) in southern Sweden. *Viltrevy* **8**: 83—114.

Hill, G. (1964). Wild mink in west Wales. *Nature in Wales* **9**: 17—18.

Kaufmann, J. H. (1962). Ecology and social behavior of the coati (*Nasua narica*) on Barro Colorado Island, Panama. *Univ. Calif. Publs. Zool.* **60**: 95—222.

King, C. M. (1975). The home range of the weasel (*Mustela nivalis*) in an English woodland. *J. Anim. Ecol.* **44**: 639—668.

Lever, C. (1977). *The naturalized animals of the British Isles*. London: Hutchinson.

Lever, C. (1978). The not so innocuous mink. *New Scient.* **78**: 812—814.

Linn, I. J. & Birks, J. D. S. (In press). Observations on the home ranges of feral American mink (*Mustela vison*) in Devon, England, as revealed by radiotracking. *Proceedings of the Worldwide Furbearer Conference*. Maryland, U.S.A. 1980.

Linn, I. & Chanin, P. (1978a). Are mink really pests in Britain? *New Scient.* **77**: 560—562.

Linn, I. & Chanin, P. (1978b). More on the mink 'menace'. *New Scient.* **80**: 38—40.

Linn, I. & Stevenson, J. H. F. (1980). Feral mink in Devon. *Nature in Devon* **1**: 7—27.

Lockie, J. D. (1966). Territory in small carnivores. *Symp. zool. Soc. Lond.* No. 18: 143—165.

Macdonald, D. W. (1977). The behavioural ecology of the red fox. In *Rabies: The facts*: 70—90. Kaplan, C. (Ed.). Oxford: University Press.

MacLennan, R. R. & Bailey, E. D. (1969). Seasonal changes in aggression, hunger and curiosity in ranch mink. *Can. J. Zool.* **47**: 1395—1404.

Marshall, W. H. (1936). A study of the winter activities of the mink. *J. Mammal.* **17**: 382—392.

Sanderson, G. C. (1966). The study of animal movements — a review. *J. Wildl. Mgmt* **30**: 215—235.

Sargeant, A. B., Swanson, G. A. & Doty, H. A. (1973). Selective predation by mink (*Mustela vison*) on waterfowl. *Am. Midl. Nat.* **89**: 208—214.

Taylor, K. D. & Lloyd, H. G. (1978). The design, construction and use of a radio-tracking system for some British mammals. *Mammal Rev.* **8**: 117–141.

Thompson, H. V. (1962). Wild mink in Britain. *New Scient.* **13**: 130–132.

Thompson, H. V. (1968). British wild mink. *Ann. appl. Biol.* **61**: 345–349.

Wise, M. H. (1978). *The feeding ecology of otters and mink in Devon.* Ph.D. thesis: University of Exeter.

DISCUSSION

D. Chivers (Cambridge) — I am not clear of the extent to which the home ranges you describe are in fact territories in the sense in which there is no overlap.

J. D. S. Birks — Mink are definitely territorial in that intra-sex overlap does not occur, though inter-sex overlap does sometimes occur. I use the term home range here to emphasize the point that we were studying isolated individuals, not in their social context.

McBride (UCL) — Is there an explanation for the fact that most den sites are on one side of the river?

Birks — This is probably due to the fact that the flood plain with suitable tree den sites lies on one side of the river, the other side is rocky.

M. J. Delany (Bradford) — Have you been able to quantify the amount of time mink spend foraging on land as against in the water?

Birks — No, due to the discontinuous collection of data and the difficulty of making close observations.

M. Ridley (Oxford) — Has your study shed any light on the efficiency of mink control methods?

Birks — It would appear that mink hunting is very inefficient. I followed one day's hunting and the hunt found three of the six residents on the stretch of river covered and killed one of these. This represents a one in six success rate. Casual localized control, such as that practised by one study area resident who claims to have killed 120 minks in six years, has practically no effect on the resident population as most of the animals killed are transients.

Symp. zool. Soc. Lond. (1982) No. 49, 259–289

Some Comparisons Between Red and Arctic Foxes, *Vulpes vulpes* and *Alopex lagopus,* as Revealed by Radio Tracking

P. HERSTEINSSON and D. W. MACDONALD

Animal Behaviour Research Group, Department of Zoology, University of Oxford, Oxford, England

SYNOPSIS

A comparison of red and Arctic foxes, *Vulpes vulpes* and *Alopex lagopus*, based on two radio tracking studies and a review of the literature, indicates that the two species are very similar. Both appear to have a flexible social organization which, in the case of our field studies, includes resident animals maintaining territories that support more than an adult pair. For both species, groups comprised only one adult male together with several adult females. Some, apparently subordinate, adult vixens within the social groups of both species did not breed, and some of these acted as helpers, feeding the cubs of the more dominant vixens, who were often their kin.

Although home range sizes were different for the populations of red and Arctic foxes in question (red = 0.45 km^2, Arctic = 12.5 km^2), there was evidence that the same hypothesis might explain the social organization of both species; i.e. the abundance and dispersion of patches of available food independently influence group and territory sizes. In each case, this hypothesis gained support from the presence of almost constant amounts of key habitats within neighbouring territories of widely differing sizes. In the case of red foxes this key habitat was "residential land" (10.4 ha per territory), whereas for the Arctic fox it was "productive coastline" (5.7 km per territory). The similarity between red and Arctic foxes suggests that they would be direct competitors where their geographical distributions overlap.

INTRODUCTION

Red foxes, *Vulpes vulpes* L., and Arctic foxes, *Alopex lagopus* L., are believed to be descendants of a common ancestor, *Vulpes alopecoides* Campana, which existed in southern Europe in the late Pliocene and early Pleistocene (Beneš, 1970; Kurtén, 1965). The two present day species diverged during the Pleistocene and now *Alopex lagopus* occupies the circumpolar tundra zone, whereas *Vulpes vulpes* spans most of the northern hemisphere north of 30°N. Generally the northern limits of the red fox's range are in the tundra zone, although in places it reaches the Arctic Ocean.

There is no evidence of interbreeding between red and Arctic foxes in the wild, although male Arctic foxes readily mate with red vixens in captivity (Cole & Shackelford, 1946). The resulting hybrids are sterile. The red fox has a diploid chromosome number of 34, together with 6—7 microsomes (Rausch & Rausch, 1979). The Arctic fox, in contrast, has a diploid chromosome number of 52. The hybrid has an intermediate number of 43 (Wipf & Shackelford, 1949). The smaller size of the Arctic fox's chromosomes may indicate that the difference in diploid numbers between the two species derives from fusion (or dissociation) of the ancestral chromosomes during evolutionary divergence.

This chapter reviews the literature on adaptations shown by each species and then, on the basis of radio tracking studies, develops a hypothesis which seeks to explain their social systems in terms of the dispersion of their food supplies. Radio tracking and direct observations have been critical to these studies so techniques and specifications of the equipment will also be summarized.

SIMILARITIES AND DIFFERENCES

Morphology

The dimensions of red and Arctic foxes are compared in Table I, which indicates that although the red fox is about 60% heavier than the Arctic species at comparable latitudes, the two species are nevertheless similar anatomically. The most striking difference in their

TABLE I

Morphological characteristics of north European Foxes

	Head-body length (mm)		Tail length (mm)		Weight (g)		
	♂♂	♀♀	♂♂	♀♀	♂♂	♀♀	Source
Red fox	681	663	437	418	5897	5193	Lund (1959)
Arctic fox	554	526	305	299	3816	3091	Pulliainen (in press)
Red fox: Arctic fox ratio	1.23	1.26	1.43	1.40	1.55	1.68	

morphology is the great density of the Arctic species' under-fur (about 70% of the Arctic fox's fur fibre consists of fine under-fur, compared to 20% in *Vulpes* (Cole & Shackelford, 1946)). Scholander *et al.* (1950) found that under controlled conditions the metabolic rate of Arctic foxes began to rise only when temperatures dropped to between -45 to $-50°C$, and these foxes began to shiver only after almost an hour at $-70°C$ (See also Henshaw, Underwood & Casey, 1972). In contrast, the metabolic rate of an Alaskan red fox, similarly in winter pelage, began to rise at $-13°C$ and at $-50°C$ it had almost doubled (Irving, Krog & Monson, 1955).

Other differences in morphology between the species can also probably be accounted for as adaptations to the cold temperatures of the northern tundra; for example, the Arctic fox has smaller, more rounded ears than the red fox. The tail and neck of Arctic foxes are relatively shorter than those of red foxes, the braincase is more rounded, and the muzzle is slightly shorter and broader (Clutton-Brock, Corbet & Hills, 1976). However, the two species have relatively the same length legs, contrary to outward appearances (Hildebrand, 1952).

The most obvious difference between the two species is in their colour. The red fox is largely russet coloured, tinged to varying extents with golden lights (especially around the neck and shoulders); its lower limbs are normally black, as are the backs of the ears. A triangular black mark on either side of the muzzle varies between being prominent or absent, and the tail may be tipped with white. The throat and underside are normally grey, but vary from almost white to almost black. Guard hairs are banded with black and since the melanin from which this colour derives shares a synthetic pathway with adrenalin, there is a relationship with coat colour and temperament (Keeler, 1975).

A minority of red foxes also occur in other colour morphs, notably silver (i.e. black) and "cross" foxes (red, with a band of black running up the back and forming a cross at the neck). These colours are under genetic control at three loci (Keeler *et al.*, 1970). The Arctic fox occurs in two distinct colour morphs, known as "blue" and "white". Each colour phase also changes seasonally, the "blue" foxes moulting from chocolate brown in summer to a winter coat of a slightly lighter shade of brown tinged with a blue sheen due to varying densities of grey guard hairs (Boitzov, 1937). In Iceland some blue foxes have a white blaze on the chest and chin, and white toes are not uncommon (P. Hersteinsson, unpublished). The "white" variety is almost pure white in winter. Their summer coat is grey to brownish-grey dorsally and light grey to white below. The colour morphs of Arctic foxes are genetically determined at a single locus (Slagsvold, 1949), the white being recessive. Heterozygotes are indistinguishable from homozygous blue foxes.

The proportions of blue to white Arctic foxes vary greatly throughout the species' range. Throughout most of their continental range, i.e. mainland Alaska, Canada and Eurasia, the "blue" phase amounts to less than 1% of the population. However, the proportion of blue foxes increases to the west in Alaska (Anderson, 1934); and to the east in Canada (reaching 4.4% on eastern Baffin Island) (Fig. 1; Fetherston, 1947). The proportions are quite different on islands. For example, in northern Greenland they are present in roughly equal proportions, whereas in the south and west of Greenland the blue form predominates (Braestrup, 1941). In fact, Elton (1949) suggests that the increased proportion of blue foxes in eastern Canada is due partly to emigration (by sea-ice) from western Greenland. In Iceland "blues" comprise two-thirds of the population overall, and considerably more on the coast (P. Hersteinsson, unpublished). Amongst the small islands of the Pribilov and Commander groups in the Bering Sea the "blue" fox comprises 90.0—99.85% of the population (Boitzov, 1937). Hence, the blue morph is associated with coastal habitats and thus with smaller islands where the ratio of coastland to mainland is greater.

Against this array of adaptations to cold climates, it is noteworthy that the Arctic fox is considerably smaller than the red fox, and thus presents an interspecific discontinuity in Bergmann's rule, to which the red fox conforms intraspecifically (Davis, 1977).

Habitat and Diet

The geographical ranges of the two species of fox both encompass various habitats, but in general that of the Arctic fox is both less varied and less productive than that of the red fox. The Arctic and Alpine tundra biome of the treeless north encompasses maritime and continental climates and landscapes as varied as dry heathland and marshland (Järvinen and Väisänen, 1977). However, this variation is small in comparison to that reflected in the latitudinal changes in habitat types and with what is, broadly speaking, a northerly decrease in primary productivity. Temperate woodlands have an average annual productivity of $1300 \, g \, m^{-2} \, year^{-1}$ (range 600—2500 $g \, m^{-2} \, year^{-1}$) in comparison to an average of $800 \, g \, m^{-2} \, year^{-1}$ (range 400—2000 $g \, m^{-2} \, year^{-1}$) for boreal forests and $140 \, g \, m^{-2} \, year^{-1}$ (range 10—400 $g \, m^{-2} \, year^{-1}$) for tundra (Whittaker & Likens, 1975) (see Table II). The consequences of this decline in productivity

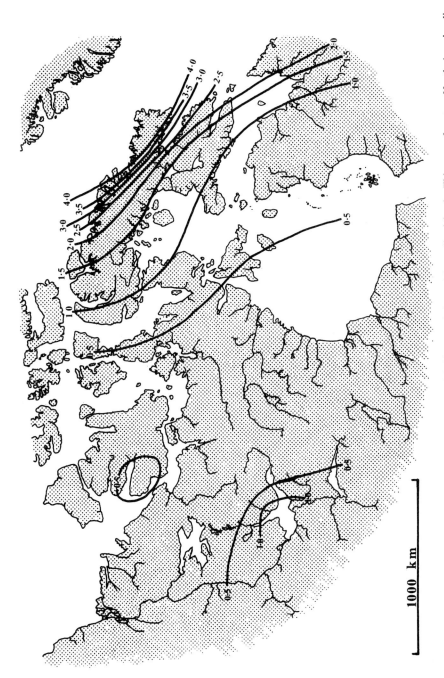

FIG. 1. The blue morph of Arctic foxes is associated with coastal habitats and thus with smaller islands. This map of eastern North America displays isoclines of given percentages of the blue morph in samples of Arctic foxes killed by fur trappers. Drawn from data in Fetherston (1947).

1000 km

TABLE II

Comparison of primary productivity and biomass of a selection of ecosystems (after Whittaker & Likens, 1975)

Ecosystem type:	Net primary production (dry matter)		Mean biomass (dry matter) (kg m^{-2})
	Normal range (g m^{-1} year^{-1})	Mean (g m^{-2} year^{-1})	
Tropical rain forest	1000–3500	2200	45
Tropical seasonal forest	1000–2500	1600	35
Temperate forest:			
evergreen	600–2500	1300	35
deciduous	600–2500	1200	30
Boreal forest	400–2000	800	20
Woodland and shrubland	250–1200	700	6
Savanna	200–2000	900	4
Temperate grassland	200–1500	600	1.6
Tundra and alpine	10–400	140	0.6
Desert and semidesert scrub	10–250	90	0.7
Extreme desert — rock			
sand and ice	0–10	3	0.02
Cultivated land	100–4000	650	1
Swamp and marsh	800–6000	3000	15
Lake and stream	100–1500	400	0.02

at higher latitudes are reflected in reduced diversity of species of birds towards the north (Kauri & Østbye, 1975; Fischer, 1960; Mac-Arthur, 1972: 212). Even between adjacent habitats productivity of more typical Arctic fox habitat is less than that of red fox habitat; Wielgolaski (1975) recorded an annual primary productivity in Norwegian birch forest of 600–900 g m^{-2} year^{-1} whereas snowbeds and shrub or lichen heaths produced 100–400 g m^{-2} year^{-1}. In Finland birch forest, pine forest and low alpine heath supported, respectively 50, 40 and 27 species of vertebrates (Haukioja & Koponen, 1975). Similarly, in the tundra of mainland Canada and Alaska the number of breeding bird species is about one-third of what it is in the rest of these countries per unit area (calculated from MacArthur, 1972: 212) and the number of rodent species in northern boreal forests of Canada and Alaska is about three times higher than in the tundra of North America (Hagmeier & Stults, 1964). The greater diversity of food taken by the red fox as compared to the Arctic fox can thus be ascribed to its access to a more diverse fauna on which to prey, rather than necessarily to any difference in the hunting behaviour of the species.

The red fox epitomizes a generalist in its feeding behaviour. Its animal prey range from young chamois, *Rupicapra rupicapra*, (Leinati *et al.*, 1960) to earthworms, *Lumbricus terrestris*, (Macdonald, 1980a) and it also eats fruits, berries and tubers together with varied forms of carrion and refuse. Following the trilogy of diet studies reported by Scott (1943) there have been over 100 studies of the diet of red fox (reviewed by Sequiera, 1980), all of which indicate its catholic tastes.

In comparison to that of the red fox, the diet of the Arctic fox seems restricted. Over most of its range in Arctic North America and the USSR rodents comprise the major prey. Indeed, cycles in number of Arctic rodents, particularly lemmings, *Lemmus* and *Dicrostonyx*, are mirrored in Arctic fox populations. However, even in the tundra there are areas where geese, *Anser* and *Branta* spp., and reindeer, *Rangifer tarandus*, (as carrion) are important, while in Scandinavia and Greenland the rock ptarmigan, *Lagopus mutus*, is a major prey (Collett, 1912; Braestrup, 1941). Lavrov (1932) reported that on the Pribolov and Commander Islands sea-birds, fish, marine invertebrates and seaweeds were eaten. In western and southern Greenland and Spitzbergen cliff-nesting birds, mainly alcids, are important food. On Kildin Island (USSR) (Lavrov, 1932) and Iceland (see below), Arctic foxes are accused of killing sheep.

Although the diet of the Arctic fox is restricted in comparison to that of the red fox it nevertheless includes diverse prey. We conclude that the Arctic species is just as opportunistic a predator as the red one, but that its geographical range incorporates a more restricted prey-base. One apparent anomaly is the case of the Arctic foxes on St George (one of the Pribilov group) where there is an endemic species of lemming. According to Osgood, Preble & Parker (1915) (cited in Braestrup, 1941) they are not eaten by the foxes even when very abundant.

The similarities in the two species' diets and a consensus of anecdotal reports suggest that Arctic and red foxes forage in broadly similar ways, and this view is supported by observations in captivity (Fox, 1971). One tactic adopted by both species is the trailing of larger predators in order to pilfer their food. Arctic foxes follow polar bears, *Thalarctos maritimus*, onto pack-ice and scavenge from the carcasses of seals killed by the bears (Braestrup, 1941). Similarly, in the desert of Israel red foxes regularly follow striped hyaena, *Hyaena hyaena*, and steal scraps of food (D. W. Macdonald, unpublished).

Movements and Social Organization

Much less is known about the Arctic fox's movements and social system than about the red fox's, and even the red fox is at best poorly understood. Red foxes have been found in broadly territorial organization (Sargeant, 1972; Ables, 1975; Macdonald, 1981) and in less clear-cut spatial systems (Niewold, 1974; Harris, 1980). Young foxes disperse from their natal territories over distances from a few kilometres to over 200 km (Lloyd, 1975; Storm *et al.*, 1976; Englund, 1970; Macdonald, 1980b) with males generally dispersing over greater distances than females. In some habitats additional females, probably offspring or close relatives, may share the home range of the pair during their first winter (Ables, 1975) or for several years (Macdonald, 1980b). Perhaps associated with larger groups are great variations in vixen productivity from one habitat to the next (Englund, 1970; Harris, 1979; Macdonald, 1979). There is considerable variation in the social system of red foxes, most clearly reflected in their occupancy of habitats ranging from wilderness to city centre (Harris, 1977, 1980; Macdonald & Newdick, in press), and the variation in the sizes of their home ranges from 0.1 to 20 km^2.

There are few data on the home range sizes of Arctic foxes. Longstaff (1932) estimated that one pair occupied 20 km^2 in Greenland, and in similar habitat D. Stroud (personal communication) saw one identifiable individual within at least 30 km^2. A calculation of the home ranges pictured by Kaikusalo (1971) shows that the home range of an Arctic fox pair in North Finland varied between 3 km^2 and 60 km^2 within one year, with changes in food availability. Speller (1972) estimated 2.9 km^2 as the mean hunting range of a pair of Arctic foxes at Aberdeen Lake (North West Territories) in 1970, a good lemming year. Arctic foxes are known to travel large distances, behaviour perhaps comparable to the emigration and dispersal of young red foxes, but sometimes on an apparently greater scale and involving a large proportion of the population. Large-scale emigrations may result from drastic reductions in food supply, such as a population crash amongst lemmings. Emigrations have been recorded in Canada (Wrigley & Hatch, 1976), Fennoscandia (Collett, 1912; Pulliainen, 1965) and the USSR (Lavrov, 1932; Shilyaeva, 1967). Ear-tagging studies have revealed journeys that exceed the longest recorded for red foxes, amounting to 1000 km or more (Skrobov, 1961; Schwartz, 1967; Shilyaeva, 1967; Chesemore, 1968; Eberhardt & Hanson, 1978).

Seasonal migrations and large-scale emigrations have not been documented for red foxes (but see Novikov, 1956: 62–63).

Macdonald (1980b: 56) has argued that red foxes disperse over greater distances in areas where their home ranges are larger. As Arctic fox home ranges generally seem to be larger than the largest documented red fox ranges, and the latter probably increase in size with increased latitude (Hersteinsson & Macdonald, in preparation), wide dispersal in Arctic and red foxes may be governed by the same ecological principles.

In addition to emigrations, some Arctic fox populations may undertake seasonal migrations. On mainland Alaska and some of the larger Arctic islands (Banks Island, Baffin Island and Novaya Zemlya), Arctic foxes leave their breeding grounds in the autumn to travel to the coast and even out onto the sea-ice; in late winter or early spring they return (Seton, 1929; Chitty & Elton, 1937; Dubrovskii, 1937; Soper, 1944; Chitty & Chitty, 1945; Chesemore, 1968). These migrating foxes are reputed to travel singly or in pairs, often along well defined "runs", particularly along coastlines and river valleys (Maksimov, 1945).

Although generally regarded as monogamous, Boitzov (1937) reports that under the semi-natural conditions of island fur farms, some males will mate with more than one female. Arctic foxes have been seen in numbers of up to 40 around large carcasses (Chesemore, 1975). Indeed, Niewold (1974) has found that red foxes from different territories will congregate around large concentrations of carrion.

Communication amongst members of the two species appears to be indistinguishable, in terms of postures and expressions (Tembrock, 1957; Fox, 1971) and olfactory and vocal behaviour are very similar (Tembrock, 1977; Macdonald, 1980d).

General Behaviour

On the basis of travellers' accounts from remote areas, the Arctic fox is often typified as a much tamer animal than the red fox. However, in Iceland where the Arctic fox is hunted intensively its flight distance from man and guile at evading hunters is equivalent to that of the red fox. Similarly, in Israel where foxes are largely protected, red foxes become very tame, although remaining shy in areas where Beduin eat them (D. W. Macdonald, personal observation). Consequently we suspect that the supposed difference is not between the species as much as between local circumstances.

Another difference that appeared to distinguish red and Arctic foxes was in their food caching behaviour. Macdonald (1976) reviewed the evidence which indicated that red foxes invariably scatter-hoarded surplus prey, limiting the contents of each cache to one mouthful of food; in contrast Arctic foxes sometimes placed dozens of prey in one cache. Recently however, W. Montevecchi (personal communication) has found an area on Baccalieu Is. (Newfoundland) where red foxes occasionally use huge hoards. One active den and adjacent excavation contained 319 prey items. Conversely, Arctic foxes watched in Iceland invariably scatter-hoarded (P. Hersteinsson, unpublished).

TWO CASE STUDIES

Methodology

Red and Arctic foxes have both been studied using radio tracking not only to pinpoint each animal's location, but also to predict where they could be found and watched (Macdonald, 1978). Techniques of equipment construction, usage, and data analysis are described in Amlaner & Macdonald (1980).

Equipment

Foxes were equipped with collar-mounted, two-stage transmitters fitted with whip aerials. Each transmitter incorporated a motion-sensitive mercury switch, so that not only the animal's position, but also whether it was active or resting, was recorded. The transmitter circuitry used most recently followed Macdonald & Amlaner (1980) and could be adapted to any carrier frequencies between 30 and 250 MHz. The appropriate tuning of the inductor ($0.2-1.0 \,\mu$H) and the trimmer capacitor ($5-25 \,$pF) for oscillation at a particular fundamental crystal frequency (F_{MHz}) is calculated from:

$$F_{(MHz)} = \frac{10^3}{2\,LC} \tag{1}$$

where L = inductance in microhenries and C = capacitance in picofarads. Figure 2 illustrates the circuitry, modified slightly from Macdonald & Amlaner (1980) as necessitated by changing component availability.

Each transmitter was fitted with a reed switch connected between the base transistor and earth. A small magnet could be taped to the encapsulated radio package and would close the reed switch, inactivating the circuit (except for an approximately $4 \,\mu$A current through the 820 K and 471 K resistors) until it was needed. This conserved the battery and avoided confusion between signals from

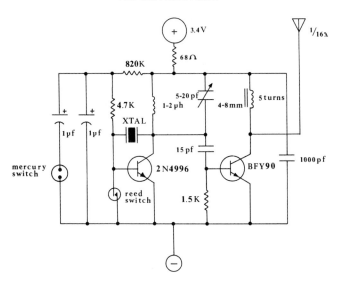

FIG. 2. Circuitry of two-stage radio transmitter used during the fox studies. The circuit is based on that described by Macdonald & Amlaner (1980) but slightly modified according to component availability.

transmitters already attached to animals and those awaiting attachment.

The transmitters were powered by lithium cells which have roughly twice the mAh capacity per unit weight of mercury cells (Ko, 1980). They also have a long shelf life, especially if stored in a refrigerator. To avoid damage to the battery through heat, tags welded to their terminals by the manufacturer are strongly recommended; connections can then be safely soldered to these tags.

Two pulse rates, of about 60 and 100 per min, were used and pulse width was set to about 40 ms. Depending on the landscape and frequency, signals could be detected at up to 10–15 km and a transmitter kept in the laboratory ran for 14 months. The transmitter (wt = 5 g) is encased in beeswax and, together with a 3.4 V battery (wt = 48 g) encapsulated in transparent plastic, weighs in total 130 g. The collar, of braided nylon ribbon, is threaded with a light-weight buckle and finally riveted after adjustment to suit the individual animal's neck (Fig. 2). It is important to use non-corrosive rivets to avoid damage to the animal. The collar should not be too loose (but should permit two fingers to be easily slipped between it and the fox's neck) or else there is a risk of the fox forcing a front leg into the gap.

Two brands of receiver have been used: the AVM model LA12 (AVM Instrument Company, 3101 West Clark Road, Champaign,

FIG. 3. Red fox equipped with radio transmitter attached to nylon collar. Numbered ear-tags bear the university's telephone number.

Illinois, 61820, USA) and the Telemetry Systems model RT-20A (Telemetry Systems Inc., Mequon, Wisconsin, USA). Receiving aerials included dipoles, Adcock, and Yagi designs (Amlaner, 1980).

At higher frequencies a three-element Yagi was small enough to be portable, but for lower frequencies a Yagi was mounted on a car and a dipole (or collapsible Adcock with telescopic elements) was used on foot.

Red foxes were watched by night using infrared binoculars (Old Delft Optics, Old Delft, Holland) and image-intensifiers (Davin Optics Ltd, Hertfordshire, UK). In summer, Arctic foxes could be watched throughout the night through conventional 10 x 50 binoculars by the light of the midnight sun.

To facilitate observation, radio collars were fitted with reflective markers and beta lights (Saunders Roe Developments, Hayes, Middlesex, UK) as described in Macdonald (1978).

Techniques

The barren landscape of north-west Iceland and the lush country-side of Oxfordshire posed different problems for radio tracking and subsequent observation. However, in both cases a detailed know-ledge of the landscape was indispensable for accurate tracking, which necessitated the rejection of signals whose direction resulted from reflections and other topographical factors. Either on foot, or by car, fixes were rarely the conventional intersection between two bearings. Instead the indicated direction of the animal was monitored con-tinuously both on a map and with respect to known landmarks as the observer moved. Each fix thus represented the intersection of a running average of bearings. For example, in Oxford a network of roads normally permitted a tracker working from a car to remain within 100–200 m of the fox. The Yagi aerial was mounted on a series of gears which enabled it to be rotated from within the car while driving. Consequently, as the driver toured around the fox, a continuous check could be kept on its direction with respect to the car and in relation to well known landmarks. Simultaneously note was taken of variations in signal characteristics that were diagnostic of the fox's position.

The same principle was applied in Iceland, where tracking was necessarily on foot. A tracker listened continuously to the radio signal, and selected routes, based on where the fox seemed to be travelling, which led to known vantage points. The fox was thus kept intermittently in sight and, between times, its route was fairly accurately known. However, working on foot it could take many hours to locate a fox.

The Red Fox on Boars Hill

Boars Hill is an area of rural-suburbia on the outskirts of Oxford. The landscape is typified by small fields divided by hedgerows, and both deciduous and coniferous woodlands, amongst which are dispersed detached houses and large gardens. Foxes have been studied there through radio tracking and direct observation since 1972. In the early phase of this project, from 1973 until 1976, 16 resident foxes were equipped with radio transmitters, within an area of just over 2 km^2, which was thought to support 26 adult resident foxes (Macdonald, 1981). Other individuals, classed as itinerant, travelled through the area. Subsequently, additional foxes have been studied in the same and adjacent habitats (Macdonald, 1980b and unpublished).

Social organization

Home ranges of resident radio equipped foxes tracked on Boars Hill between 1973 and 1976 varied between 18.6 and 72.0 ha. The foxes lived in social groups composed of one adult male and two to five adult vixens (mean group = 4.4 ± 1.1 s.d., $n = 5$) together with cubs of the year. The ranges of group members overlapped widely, and in the majority of cases completely (mean overlap (24 combinations) for 62.3% confidence ellipses (i.e. core areas), of eight home ranges = 65.0%). There was little or no overlap between the ranges of foxes belonging to adjacent neighbouring groups (mean overlap for 62.3% confidence ellipses = 6.1%). In many places the borders of group ranges were clearly defined and precisely tessellated, often along a topographical feature such as a road, ditch or garden fence (Fig. 4). Where neighbouring ranges did overlap there were indications that it was in areas which were of less importance for foraging. These group ranges were termed territories because of their non-random distribution with respect to each other and observations indicating that intruding foxes are at least sometimes attacked and expelled. The average territory size, measured from hand-drawn boundaries, was 45.2 ha (± 16.5, s.d., $n = 7$). There was no relationship between territory size and group size (Spearman Rank Correlation, $r_s = 0.07$, $P > 0.1$).

Encounters between foxes, watched either at dusk and dawn, or through night-vision equipment, indicated that group members met frequently, but most often fleetingly. However, there were occasions, especially soon after dawn in the late summer, when several adults, together with maturing cubs, could be seen lazing and playing together in the morning sun. Adult group members generally behaved

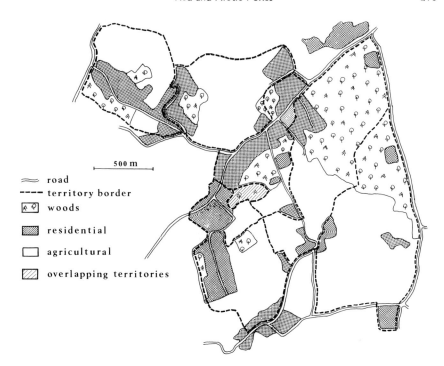

FIG. 4. Map of the territory borders of seven groups of red foxes on Boar's Hill, Oxford-shire, showing principal habitat types.

amicably towards each other, although there was a discernible hierarchy within each group, with vixens generally being subordinate to dog-foxes. For example, in the absence of food, of 32 interactions between adults only three involved even mild aggression (41 out of 68 encounters between the members of one group at a feeding site involved some aggression). In contrast, serious attacks were seen against foxes intruding into a group's territory. Full accounts of a selection of interactions are given in Macdonald (1977, 1981).

Radio tracking studies of maturing foxes indicated that most, perhaps all, young males disperse from their natal territory within one year, whereas some young vixens become incorporated into the adult group. Furthermore, vixens within a group often shared unusual morphological characteristics (e.g. white toes). Consequently these groups are at least partially composed of related animals. In one case where relationships were fairly well known the coefficient of relatedness was estimated at $r = 0.38$ (Macdonald, 1980c).

Amongst these family groups, at least in some years, only a pro-portion of the adult resident vixens reared their own cubs. In one sample only six of 15 vixens reared cubs. In effect this meant that only one or two vixens reared cubs in each group. In one case where two vixens within one group reared cubs, both litters occupied the same earth and both mothers suckled, guarded and played with all the cubs and provisioned them with prey, as did a third, non-breeding vixen. In other groups in which only one vixen bred, at least some of the non-breeding vixens frequently visited the cubs and occassionally fed them. Studies in captivity indicated that such "helpers" may contribute as much food to cubs as does the real mother (Macdonald, 1979, 1980c).

Diet and foraging

The diet of foxes on Boars Hill varied between neighbouring territories, and very probably between some individuals within each group. This variation depended partly on the distribution of prey; for instance, rabbit, *Oryctolagus cuniculus*, only featured significantly in the diets of foxes whose territory encompassed a rabbit warren. Overall, ranked in order of volume of remains in faeces, the diet was largely composed of offal scavenged from people, earthworms, *Lumbricus terrestris*, and fruit (mainly blackberries, *Rubus ulmifolius*, and apples, plums and cherries). Birds, especially passerines, and rodents, especially voles, were also eaten. Earthworm chaetae were the single most frequently found prey remains in 1405 faeces. The scavenged food came largely from feeding sites, provisioned by householders with an interest in foxes, or from bird-tables and compost heaps. Foxes were consuming all the available offal (Macdonald, 1981). Each of these food types was spatially and/or temporally variable in its availability to foxes. For instance, earthworms were only available on the surface under certain weather conditions and at locations which depended on, for example, prevailing wind direction (Kruuk, 1978; Macdonald, 1980a). Worms were most available to foxes on mild, humid nights and indeed foxes responded to this variation in availability by foraging for them more often on wet nights than on dry nights. Even on nights when worms were available, their numbers varied from field to field and the foxes foraged most frequently in the fields with greatest worm abundance. Within a field of high worm abundance various factors, including wind direction, affected their availability to foxes. Worms were more inclined to surface in the lee of windbreaks and foxes responded to this by concentrating their foraging in patches of ground in such positions. This prey, like several others, was spatio-temporally heterogeneous in its availability.

Description of territories

Seven group territories were composed on average of 14.6 ha of woodland, 9.6 ha arable, 10.5 ha pasture, and 10.4 ha human residential habitat (i.e. houses, gardens, and adjoining orchards and scrubland). With the exception of residential habitat, the areas of each habitat type varied widely from territory to territory. Even more constant than the area of residential land in each of the seven territories was the number of houses, at 23.7 ± 4.8 s.d. Irrespective of its overall size, each territory has approximately the same area of residential habitat, indeed six of the seven territories embrace 8–13 ha of residential habitat, although varying fourfold in total size. Figure 4 shows the dispersion of residential habitat in Boars Hill and illustrates clearly that the configuration of territorial borders is significantly influenced by this pattern.

The Arctic Fox at Ófeigsfjördur

Ófeigsfjördur is one of a series of fjords on the north-west coast of Iceland (Fig. 5), from which the rocky coast (approximately 350 m wide) rises swiftly to barren (vegetation cover = 0–20% of surface) fells at 200–400 m. Arctic foxes were studied there through radio tracking and direct observation in 1978 and 1979. Five resident adult foxes were equipped with radio transmitters, out of the total of nine adult foxes thought to occupy the 40 km² study area.

Arctic foxes have inhabited Iceland since the end of the last Ice Age (about 10,000 years B.P.). They have probably not been genetically isolated throughout that period since Arctic foxes from Greenland almost certainly reach Iceland intermittently aboard drift ice (Saemundsson, 1939). The Hansa Expedition in 1869 (Saemundsson, 1939) noted an Arctic fox on drift ice mid-way between the two islands. In addition, Alaskan "blue" foxes were imported to Icelandic fur farms during the 1920s and some of these escaped and doubtless bred with the local foxes. There are no red foxes in Iceland, but since the climate is relatively mild (64–66°N) and the habitat comparable to that in some more northerly parts of its range, the red fox's absence seems likely to be an accidental consequence of geographical isolation.

Like the red fox in Britain, the Arctic fox in Iceland is heavily persecuted. Since at least 1295 Icelandic legislation has promoted its extermination, because of its reputed depredation of sheep and lambs (Hersteinsson, 1980). State-subsidized hunting has been so efficient that in spite of ever-mounting hunting pressure the annual tallies have dropped over the last 21 years (1958–78) from 1590 to

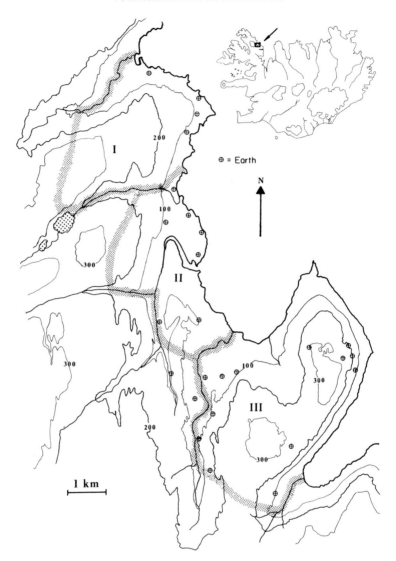

FIG. 5. Map of Ófeigsfjördur, showing its location in Iceland and delineating the borders between three territories of Arctic fox groups, and showing principal rivers and the contour-lines at 100 m intervals. Earthsites, like prey, are largely confined to the lower, coastal slopes.

456 (Fig. 6). Consequently, for the first time, large tracts of land are apparently devoid of foxes, as indicated by the absence of any field signs. At Ófeigsfjördur the hunters agreed not to kill foxes during the study.

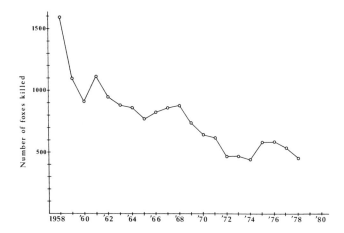

FIG. 6. Over the past 22 years the number of Arctic foxes killed each year in Iceland has consistently fallen in spite of a mounting hunting effort. This decline almost certainly reflects a serious reduction in numbers, as a consequence of intensive persecution, utilizing motor sleds and high velocity rifles.

Social organization

The home ranges of Arctic foxes at Ófeigsfjördur varied between 8.6 and 18.5 km^2. The foxes lived in social groups comprising one male and two adult vixens together with cubs of the year. In 1978 and 1979 three groups were studied and, as with the red foxes of Boars Hill, the home ranges of members of a group overlapped widely with each other and very little, if at all, with those of neighbouring groups. Observations suggested that intruders were at least sometimes attacked and expelled, and consequently that these group ranges constituted territories from which their occupants rarely strayed. During the two years of study the territory borders remained very similar, bounded to the east by the Arctic Ocean, and largely partitioned by rivers along their inland boundaries. Rivers that were impassable to humans were, arguably, also impassable to the foxes and on this criterion the territorial boundaries were divided into proportions which could or could not be forded. At least half of the rivers bordering each territory were passable and thus were boundaries rather than barriers. Nevertheless, most territorial transgressions (16 out of 20) were seen at those borders without clear-cut landmarks or along river boundaries between November and April, when they were frozen.

Dog-foxes frequently (17 times out of 42 approaches observed) barked as they approached within 500 m of the boundaries of their territories. On four occasions a fox barked back from the neighbouring territory, and twice the neighbouring dog-fox rushed to the scene. The two neighbours then moved stiff-leggedly, with tails held vertically upwards, along a section of the boundary, separated by 100—200 m, token-urine marking every 5—10 m and barking very frequently (6—15/min) for 28 min and 35 min on each occasion respectively. They then went their separate ways, away from the boundaries, the frequency of barking falling gradually and ceasing altogether within 5 min. Once, a dog-fox was seen at a territory border 60 min after a vixen of the neighbouring territory was seen token-urinating there. He followed her tracks along the boundary for about 50 m and urine-marked on most of the spots the vixen had urine-marked. On two occasions a vixen was seen at the boundary 10 and 30 min respectively after the neighbouring dog-fox had token-urine marked there. On both occasions she followed in the tracks of the dog-fox for about 10—15 m, and urine-marked where he had urine-marked. In contrast, five days later, when this same vixen appeared on the boundary 75 min after the breeding vixen from the neighbouring territory had been seen token urine-marking there, she reacted by violently barking after sniffing this vixen's urine-marks. She then followed the other vixen's tracks for about 150 m, apparently urinating at each spot the other one had urine-marked. She then moved about 200 m away from the border, where she sat and barked intermittently for the next 13 min.

On one occasion when a strange fox was seen in the study area it was spotted running hard from the southern boundary of Territory II, and it continued to run for 400 m, at which point it stopped for less than one second to look back. It then continued at a fast gallop, with a few stops to sniff and look back in the direction it had come, for another 2 km when it was lost from sight. Smeared mud across this fox's neck and shoulders, together with its glances in the direction it had come from, were interpreted as signs that it was fleeing after being attacked by a resident fox.

In the spring and early summer of 1978 each of the three territories was occupied by a dog-fox and two vixens. The same was true of Territories I and III for 1979, but this could not be confirmed for Territory II. In 1978 only one vixen of the two in each territory reared cubs. Similarly, in 1979 only one vixen bred in Territory I, whereas in Territory III either the cubs all died or none were born.

Aggressive encounters between adult group members were never observed. During the mating season (March—April) the breeding pairs spent much time together, with the vixens behaving submissively towards the male. The non-breeding vixens behaved submissively towards both the dog-fox and the breeding vixen. Thus there appeared to be dominance hierarchies within the groups.

In each case the non-breeding vixens not only occupied the same range as the breeding pair, but also visited the vicinity of the earth and cubs. From 12 June and 15 June 1979 the earth in Territory I was watched continuously for 68 h. During this time the non-breeding vixen brought food to the cubs at least three times and remained with them for 16 h in total. She disappeared from the area the following month. Indeed each of the five non-breeding vixens that were studied disappeared when the cubs of their group reached 6–8 weeks of age (both species of foxes begin to eat solid food at about three weeks and are largely weaned at eight weeks). Two of these non-breeding vixens were fitted with radio collars in the spring of 1978 and tracked within their group territories until they emigrated. A year later both were shot at separate earths outside the study area, where each then had a litter of its own. They had travelled respectively 25 and 27 km in straight line distances, in opposite directions. Examination of cementum annuli (Jensen & Nielsen, 1968; Grue & Jensen, 1976) confirmed that both vixens died when they were two years old. Consequently they had been yearlings when they acted as helpers. The morphological characteristics of non-breeding vixens strongly suggested that they were relatives, presumably off-spring, of the breeding pair.

Diet and foraging

The analysis of 259 faeces revealed that from season to season 60–80%, by volume of faeces, of the prey remains were of food most probably foraged at the sea-shore. This consisted largely of seabirds (mainly as carrion) especially guillemot, *Uria aalge*, and Brunnich's guillemot, *U. lomvia*, together with seal, *Phoca vitulina*, carcasses. In addition the foxes ate lumpsuckers, *Cyclopterus lumpus*, and various invertebrates from the shore. Eider ducks, *Somateria mollissima*, and a variety of waders (Charadriidae and Scolopacidae) were killed by the foxes, and in summer the eggs and chicks of both passerines and waders were taken from the narrow (100–500 m) lowland coastal strip rising from the beach. The highland plateaux were almost devoid of prey for these foxes, save for the ptarmigan, *Lagopus mutus*, which never amounted to more than 10% of prey remains.

In the autumn black crowberries, *Empetrum nigrum*, were eaten in large numbers. These grew on the slopes above the shore. The only rodent in the area, *Apodemus sylvaticus*, provided approximately 2% of the diet.

Amongst invertebrates, mussels (*Mytilus edulis*) and the larvae and pupae of *Coelopa frigida* predominated in the diet. The maggots occurred in very large numbers in clumps of decomposing seaweed

between July and early September. The Arctic foxes scraped the dried
crust off a clump of decomposing seaweed to expose the maggots in
the same way that red foxes gather blowfly larvae from cow dung.

Since the bulk of the diet was derived from gleaning the shoreline the rhythm of the tides greatly affected food availability and
hence fox foraging behaviour. The availability of carrion and birds on
the shores was obviously influenced by the tides. Indeed, this was
also true for other prey such as mussels. Observations of the foxes
foraging inland and along the coast indicated that over the 12.3 h
tidal cycle they concentrated their visits to the shoreline to the three
hours before low tide (Fig. 7). This is the best time for beachcombing.

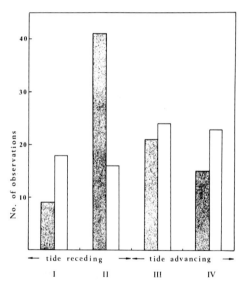

FIG. 7. The availability of carrion on the shore was determined by the tides and this was
reflected in the movements of the foxes along the coast and inland. The figure shows the
number of foraging incidents on the shore (shaded columns) and inland (unshaded columns)
during the four 3-h periods of the tidal cycle. Significantly more foraging was observed on
the shore in the 3 h before Low Tide (Period II), than while the tide was advancing (Periods
III and IV) or in the 3 h after High Tide (Period I) ($\chi^2 = 26.930$; $P < 0.001$; $n = 86$). No
significant difference was found in inland foraging with respect to the tidal cycle.

A fox arriving at the sea-shore would almost invariably do so
furtively, along a gully. It would creep forward until it had
apparently ascertained whether any birds were resting or feeding on
the beach. If there were birds there the fox would stalk them, if not
it would begin gleaning for carrion and invertebrates.

TABLE III

Anatomy of territories in Ófeigsfjördur

	Territory I	Territory II	Territory III	Average
Territory size	10.3 km^2	8.6 km^2	18.5 km^2	12.5 km^2 ±5.3 km^2 S.D.
Length of coastline	5.6 km	5.4 km	10.5 km	7.2 km ±2.9 km S.D.
Length of productive coastline	5.6 km	5.4 km	6.0 km	5.7 km ±0.3 km S.D.
Coastal productivity index*	1800	1800	2100	1900±44 S.D.
Length of territory boundaries	9.6 km	9.4 km	8.0 km	9.0 km ±0.9 km S.D.
Type I boundary	2.7 km	2.2 km	θ	1.6 km ±1.4 km S.D.
Type II boundary	2.7 km	2.2 km	3.1 km	2.7 km ±0.5 km S.D.
Type III boundary	4.2 km	5.0 km	4.9 km	4.7 km ±0.5 km S.D.
Type II and III boundary	6.9 km	7.2 km	8.0 km	7.4 km ±0.6 km S.D.

* The coastal productivity index is the number of fence posts extracted annually by farmers from the driftwood in each territory.
Type I boundary: Boundary along impassable rivers.
Type II boundary: Boundary along passable rivers.
Type III boundary: Boundary without clear-cut landmarks.
Type II and III boundaries: Boundaries which had to be defended against intruding foxes.

Anatomy of territories

The three territories each contained two broad habitats: the coastal strip and the rest. Table III shows that the areas (and lengths of coastline) varied widely. Just as much of the food of red foxes on Boars Hill came from residential habitat, so, by analogy, the coastal strip was crucial to the Arctic foxes at Ófeigsfjördur. Furthermore, the availability of that food is not uniform temporally even within the coastal strip, as constrained by phases of the tide (see above) nor even uniform spatially at low tide; beachcombing revealed areas of greater and lesser abundance of carrion, invertebrate prey, and bird flocks. The potential productivity of a strip of shore is partly determined by the flow of the tides. The quantity of driftwood found along each section of shoreline can be used as an indication of the amount of debris, including carrion, washed up there over a long period of time. Defining "productive coastline" as that where driftwood was washed ashore, the relationship between territory size and total coastline length disappears since each territory has almost exactly the same mean length (5.7 km ± 0.3 s.d.) of productive coastline. The productive coastlines of territories I to III respectively yielded comparable quantities of driftwood (1800, 1800 and 2100 fenceposts extracted annually from the driftwood in each fox territory by farmers). Observation showed that the foxes were familiar with the spatial pattern of productive bays; having combed one stretch of productive coastline a fox would generally trot quickly along the beach to the next such stretch, or even take a shortcut across a peninsula to reach it.

The area of inland habitat varies widely between each territory, and the higher fells are unoccupied by resident foxes. To that extent, defining the inland border is difficult; snow tracking in winter indicated that sporadic forays were made up to 6.5 km inland, but these were never found while radio tracking.

An Ecological Basis of Red and Arctic Fox Societies

To the extent that habitat types mirror food supplies, the foregoing analyses of the habitat components of red and Arctic fox territories are compatible with a hypothesis linking social organization and territory sizes with the abundance and dispersion of available food. This hypothesis is more fully considered by Macdonald (1981). Briefly, the patterns of prey availability for red foxes on Boars Hill and the Arctic foxes of Ófeigsfjördur were spatially and temporally heterogeneous both daily and seasonally. As a consequence, the smallest home range which will reliably support a pair of foxes whilst having an economically defensible (Brown, 1964) configuration may, at least sometimes, support additional foxes too. Thus range size and configuration may be determined by the dispersion of food patches (and the influence of topography on the economics of defence), whereas group size may be determined by the richness of these patches. Additional foxes will be tolerated in numbers and at times when any costs to the basic pair due to their presence are outweighed by the benefits. In more homogeneous habitats territory sizes which support a pair may not support additional adults. A similar explanation is advanced by Kruuk (1978) to explain the social system of another carnivore, the European badger, *Meles meles*.

In each case the argument is applicable in both the short and long terms. For the red fox a change in wind direction during one night will alter the dispersion of available worms, which surface in patches leeward of a windbreak (Macdonald, 1980a), and similarly, on one night a householder at one extreme of a territory may provide scraps, while on the next night one at another extreme does so. To ensure adequate food on both nights a pair of foxes must monopolize both households, and sufficient faces of a hill to guarantee that some slopes will be sheltered irrespective of wind direction. However, since some patches contain more food than the pair need and sometimes several patches may be available simultaneously, the dispersion of prey together with the costs of defending a range with an irregularly shaped perimeter may make it economic to tolerate, or even encourage, additional group members who may bring with them other benefits to the inclusive fitness of the pair. Similarly, for the Arctic fox even the productive coastline provides a shifting patchwork of available carrion from one tide to the next with the fruits of beachcombing any given cove depending on its geological structure, the run of the tide, direction and strength of wind, and chance deaths at sea. A raft of eider duck may settle in one of several coves, and a seal carcass may be washed ashore anywhere along the beach. To ensure that food is always available, several potentially productive coves (patches) must be defended and we suggest that the minimum number that will support a pair will also sometimes support others too. The benefit of having rivers as borders may further make it advantageous to expand the minimum range to the nearest river, and other factors may also favour larger ranges than the minimum.

Analogous reasoning could be applied on a seasonal (or yearly) scale. To ensure access to ptarmigan in winter and eider duck in summer the Arctic fox must encompass within its territory both coast and fell. To ensure access to both apples in autumn and baby rabbits in spring, the red fox's territory must embrace both garden orchard and woodland warren. Once the orchard, or the eiders' nursery, has been secured it may then support more than a pair.

We argue that an ecological factor, namely the pattern of food availability, sets fundamental constraints on territory and group sizes and may, in particular, determine that the minimum range required by a pair can sometimes also support additional foxes. Of course, pairs occupying a range larger than the local minimum could also afford additional group members. The indications are that the social systems of these two species are so similar that the consequences of group membership to the fitness of both the basic pair and additional foxes are similar for red and Arctic foxes; for both species we have reported female helpers, evidence of hierarchies and the involvement of both sexes in territorial defence. It also seems likely that individuals of both species maintain long-term pair bonds and that where groups develop they are at least often composed of close kin. The behavioural similarities between red and Arctic foxes are perhaps more striking than their morphological differences.

The only difference between the societies of the two species of fox was in the consistency with which non-breeding female Arctic foxes at Ófeisgsfjördur dispersed after acting as helpers for the first 6—8 weeks of the cubs' lives. There are few data on the emigration of adult female red foxes, although it does occur. It seems more likely that the behaviour of these vixens at Ófeigsfjördur was an adaptation to their particular surroundings rather than a specific difference between red and Arctic foxes. For example, one might speculate that dispersing after only 6—8 weeks of provisioning for cubs is a compromise for the inclusive fitness of these young vixens. At this time, prey availability is at its maximum and the cubs are about to begin foraging to some extent on their own in the littoral zone. In addition, the barren vixens have acquired maternal experience and the timing of their dispersal may also be influenced by the fact that most of the persecution by man of foxes at earths is over at the end of June, leaving surrounding countryside well below carrying capacity.

CONCLUSION

Both the red and Arctic fox have a very flexible social system; within each species there are indications that range sizes, extent of territoriality, social dynamics and population structure may all vary between broad limits. However, there are no indications that these limits differ in their nature for the two species, and under the particular ecological circumstances of our two case studies the red and Arctic foxes were striking in their similar behaviour. These similarities lead us to suggest that the potential for complete direct competition between these species underlies their present distribution (Hersteinsson & Macdonald, in preparation).

REFERENCES

Ables, E. D. (1975). Ecology of the red fox in America. In *The wild canids:* 216—236. Fox, M. W. (Ed.). New York: Van Nostrand Reinhold.
Amlaner, C. J. (1980). The design of antennas for use in radio telemetry. In *A handbook on biotelemetry and radio tracking*: 251—262. Amlaner, C. J. & Macdonald, D. W. (Eds). Oxford: Pergamon Press.
Amlaner, C. J. & Macdonald, D. W. (Eds). (1980). *A handbook on biotelemetry and radio tracking*. Oxford: Pergamon Press.
Anderson, R. M. (1934). The distribution, abundance, and economic importance of the game and furbearing mammals of western North America. *Proc. Pacif. Sci. Congr.* 5: 4055—4075.

Beneš, J. (1970). Die Füchses des mitteleuropäischen Pleistozäns. *Lynx* 11: 61—65.

Boitzov, L. V. (1937). [The Arctic fox: biology, food habits, breeding.] *Trudy Arkt. Nauchno-Issled Inst.* 65: 7—144. [In Russian.]

Braestrup, F. W. (1941). A study on the Arctic fox in Greenland. *Meddel. Grönland* 131: 1—101.

Brown, J. L. (1964). The evolution of diversity in avian territorial systems. *Wilson Bull.* 70: 160—169.

Chesemore, D. L. (1975). Ecology of the Arctic fox (*Alopex lagopus*) in North America — A review. In *The wild canids. Their systematics, behavioral ecology and evolution*: 143—163. Fox, M. W. (Ed.). New York: Van Nostrand Reinhold.

Chitty, D. & Elton, C. (1937). Canadian Arctic wildlife enquiry, 1935—36. *J. Anim. Ecol.* 6: 368—385.

Chitty, H. & Chitty, D. (1945). Canadian Arctic wildlife enquiry, 1942—43. *J. Anim. Ecol.* 14: 37—41.

Clutton-Brock, J., Corbet, G. B. & Hills, M. (1976). A review of the family Canidae, with a classification by numerical methods. *Bull. Br. Mus. (Nat. hist.) (Zool.)* 29 (3): 1—199.

Cole, I. J. & Shackelford, R. M. (1946). Fox hybrids. *Trans. Wis. Acad. Sci.* 38: 315—332.

Collett, R. (1912). *Norges Pattedyr*. Kristiania: H. Aschehaug & Co. (W. Nygaard).

Davis, S. (1977). Size variation of the fox, *Vulpes vulpes*, in the palaearctic region today, and in Israel during the late Quaternary. *J. Zool., Lond.* 182: 343—351.

Dubrovskii, A. N. (1937). [The Arctic fox (*Alopex lagopus* (L)) and Arctic fox trapping in Novaya Zemlya.] *Trans. Arctic Inst. Leningrad* 77: 7—31. [In Russian.]

Eberhardt, L. E. & Hanson, W. C. (1978). Long-distance movements of Arctic foxes tagged in Northern Alaska. *Can. Fld Nat.* 92: 386—389.

Elton, C. (1949). Movements of Arctic fox populations in the region of Baffin Bay and Smith Sound. *Polar Rec.* 5: 296—305.

Englund, J. (1970). Some aspects of reproduction and mortality rates in Swedish foxes (*Vulpes vulpes*), 1961—63 & 1966—69. *Viltrevy* 8: 1—82.

Fetherston, K. (1947). Geographic variation in the incidence of occurrence of the blue phase of the Arctic fox in Canada. *Can. Fld Nat.* 61: 15—18.

Fischer, A. H. (1960). Latitudinal variations in organic diversity. *Evolution* 14: 64—81.

Fox, M. W. (1971). *Behaviour of wolves, dogs and related canids*. London: Jonathan Cape.

Grue, H. & Jensen, B. (1976). Annual cementum structures in canine teeth in Arctic foxes (*Alopex lagopus* (L)). from Greenland and Denmark. *Dan. Rev. Game Biol.* 10 (3): 1—12.

Hagmeier, E. M. & Stults, C. D. (1964). A numerical analysis of the distributional patterns of North American mammals. *Syst. Zool.* 13: 125—155.

Harris, S. (1977). Distribution, habitat utilisation and age structure of a suburban fox (*Vulpes vulpes*) population. *Mammal Rev.* 7: 25—39.

Harris, S. (1979). Age-related fertility and productivity in Red foxes, *Vulpes vulpes*, in suburban London. *J. Zool., Lond.* 183: 91—117.

Harris, S. (1980). Home ranges and patterns of distribution of foxes, *Vulpes vulpes*, in an urban area, as revealed by radio tracking. In *A handbook on biotelemetry and radio tracking*: 685—690. Amlaner, C. J. & Macdonald, D. W. (Eds). Oxford: Pergamon Press.

Haukioja, E. & Koponen, S. (1975). Faunal structure of investigated areas at Kevo, Finland. In *Fennoscandian tundra ecosystems* Part 2: 19—28. Wielgolaski, F. E. (Ed.). Berlin: Springer Verlag.

Henshaw, R. E. Underwood, L. S. & Casey, T. M. (1972). Peripheral thermoregulation: foot temperature in two Arctic canines. *Science, N.Y.* 175: 988—990.

Hersteinsson, P. (1980). [Arctic fox.] *Rit Landverndar* 7: 65—79. [In Icelandic with English summary.]

Hildebrand, M. (1952). An analysis of body proportions in the Canidae. *Am. J. Anat.* 90: 217—256.

Irving, L., Krog, H. & Monson, M. (1955). The metabolism of some Alaskan animals in winter and summer. *Physiol. Zool.* 28: 173—185.

Järvinen, O. & Väisänen, R. A. (1977). Long-term changes of the North European land bird fauna. *Oikos* 29: 225—229.

Jensen, B. & Nielsen, B. L. (1968). Age determination in the Red fox (*Vulpes vulpes* (L.)) from canine tooth sections. *Dan. Rev. Game Biol.* 5 (6): 1—15.

Kaikusalo, A. (1971). Naalin pesimisestä Luoteis — Enontekiössä. *Suomen Riista* No. 23: 7—16.

Kauri, H. & Østbye, E. (1975). Introduction. Fennoscandian tundra ecosystems. Part 2. *Ecol. Stud.* 17: 2—7.

Keeler, C. (1975). Genetics of behaviour variations in color phases of the red fox. In *The wild canids. Their systematics, behavioural ecology and evolution*: 399—413. Fox, M. W. (Ed.). London: Van Nostrand Reinhold.

Keeler, C., Mellinger, T., Fromm, E. & Wade, L. (1970). Melanin, adrenalin and the legacy of fear. *J. Hered.* 61: 81—88.

Ko, W. H. (1980). Power sources for implant telemetry and stimulation systems. In *A handbook on biotelemetry and radio tracking*: 225—246. Amlaner, C. J. & Macdonald, D. W. (Eds). Oxford: Pergamon Press.

Kruuk, H. (1978). Foraging and spatial organization of the European badger, *Meles meles* L. *Behav. Ecol. Sociobiol.* 4: 75—89.

Kurtén, B. (1965). The carnivores of the Palestine caves. *Acta zool. fenn.* 107: 1—74.

Lavrov, N. P. (1932). *The Arctic fox.* (Scientific Popular Library of Fur Animals of the U.S.S.R. Zhitkov, G. M. (Ed.).) Moscow: Board of Foreign Trade.

Leinati, L., Mandelli, G., Videsott, R. & Grimaldi, E. (1960). Indagini sulle abitudini alimentari della volpe (*Vulpes vulpes* (L.)) del parco Nazionale Gran Paradiso. *Clin. Vet.* 83: 305—328.

Lloyd, H. G. (1975). The Red fox in Britain. In *The wild canids*.: 207—215. Fox, M. W. (Ed.). London: Van Nostrand Reinhold.

Longstaff, T. G. (1932). An ecological reconnaisance in West Greenland. *J. Anim. Ecol.* 1: 119—142.

Lund, Hj. M. K. (1959). The Red fox in Norway I. Survey of 551 Red foxes collected, their size and sex ratio. *Pap. Nor. St. Game Res.* (2) 5: 57.

MacArthur, R. H. (1972). *Geographical ecology: Patterns in the distribution of species.* New York: Harper & Row.

Macdonald, D. W. (1976). Food caching by red foxes and some other carnivores. *Z. Tierpsychol.* 42: 170—185.

Macdonald, D. W. (1977). *The behavioural ecology of the Red fox*, Vulpes vulpes, *a study of social organisation and resource exploitation*. D. Phil. thesis: University of Oxford.

Macdonald, D. W. (1978). Radio-tracking: some applications and limitations. In *Recognition marking of animals in research*: 192—204. Stonehouse, B. (Ed.). London: Macmillan.

Macdonald, D. W. (1979). 'Helpers' in fox society. *Nature, Lond.* 282: 69—71.

Macdonald, D. W. (1980a). The Red fox, *Vulpes vulpes*, as a predator upon earthworms, *Lumbricus terrestris*. *Z. Tierpsychol.* 52: 171—200.

Macdonald, D. W. (1980b). *Rabies and wildlife: a biologist's perspective*. Oxford: Oxford University Press.

Macdonald, D. W. (1980c). Social factors affecting reproduction by the Red fox, *Vulpes vulpes*, L., 1758. *Biogeographica* 18: 123—175.

Macdonald, D. W. (1980d). Patterns of scent marking with urine and faeces amongst carnivore communities. *Symp. zool. Soc. Lond.* No. 45: 107—140.

Macdonald, D. W. (1981). Resource dispersion and the social organisation of the red fox, *Vulpes vulpes*. In *Proceedings of the Worldwide Furbearer Conference*. Chapman, J. & Pursely, D. (Eds).

Macdonald, D. W. & Amlaner, C. J. (1980). A practical guide to radio-tracking. In *A handbook on biotelemetry and radio-tracking*: 143—160. Amlaner, C. J. & Macdonald, D. W. (Eds). Oxford: Pergamon Press.

Macdonald, D. W. & Newdick, M. T. (In press). The distribution and ecology of foxes, *Vulpes vulpes*, in urban areas. In *Proceedings of European Ecological Symposium on Urban Ecology*.

Maksimov, A. A. (1945). Migrations of the Arctic fox in the north of the European part of the U.S.S.R. *Bull. Soc. Nat. Mosk.* (Biol.) 50 (5—6): 45.

Niewold, F. J. J. (1974). Irregular movements of the Red fox (*Vulpes vulpes*), determined by radio tracking. *Int. Congr. Game Biol.* 11: 331—337.

Novikov, G. A. (1956). Carnivorous mammals of the fauna of the U.S.S.R. *Fauna USSR* No. 62: 1—294. (Israel Program for Scientific Translations, Jerusalem, 1962.)

Osgood, W. H., Preble, E. A. & Parker, G. H. (1915). The fur seals and other life of the Pribilof Islands, Alaska, in 1914. *Bull. Bur. Fish. U.S. Dept. Comm.* 34: 1—172.

Pulliainen, E. (1965). On the distribution and migrations of the Arctic fox (*Alopex lagopus* (L.)) in Finland. Aquilo (Zool.) 2: 25—40.

Pulliainen, E. (In press). *Alopex lagopus* (Linnaeus, 1758) — Der Eisfuchs. *Säugetiere Europas*.

Rausch, V. R. & Rausch, R. L. (1979). Karotype of the Red fox (*Vulpes vulpes* (L)) in Alaska. *Northw. Sci.* 53: 54—57.

Saemundsson, B. (1939). Mammalia. *Zool. Iceland* 4 (76): 1—52.

Sargeant, A. B. (1972). Red fox spatial characteristics in relation to waterfowl predation. *J. Wildl. Mgmt* 36: 225—236.

Scholander, P. F., Hock, R., Walters, V., Johnson, F. & Irving, L. (1950). Heat regulation in some arctic and tropical mammals and birds. *Biol. Bull. mar. biol. Lab. Woods Hole* 99: 237—258.

Schwartz, S. S. (1967). Biological foundations of the hunting and trapping industries. *Probl. N.* 11: 1—11.

Scott, T. G. (1943). Some food coactions of the northern plains Red fox. *Ecol. Monogr.* 13: 427—429.

Sequiera, D. (1980). Comparison of the diet of the red fox (*Vulpes vulpes* (L)) in Gelderland, Holland, Denmark and Finnish Lapland. *Biogeographica* 18: 35—52.

Shilyaeva, L. M. (1967). Studying the migration of the Arctic fox. *Probl. N.* 11: 103—112.

Skrobov, V. (1961). *Pesets: Okhota i okhotnich'e Khozyaistvo, U.S.S.R.* [The arctic fox] No. 1: 17—20. Moscow: Ministry of Agriculture.

Slagsvold, P. (1949). Nedarvning av den blå og hvite farge hos polarreven (*Alopex lagopus*). Nord. Vet. -Med. 1: 429—441.

Soper, J. D. (1944). The mammals of southern Baffin Island, Northwest territories, Canada. *J. Mammal.* 25: 221—254.

Speller, S. W. (1972). *Food ecology and hunting behavior of denning Arctic foxes at Aberdeen Lake, Northwest Territories.* Ph.D. thesis: University of Saskatchewan.

Storm, G. L., Andres, R. D., Phillips, R. L., Bishop, R. A., Siniff, D. B. & Tester, J. R. (1976). Morphology, reproduction, dispersal and mortality of mid-western Red fox populations. *Wildl. Monogr.* 49: 1—82.

Tembrock, G. (1957). Zur Ethologie des Rotfuchses (*Vulpes vulpes* (L.)) unter besondere Berücksichtigung der Fortpflanzung. *Zool. Gart. Lpz* 23: 289—532.

Tembrock, G. (1977). Canid vocalisations. *Behav. Proc.* 1: 57—75.

Whittaker, R. H. & Likens, G. E. (1975). The biosphere and Man. In *Primary productivity of the biosphere*: 305—328. Lieth, H. & Whittaker, R. H. (Eds). Berlin: Springer Verlag.

Wielgolaski, F. E. (1975). Primary productivity of alpine meadow communities. In *Fennoscandian tundra ecosystems* Part 1: Plants and microorganisms: 121—128. Wielgolaski, F. E. (Ed.). New York: Springer Verlag.

Wipf, L. & Shackelford, R. M. (1949). Chromosomes of a fox hybrid (*Alopex — Vulpes*). *Proc. nat. Acad. Sci., Wash.* 35: 468—472.

Wrigley, R. E. & Hatch, D. R. M. (1976). Arctic fox migrations in Manitoba. *Arctic* 29: 147—158.

DISCUSSION

P. J. Garson (Newcastle) — How do you explain the apparent adjustment of fox group territory size to the availability of productive foraging habitat in those territories alongside the fact that fox group size varies?

D. W. Macdonald — For the seven red fox territories and three Arctic fox territories which we know well, there are no correlations between group sizes and territory size. Territory size varies greatly, yet each territory contained the same amount of critical habitat — in one case "residential" land, in the other "productive" coastline. The idea is that the dispersion of patches of available prey determines the minimum size and configuration of a territory that will support a pair of foxes, whereas patch richness therein sets the maximum possible group size. Group size and territory size could thus be determined quite independently of each other.

B. C. R. Bertram (Zoological Society of London) — If the red fox northerly distribution is limited by body size one would expect smaller social groups further north.

Macdonald — One would expect smaller groups of red fox in more northern latitudes in as much as one expects such habitats to harbour more homogeneously dispersed food. In that case we argue that the minimum territory size that will support a pair of foxes cannot (in contrast to a patchy food supply) support others at no net cost to the original pair. However, I know of no data to test your suggestion. Our argument would be that while primary productivity may limit the northerly distribution of the red fox (through the costs of larger body size), it would be the dispersion of that productivity in available prey that would determine the group structure at these northern limits (and elsewhere) of the species' range. Of course, the idea that northern habitats have a more homogeneous prey availability overall is too simple an assumption — one might think of northern colonies of nesting seabirds, for example, which could be highly clumped and favour the development of large groups.

Symp. zool. Soc. Lond. (1982) No. 49, 291–299

The Uses of Radio Tracking Combined with other Techniques in Studies of Badger Ecology in Scotland

T. PARISH and H. KRUUK

Institute of Terrestrial Ecology, Banchory Research Station, Hill of Brathens, Glassel, Banchory, Kincardineshire, Scotland

SYNOPSIS

The chapter gives a brief description of radio tracking equipment used in a study of the ecology and social organization of the European badger *Meles meles* in Scotland. The equipment is used in conjunction with other techniques (bait colour-marking, isotope recovery, use of night-viewing equipment and automatic activity recording) to assess home range, numbers of animals per group, feeding behaviour and various activity patterns. Examples are given of the kind of information obtained. The importance is stressed of combining or comparing results of several different methods.

INTRODUCTION

This chapter describes ways in which radio tracking equipment is used in conjunction with other techniques in a study of the ecology and social organization of the European badger *Meles meles* in Scotland. The methods were used to identify the size of group range, number of animals per group, and habitat utilization and behaviour of individuals under various environmental conditions. Radio tracking techniques were used in the following aspects of the study:

(i) To locate individuals for direct, visual observations using infra-red night glasses, electronic image-intensifier or ordinary 7 × 50 binoculars.

(ii) To locate and map the range of movements by individuals for establishing range-size and habitat utilization.

(iii) To identifiy ranges of individuals injected with a radio isotope in order to assess the number of badgers using a particular range (Kruuk, Gorman, & Parish, 1980).

(iv) To establish, with a continuous activity recorder, the activity of different individual badgers under various environmental conditions.

METHODS

In a long-term study, begun in 1976, in a study area of approximately 12 km² at Aviemore, Inverness-shire, Scotland, badgers were trapped in large cage traps (1.75 × 0.9 × 0.7 m) of welded steel mesh (7.5 cm) with a single hinged drop door and locking catches. The traps had a trigger or treadle release and peanuts were used as bait. Captured badgers were immobilized with a blowpipe and dart, injecting a dose of approximately 20 mg/kg body weight of ketamine hydrochloride. Trapping success was approximately one animal per five trap nights. Badgers were fitted with a transmitter and weighed, tooth-wear was described, samples of sub-caudal and anal gland secretions were collected and the animal was tattooed inside the upper part of the hind leg. In the early part of the study the transmitter was housed in a back-pack on a harness comprising a neck strap, chest strap and a linking strap along the sternum (Kruuk, 1978a) but this was later replaced by a collar (Cheeseman & Mallinson, 1980). The transmitter was based on the Ashwell circuit (Taylor & Lloyd, 1978) with a ferrite rod aerial and was housed with a lithium AA cell in threaded chambers in a shaped block of rigid PVC. Silica gel crystals were placed in the cell chamber to prevent internal condensation and corrosion, and the threaded bungs were welded in place with a PVC solvent. One of the threaded bungs could be drilled out for replacement of the cell.

The transmitter hung under the neck of the animal and a suitably embedded 2 cm diameter luminous beta-light was mounted on the top of the collar to aid visual location at night.

The equipment operated on 102.2 MHz. AVM LA 12 receivers with 12 channels were used, with a capacity for at least two different frequencies on each channel. The receiving aerials were hand-held rigid and collapsible dipoles (Parish, 1980), a vehicle-mounted, non-directional whip-aerial and an H-Adcock array (Taylor & Lloyd, 1978), also vehicle mounted.

The transmitter gave line-of-sight reception ranges in excess of 2.5 km but this was often considerably reduced by topography and the poor output of some transmitters. The signal consisted of 40 to 70 pulses per minute, and a pulse duration of 80 ms as the best compromise between audibility and extended cell life (usually nine months to one year).

For observing the animals we used 7 × 50 binoculars or Old Delft infrared binoculars ("hot-eye", Kruuk, 1978a) or an image intensifier based on a Mullard XX1306 microchannel plate intensifier tube (second generation type). This last instrument was housed in a purpose made pistol-grip casing, incorporating a battery chamber and fitted with a × 6 eye-piece and a 135 mm F2. camera lens. The magnification of this equipment could be doubled using a × 2 teleconverter bringing it to approximately × 7; and the background illumination and hence the definition of the image can be increased by using a small low-power narrow beam torch with a red filter.

The bait marking was carried out at all main badger setts or probable main setts by baiting with a mixture of peanuts and treacle, with disks of polythene of various colours and sizes added as indigestible markers. The polythene was subsequently replaced by small coloured alkathene pellets (ICI) used in plastic industries. Up to ten dessert-spoons of the mixture were put out at each sett every day, usually under slates or stones, or in tunnel entrances to prevent the bait being eaten by birds. Badger latrines were checked for the presence of colour markers as often as possible, at least once a week, until an adequate boundary map was produced (four to six weeks).

Although location and movements of badgers and range boundaries were plotted in as much detail as possible, part of the range-use analysis was done using a grid system with 4-ha squares, evaluating habitat utilization on a PDP 11 computer against season, weather, human land use, sex and status of the badger, and other parameters.

After badgers had been injected, for group size assessment, with the isotope ^{65}ZnCl (0.1 mCi/animal, intramuscular; Kruuk et al., 1980) and fitted with a radio transmitter, all latrines in the area were cleared of dung, so that only faeces produced after all animals were injected were subsequently collected. All dung produced in latrines within group ranges and on boundaries common to groups with injected animals was collected and tested for radio-activity; the proportion of radio-active faeces was indicative of the group size. Each sample was scanned using a Mini-monitor counter with a gamma-tube which was inserted into a hole in a lead castle containing the whole fresh sample. This method was found to be more efficient for a presence or absence assessment than the use of wet or dry sub-samples in a scintillation counter. Adequate results could be obtained for about four months from the date of injection. All injected animals had radio collars to confirm that for the duration of the dung collections they lived in the groups under study.

For continuous recording of badger presence or absence in the sett a Rustrak 100 micro-amp chart recorder was used, operating at a chart speed of 2.5 cm h^{-1}, connected to an AVM receiver and vertical dipole aerial. The whole system was powered by a 12-V car battery and operated for up to 30 days without attention, located close to the relevant badger sett.

RESULTS

A brief indication of some of the results that are being obtained with the above techniques follows.

Direct observations of radio collared individuals using night-viewing devices allowed the collection of data on interactions between badgers and their prey over periods of one hour or more. This made it possible to identify food "patches" (Kruuk, 1978b) and to measure the foraging efficiency with respect to the main foods such as earthworms, pignuts (*Conopodium majus*) and others, as well as to observe interactions between badgers.

TABLE I

Range sizes of badger groups, 1980 in 4-ha squares, as determined by colour-marked bait recovery and radio tracking; a comparison

Group	Range size by colour mark recovery	Range size by radio tracking	Colour-mark range outside radio range (%)	Radio range outside of colour-mark range (%)
FF	63	49	22	4
P	29	29	0	0
LL	33	33	0	0
M/S/WCM	67	60	10	3
ML/A	57	59	0	3
G	28	30	0	13

Using the relatively quick assessment of badger ranges with colour marked bait and radio tracking (which give good agreement; Table I and Fig. 1), differences between areas and changes in the range system from year to year were measured (Fig. 2), and radio tracking provided information on the way in which individual badgers fit into this system. In the study area badgers appeared to use exclusive group ranges with virtually no overlap between neighbours; however, there were variations on this general theme. Usually, the ranges of radio collared individuals were approximately the same as the ranges of the whole group to which they belonged (Fig. 1), but some individuals used only part of that group territory; most of these were females, but not all. Over prolonged periods (one year or more) these individual ranges sometimes changed, within the group territory. Some individual badgers would habitually cover the ranges of two otherwise separate groups; these were mostly males, but not always. Brief intrusions into neighbouring ranges were also observed.

Table II presents estimates of the number of badgers in groups to

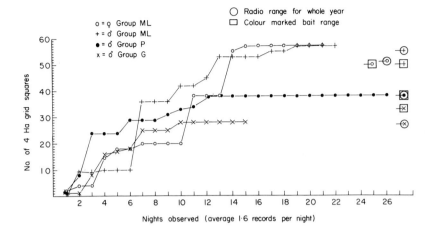

FIG. 1. Examples of the rate at which individual home ranges can be assessed (for the period May—August 1979). Range size as the number of grid squares included after each night when used squares are joined to the nearest two peripheral previously used squares.

Group	1977	1978	1979	1980
FF		45	50	63
	78			
P		33	38	29
LL	44	44	39	33
S	31	25		
			57	(57)
M		(25)		
	38	36		67
WCM		(11)	10	(10)
ML	20			
		(64)	(50)	59
A	52			
		103	83	
G	21	(39)	(33)	30
Total no. of grid squares	284	286	277	281

FIG. 2. Group range size as shown by radio tracking and colour marker returns 1977—80 in Speyside study area. Units used are 200 × 200 m grid squares (4 ha) entirely or more than half within the range as drawn by linking boundary latrines, i.e. those with colour markers from at least two groups. Figures in brackets ( ) were possible subdivisions suggested by the colour marker results which were subsequently shown by radio location to be valid boundaries for the majority of animals in the groups.

TABLE II

Group size estimates from isotope recovery from the faeces, 1980. Number of faeces in brackets

Group	No. of injected badgers	26 May	8 June	9 July	22 July	10 August	26 August	Total ± s.e.
FF	2	5.3 (42)	4.7 (61)	5.6 (14)	7.3 (62)	5.1 (102)	4.9 (106)	5.3 ± 0.4 (387)
P	1	3.3 (13)	– (3)	– (3)	– (1)	1.8 (9)	1.3 (5)	2.4 ± 0.6 (34)
LL	1	2.1 (19)	2.4 (36)	2.6 (31)	3.0 (80)	2.0 (24)	2.1 (12)	2.5 ± 0.2 (207)
M/S/WCM	3	9.4 (47)	6.7 (40)	8.3 (72)	9.0 (6)	10.1 (61)	9.3 (102)	8.8 ± 0.5 (328)
ML/A	3	6.5 (26)	10.3 (31)	6.3 (19)	6.9 (32)	7.5 (10)	3.4 (12)	6.4 ± 0.8 (135)
G	2	4.5 (34)	7.1 (25)	3.3 (5)	4.0 (28)	4.7 (7)	– (3)	4.8 ± 0.6 (102)
Totals for whole area[a]	12	31.6 (357)	30.2 (340)	28.3 (278)	30.4 (342)	31.5 (304)	31.2 (328)	30.6 ± 0.5 (1949)

[a] The total area estimates include boundary latrines shared by two or more groups.

which radio collared animals belonged; the estimates were derived from several independent samples of faeces. The standard errors were reasonably small; the final assessment of group size also took into account direct observations when available.

Some of the results of the activity recording analysed so far suggest that the length of the activity period is related to season (it is very short in winter) but not obviously to the size of the range. Badgers fairly frequently return to the sett for brief periods at night. Males are probably active longer each night than females, except when females are foraging with their cubs.

DISCUSSION OF RESULTS

The kind of results which are being obtained with the combination of radio tracking and various other techniques are important for our understanding of the effect of the environment on populations of badgers and on carnivores in general. The observation of individual badgers with radio transmitters is essential; for instance, individual differences in range use within the group range system often appear to be associated with the previous or subsequent changes in the group ranges as illustrated in Fig. 2. The combination of the colour-marked bait technique and radio tracking is probably the most efficient way of observing such effects; an automatic location recording system (Deat et al., 1980) would give continuous range boundary and utilization data, but because of high cost, difficult topography and small-scale diversity of habitat this method was rejected in favour of visual records and manual triangulation. It would be difficult to carry out the group size estimates using radio isotopes without the detailed knowledge of the ranges of injected badgers, and group boundaries. It is only through radio tracking that it has been possible to integrate the various techniques needed to answer the main questions of this study.

ACKNOWLEDGEMENTS

We thank the various estates for their permission to carry out this research, also students Joanna Walker, Chris Brown, Alex Hutt and Roy Paton for their fieldwork in successive years, and John Morris for his technical assistance in the construction and maintenance of equipment. We are grateful to Dr David Jenkins for his comments on the manuscript.

REFERENCES

Deat, A., Mauget, C., Mauget, F., Maurel, D. & Sempere, A. (1980). The automatic continuous and fixed radio tracking system of the Chize Forest: theoretical and practical analysis. In *A handbook of biotelemetry and radio tracking*: 439–451. Amlaner, C. J. & Macdonald, D. W. (Eds). Oxford: Pergamon Press.

Cheeseman, C. L. & Mallinson, P. J. (1980). Radio tracking in the study of bovine tuberculosis in badgers. In *A handbook of biotelemetry and radio tracking*: 649–656. Amlaner, C. J. & Macdonald, D. W. (Eds). Oxford: Pergamon Press.

Kruuk, H. (1978a). Spatial organisation and territorial behaviour of the European badger *Meles meles. J. Zool., Lond.* 184: 1–19.

Kruuk, H. (1978b). Foraging and spatial organisation of the European badger *Meles meles. Behav. Ecol. Sociobiol.* 4: 75–89.

Kruuk, H., Gorman, M. & Parish, T. (1980). The use of ^{65}Zn for estimating populations of carnivores. *Oikos* 34: 206–208.

Parish, T. (1980). A collapsible dipole antenna for radio tracking on 102 MHz. In *A handbook of biotelemetry and radio tracking*: 263–268. Amlaner, C. J. & Macdonald, D. W. (Eds.) Oxford: Pergamon Press.

Taylor, K. D. & Lloyd, H. G. (1978). The design, construction and use of a radio-tracking system for some British mammals. *Mamm. Rev.* 8: 117–141.

DISCUSSION

E. G. Neal — Would the fact that badgers often defecate below ground affect the accuracy of your estimate of numbers by using ^{65}ZnCl?

T. Parish — The only way such behaviour could affect our estimates is if a particular individual habitually defecated below ground. We have no evidence that this happened. The method was tested on a captive group of badgers and gave good results. Whenever it is possible to cross-check our information the method gives consistently reliable results.

P. Chanin (Exeter) — Did the isotope method give any indication of differences in the way males and females or different age groups distribute their faeces?

Parish — We intend to look at this aspect in more detail but at present our impression is that males and females behave in a similar way in distributing their faeces.

J. White (Loughborough University) — Have you any information on mean territory size related to altitude?

Parish — No. All our territories cover similar ranges of altitude, although topography and the distribution of moorland often means that territories including a greater proportion of higher ground are in fact larger.

J. Gallagher (MAFF) — Did you find a difference between the size of the ranges used by males and females?

Parish — Individual deviations were observed in both sexes but in general there was a marked similarity in range sizes with territorial boundaries respected by all members of the social group.

Symp. zool. Soc. Lond. (1982) No. 49, 301—323

Activity Patterns and Habitat Utilization of Badgers *(Meles meles)* in Suburban Bristol: A Radio Tracking Study

STEPHEN HARRIS

Department of Zoology, University of Bristol, Bristol, England

SYNOPSIS

The distribution of badger setts on an area of 5.5 km^2 in a north-west suburb of Bristol is described; most of the larger setts were probably present before the area was developed, but in two of the seven social groups the main sett was only dug in the early 1970s.

The activity patterns of Bristol badgers are described; during the summer months emergence was delayed by about an hour when compared with a rural study, but time of return to the sett was unaltered. Periods of above-ground activity were interspersed with periods of rest; individual activity periods were longest in July to September, and shortest from November to April. Individual rest periods were shortest from May to July, and longest from December to February. During January badgers were only active above ground for 8.4% of the night, this increasing to 65.3% in July.

The food of Bristol badgers is described, based on dung and stomach contents analysis. Over the year as a whole, the badgers were not specializing in taking any one main food item, but took a variety of food. The greatest diversity of food was taken in May to July, the lowest diversity in August to November, when fruit was the most important item in the diet. Throughout most of the year, the food sources were scattered, and the badgers' rate of travel while foraging increased with the diversity of food eaten.

On the study area territorial boundaries were not well defined and the pattern of individual home ranges was complex. A group foraging range was defined for each social group; within these group foraging ranges eight habitat types were described. No group foraging range contained all eight habitat types. The badgers showed no significant selection of any one habitat for foraging, and in each social group the diversity of habitats used by the badgers for foraging bore a direct relation to the diversity of habitats available to them. Using a principal components analysis, two main types of foraging range were recognized. Smaller ranges contained a lower proportion of gardens, a greater diversity of habitat types, and the badgers ate a lower diversity of food types. Larger ranges contained a higher proportion of gardens, a lower diversity of habitat types, and the badgers ate a greater diversity of food types. These data are discussed in relation to earlier studies on badger ecology.

INTRODUCTION

The presence of large wild animals living in an urban environment is an interesting phenomenon that has been described by a number of authors, and Gill & Bonnett (1973) have discussed urban habitats and their value to wildlife, with particular reference to large carnivores. Of the wild carnivores found in British cities, foxes (*Vulpes vulpes*) are the most common; Teagle (1967) and Harris (1977) have documented their distribution in London, and more recently Harris (1981) has presented a fox population estimate for the city of Bristol. Badgers (*Meles meles*) are less common in urban areas, but can be found in some cities. Humphries (1958) described the behaviour of badgers on the edge of Cheltenham, and Stirling & Harper (1969) documented the distribution, numbers and behaviour of badgers on the southern edge of Durham city. Teagle (1969) described the distribution of badgers in London, and Cowlin (1972) recorded the distribution of badgers in the urban areas of south Essex. Middleton & Paget (1974) have noted the occurrence of setts in urban areas in Yorkshire and Humberside; these were generally rare and confined to the periphery of the conurbations. In the national badger survey, Neal (1972) reported that only 0.6% of the setts were recorded from built-up areas. More recently, Neal (1977) has discussed the distribution and behaviour of urban badgers. In the northern suburbs of Greater Copenhagen, Asferg, Jeppesen & Aaris-Sørensen (1977) found an average of one occupied badger sett per 2–3 km^2. These setts were largely located in little disturbed areas in parks, woods, or large private gardens.

In Bristol, the badgers are confined to certain parts of the city; details of individual records can be found in Symes (1972, 1973) and Jayne (1978, 1980, 1981). This chapter aims to give a preliminary account of the ecology of the badger in one part of the city of Bristol, and to describe some of their behavioural adaptations to the urban environment.

METHODS

The study area was 5.5 km^2 in size, in the north-west suburbs of Bristol. Most of the area was developed between the two World Wars, although some development has continued to the present day. The north-west part of the study area around Sea Mills is a Council housing estate, most of the eastern half consists of privately owned semi-detached housing, and the western half consists mainly of larger, often detached houses, with several blocks of luxury flats.

Between October 1977 and September 1978 all 5480 dwellings on the study area were visited, and for 5191 (94.7%) a questionnaire on the presence of badger setts, badger damage and the level of badger activity in the area was completed. In addition all vacant land was visited, and checked for the presence of badger setts and dung pits.

During 1978 and 1979 badger dung was collected each month, and badgers killed (usually by cars) in the area were collected for post mortem. The contents of 1920 dung samples and 20 stomachs were analysed, each food item being graded on a 1—6 scale of abundance devised by Kruuk & Parish (personal communication). Full details of the analysis and food items will be published elsewhere, but broad trends in food habits will be presented as they affect the behaviour of the badgers.

From May 1978 badgers were caught for radio tracking in traps baited with peanuts, in snares, and with long-handled nets while feeding in open spaces (Cheeseman & Mallinson, 1980). In addition some animals were bolted from under sheds with drain rods and caught in purse nets. The badgers were immobilized by an intramuscular injection of ketamine hydrochloride (Mackintosh et al., 1976; Hunt, 1976; Cheeseman & Mallinson, 1980). For caged badgers, the injection was either administered by hand, or with a Mini-ject blow pipe (Dist-Inject, 23/25 Marine Drive, Torpoint, Cornwall). Radio collars, frequency 173 MHz, were fitted as described by Harris (1980) for foxes. All badgers were marked with a white numbered plastic tag in each ear, these tags serving as a temporary mark; they were also more permanently marked with a number tattooed on the upper inside of each hind leg (Cheeseman & Harris, 1982). The badgers were allowed to recover from the anaesthetic either in a trap or returned to the nearest sett.

A total of 31 badgers (15 males, 16 females) were radio tracked; between three and five animals from each social group were studied. (A number of terms have been used to describe badger groups; "clans" by Kruuk & Parish (1977) and Kruuk (1978a,b); "social groups" by Ministry of Agriculture, Fisheries and Food (1976, 1977, 1979), Neal (1977) and Cheeseman & Mallinson (1980). The term "social groups" is preferred by the author and will be used thoughout this chapter.) When radio tracking, a three-element hand-held Yagi aerial and an AVM LA12 receiver were used. Only one animal was followed each night, normally from its point of emergence from the sett until it finally ceased above-ground activity in the morning. Tracking was done on foot; the discontinuous nature of the urban habitat necessitated maintaining close contact with the badger, which was normally less than 30 m away. Most of the tracking was done from the road, with forays into gardens and waste land whenever practical. Since human activity continued throughout at least half the night, the badgers took little notice of human presence on the streets, though on windy nights care had to be taken to remain downwind. Even if disturbed the badger would normally only move a few yards, either into cover or into the garden adjacent to the one in which it was disturbed. Whenever possible, all activity was observed using a pair of Oldelft Image Intensifiers (PB4DS), and data were recorded on a pocket dictaphone.

RESULTS

Distribution of Setts

Eighty badger setts were found in the study area (Fig. 1). These setts were divided amongst eight social groups; the most northerly group foraged largely outside the study area on a large golf course and in a natural wooded park, and, since this habitat was atypical of the urban area, this group was not included in the present analysis.

Many of the setts were long-established, and were probably present before the area was developed. Unfortunately this cannot be substantiated since there are no early records of badgers from this part of Bristol. Tetley (1940) recorded a badger shot in another part of Bristol (Fishponds) for killing over 20 fowl, and Bassindale (1955) said that badgers were found within the city boundaries but gave no details. However, for two social groups (Old Sneed Park and St Monicas), the main sett was only established in the early 1970s; both were originally fox earths that had been used for rearing cubs.

TABLE I

Habitats from which the 80 badger setts on the study area were recorded

In private gardens	20
In ornamental gardens, grounds of residential homes and University halls of residence	9
In disused gardens	1
Under sheds, outbuildings	2
In drains	2
In strips of woodland	28
In scrub	7
Under dense bramble	5
In horse paddocks	5
In playing fields	1

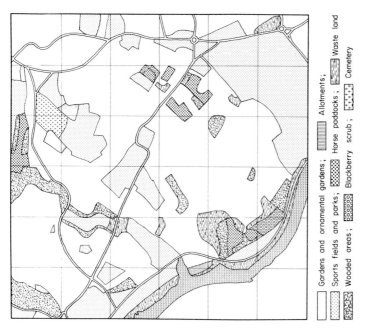

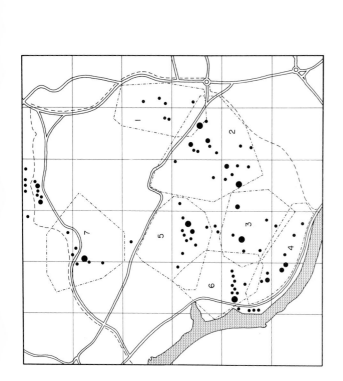

FIG. 1 (left). Distribution of badger setts on the study area. The limits of the study area are marked by the River Avon, the south-west boundary being formed by the River Avon. Only "A" and "B-class" roads are marked; the grid lines denote 500 m squares. The setts are marked by black circles as follows: ● less than five holes (including active and inactive holes); ● five to ten holes; ● eleven or more holes. The dot-dash lines mark the group foraging ranges, the numbers denoting the social groups as follows: 1. St Monicas; 2. Hiatt Baker; 3. Old Sneed Park; 4. Bishops Knoll; 5. Druid Stoke Ave; 6. Glen Avon; 7. Coombe Bridge Ave.

FIG. 2 (right). Distribution of main habitat types on and around the study area; no habitat data are supplied for the south-west side of the River Avon.

Gardens and ornamental gardens; ▓ Allotments;
Sports fields and parks; ▒ Horse paddocks; ▒ Waste land
Wooded areas; ░ Blackberry scrub; ░ Cemetery

Of the 80 setts, 40 were dug on wooded or scrub and bramble covered banks unsuitable for building development (Fig. 2). However, 25% of the setts were in private gardens, sometimes very small gardens (Table I). Two of these setts were located on small banks that formed the old field boundaries prior to development.

Activity Patterns

Although badgers are predominantly active by night, they are occasionally active during the day, particularly in quiet rural areas (Neal, 1977). During the present study animals have been observed occasionally moving around in gardens on hot, sunny days, and one animal, whose sett was under dense bramble, would sometimes lie up for the day above ground by the sett entrance. However, such activity was rare and no attempt was made to record the level of diurnal activity. Also, badgers can be very active underground; the following account only discusses activity patterns once the animals had emerged from the sett. Heavily pregnant sows, and sows with young cubs (i.e. up to June) were excluded from this analysis. No difference between sexes was then apparent, and so males and females were not analysed separately.

Times of emergence from, and final return to, the sett are shown in Fig. 3; these data are compared with those from Neal (1977) for rural areas. During the summer months mean emergence time for the Bristol badgers was delayed by about one hour, whereas the time of return to the sett was comparable to that of animals in rural areas. For the winter months the Bristol and rural data were at variance; whilst aware of the limitations of his data, Neal (1977: fig. 7.4) gave the impression that the length of time between mean emergence and final return to the sett was greatest in the winter, whereas the Bristol data did not show this. Such differences are probably attributable to the mode of data collection. Direct observations, as used by Neal, cannot normally differentiate between individual animals. Hence one animal can emerge early, return to the sett shortly, and not emerge again; it is then easy to assume that another animal that does not emerge until late in the night is the same animal re-emerging for another period of activity before dawn. With radio collared animals no such confusion occurs, and as Fig. 3 shows, during October to March, some animals did complete their nocturnal activity before others had even emerged. During January, 20.0% of the badgers failed to emerge at all during the night of study, and during December and January above-ground activity was limited to 8.6% and 8.4% of the night respectively. However, during December and January there was often considerable underground activity, with the badger repeatedly coming near to the sett entrance but not emerging.

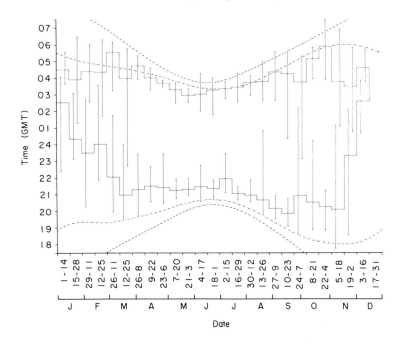

FIG. 3. Times of emergence from, and final return to, the sett by radio tracked badgers. The data are presented in two-week periods. The lower horizontal lines mark the mean time of emergence, the upper horizontal lines the mean time of the final return to the sett. Vertical lines bisecting these horizontal lines mark the range of recordings. The lower and upper dot-dash lines mark mean time of emergence from, and final return to, the sett as recorded by Neal (1977) for a rural area. The lower and upper dashed lines mark sunset and sunrise.

Once the badgers had emerged, periods of activity were interspersed with periods of rest; they either returned to a sett, or lay up under sheds or above ground, often under garden hedges or bushes (Table II), and couches, as described in Neal (1977), were built in

TABLE II

Situations in which radio tracked badgers lay up during the night between periods of activity

	No.	%
Returned to main sett	247	35.5
Went to an outlying sett	262	37.6
On surface under bushes, hedges	158	22.7
Under sheds, summerhouses, in old buildings, air-raid shelters	29	4.2
Total	696	

TABLE III

Activity patterns of badgers in Bristol

Months	Total time active during night (min) with S.D.	Number of times lay up during night with S.D.	Lengths of individual activity periods (min) with S.D.[a]	Lengths of individual rest periods (min) with S.D.[a]	% night (i.e. between sunset and sunrise) active	Distance travelled each night between foraging sites (in metres) with S.D.[c]
J	78.8 ± 77.4	1.2 ± 1.8	60.5 ± 47.3	98.1 ± 95.1	8.4	1140 ± 1320
F	139.3 ± 90.4	1.5 ± 1.0	73.0 ± 53.1	118.0 ± 80.7	16.5	2020 ± 878
M	256.4 ± 87.1	3.2 ± 2.4	69.3 ± 73.2	72.0 ± 64.6	35.1	2240 ± 1346
A	218.3 ± 69.9	2.3 ± 1.4	56.3 ± 45.4	78.2 ± 67.2	35.8	2400 ± 1731
M	287.2 ± 61.0	2.6 ± 1.6	96.2 ± 81.7	29.5 ± 21.0	56.8	3010 ± 1648
J	274.0 ± 53.2	3.0 ± 2.0	91.6 ± 85.3	27.5 ± 26.4	61.7	3950 ± 2542
J	307.1 ± 46.5	2.6 ± 1.5	107.3 ± 89.2	24.6 ± 18.2	65.3	3080 ± 1707
A	316.0 ± 102.6	2.6 ± 1.4	105.0 ± 89.3	46.8 ± 32.9	57.6	2590 ± 1277
S	389.1 ± 128.5	3.0 ± 2.4	109.3 ± 106.4	41.8 ± 30.2	57.1	2580 ± 1265
O	320.0 ± 137.0	3.5 ± 2.1	78.5 ± 82.6	53.6 ± 46.0	40.2	2320 ± 943
N	156.1 ± 149.3	3.1 ± 2.4	52.8 ± 55.7	59.7 ± 68.9	17.2	1340 ± 1016
D	83.0 ± 22.9	1.0 ± 0.7	41.5 ± 32.8	138.0[b]	8.6	1220 ± 769

[a] Individual rest and activity periods of less than 10 min duration excluded.
[b] Limited sample size; no S.D. given.
[c] Measured to nearest 10 m.

favoured sites. Individual activity periods were longest during July to September, rest periods were longest during December to February and shortest during May to July (Table III). The total time active was greatest from July to October, but even during this period the badgers were still only active from 40.2% (in October) to 65.3% (in July) of the night.

During most months there were no distinct peaks of above ground activity (Fig. 4). From November to February it was rare for more than 50% of the animals to be active above ground at any one time. During March and April there were two activity peaks; one was two to three hours after the onset of emergence, and this was followed by

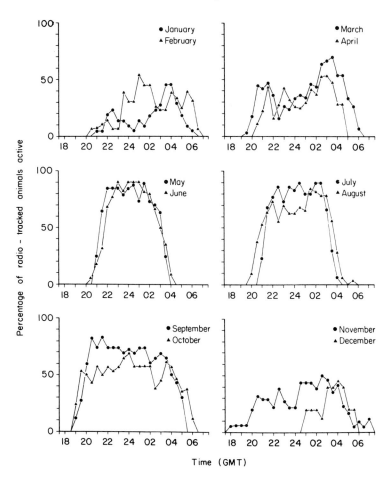

FIG. 4. Distribution of above ground activity periods during the night. The graphs show the percentage of radio tracked animals that were active at half-hour intervals throughout the night.

a larger peak about three hours before the final return to the sett. From May to August there was a high level of activity throughout the night. In September and October there was a slight decrease in the number of animals active at any one time, but the number of animals active was fairly constant throughout the night, with no distinct peaks of activity above ground.

The present results contrast with those of Bonnin-Laffargue & Canivenc (1961), who studied the activity patterns of female badgers in the Dordogne from November to June. They used implanted radio transmitters and fixed aerials which could record the animals leaving and returning to the sett. They described three distinct activity phases, with much activity around the sett soon after emergence and before the animal finally ceased above ground activity; these were interspersed with a period of foraging away from the sett. They found that this period of activity away from the sett was constant throughout the period of their study, and they suggested that the length of this period was biologically important. These results differ from the present study, where the length of the period of above ground activity varied markedly according to the time of year. Neither did these French workers make any reference to their badgers returning to the sett to lie up for parts of the night, a conspicuous feature in the present study. There are no obvious explanations for the differences between the French study and the present results. It may be that the differing results reflect adaptations to different patterns of food availability, disturbance, and other possible factors; further data from different habitats are needed to clarify this point.

Food

The food items were divided into six main categories (Table IV). "Scavenged" items included food taken from dustbins, compost heaps bird tables, and food specifically put out by householders for the badgers, since it was impossible to distinguish between the sources of the various items included under this category. The percentage importance of these six food categories each month, based on a summation of the abundance ratings, is shown in Table IV. During January to May, scavenged food was the major item, during June and July it was invertebrates other than earthworms, from August to November it was fruit, and in December it was earthworms.

The mean of these monthly figures is also given in Table IV as an approximate indication of the relative importance of these food items on an annual basis. This is probably misleading, since it considers

TABLE IV

Food of badgers in Bristol; the figures (except the bottom row) are percentages. As an indication of the relative importance of the different food items on a yearly basis, the means of the monthly percentages are given, and in the right-hand column these have been adjusted to allow for the different amount of time spent active above ground each month.

	J	F	M	A	M	J	J	A	S	O	N	D	% mean	% corrected to allow for length of time active
Earthworms	24.0	24.0	30.2	23.1	21.0	16.3	18.3	15.3	11.6	13.1	10.3	35.3	20.2	18.4
Invertebrates	22.9	16.8	19.8	21.6	26.4	30.9	36.1	21.0	15.9	12.4	11.0	20.1	21.2	21.7
Vertebrates	1.4	4.5	2.2	2.3	6.4	6.1	2.7	1.2	1.0	1.6	0.6	0.6	2.6	2.7
Vegetables	3.4	5.9	10.4	8.2	12.4	13.0	4.1	2.8	2.2	2.1	0.0	2.9	5.6	5.9
Fruit	9.8	7.5	4.1	5.9	2.5	8.3	20.3	45.5	60.0	57.0	61.1	16.6	24.9	28.4
"Scavenged" items	38.5	41.3	33.3	38.9	31.3	25.4	18.5	14.2	9.3	13.8	17.0	24.5	25.5	22.9
% occurrence of earthworms	95.1	91.0	95.6	95.9	88.6	86.5	97.6	91.0	78.3	81.5	60.0	95.2	—	—
Shannon index of food diversity	2.09	2.18	2.17	2.16	2.27	2.38	2.23	2.03	1.73	1.83	1.60	2.12	—	—

each month as being of equal importance. During December and January the Bristol badgers were only active for about 80 min each night, whereas in September they were active for nearly 390 min (Table III), when they were foraging extensively on fruit and gaining weight prior to their period of reduced activity. A more accurate indication of the annual importance of each food category may be obtained by weighting each month according to the length of time the animals were active; this is also shown in Table IV. The main change this produced was the increased importance of fruit in the diet, and the reduced importance of scavenged items and earthworms. However, this analysis did not take account of differential digestion of the various food items. In general, badgers digest animal food more completely than fruit and vegetables, and so until some allowance can be made for the nutritional value of the various food items, this analysis must be considered simply as showing the basic trends in the food of Bristol badgers.

To determine the diversity of food eaten each month, the Shannon index was used (Odum, 1971). Food diversity was greatest from May to July and lowest from August to November, when fruit was the major item in the diet. When feeding on fruit, the badgers were able to utilize an abundant but localized food source, e.g. windfall apples under one tree, a heavy fall of yew berries, or blackberries from a small clump of bushes; when utilizing such a food supply the badgers would forage for long periods in one locality. For the rest of the year the food sources were more scattered, and the badgers spent much less time foraging in any one place. Hence the rate of travel was related to the diversity of food eaten (Fig. 5); $r = 0.625, 0.05 > P > 0.01$.

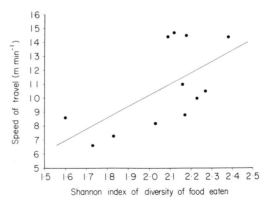

FIG. 5. Effect of diversity of food eaten on rate of travel while foraging; $y = 8.17x + 6.13$, $r = 0.625, 0.05 > P > 0.01$. Each symbol denotes a monthly mean.

Habitat Utilization

The pattern of home ranges was complex; one animal's home range would often overlap with those of animals from adjacent social groups, and the home ranges of six animals (two females, four males) included the majority of the foraging area of one or more adjacent social groups. Three of these six animals would also use the setts of adjacent social groups for lying up during the night, but normally returned to their own social groups' setts by dawn. However, these animals would occasionally lie up for the day in an outlying sett of the adjacent group. This sharing of ranges did not appear to be seasonal; of the two females, one used the foraging area of an adjacent group's range during January to March, the other used the foraging areas of three adjacent groups during March to April. Of the four males, one used an adjacent group's range during March and April, another in November, one during June to August, while the fourth used the foraging areas of two social groups during July and August, then at the end of August it moved to a third social group (from which it probably originated), and has since stayed there. Of the remaining animals studied, three males and one female permanently changed social group either before or after the period during which they were radio tracked. However, in most cases it was obvious to which social group a particular badger belonged at any one time; animals were ascribed to a social group depending on the setts they normally used to lie up in during the day.

Since the pattern of individual home ranges was so variable, a "group foraging range" was determined for each social group. For the purpose of this chapter, a group foraging range is defined as the area in which each individual member of the group spent the majority of its time; some animals never left it, whereas other members of the social group would sometimes forage beyond the group foraging range. The foraging ranges of the seven social groups are shown in Fig. 1.

In the Hiatt Baker group, the foraging range shown in Fig. 1 was used by most members of the group for the majority of the study, but in early 1980 the area was divided between two breeding groups, with an additional three adults living independently. Since this division occurred late in the study, the Hiatt Baker animals will be treated as a single social group for the purpose of this analysis. Also a sett was dug under a summerhouse on the Druid Stoke Avenue group's range by a single male cub, thought by the householder to have moved into the area in November of its first year. The animal was radio tracked from March to June the following year, after which it disappeared. During that period the cub failed to integrate

TABLE V

Distribution of habitat types within the foraging ranges of the seven social groups

		Gardens	Ornamental gardens	Allotments	Sports fields and parks	Horse paddocks	Wooded areas	Blackberry scrub	Scrub and waste land
St Monicas	: ha	25.8	8.1	—	2.9	1.2	1.0	0.6	0.4
	%	64.5	20.3	—	7.2	3.0	2.5	1.5	1.0
Hiatt Baker	: ha	42.6	21.9	—	7.7	5.5	2.5	—	0.8
	%	52.6	27.0	—	9.5	6.8	3.1	—	1.0
Bishops Knoll	: ha	14.1	—	—	4.7	1.5	3.5	2.5	4.7
	%	45.5	—	—	15.2	4.8	11.3	8.0	15.2
Druid Stoke Ave	: ha	45.3	—	—	4.2	—	1.8	—	0.7
	%	87.1	—	—	8.1	—	3.5	—	1.3
Glen Avon	: ha	24.8	—	0.6	—	4.3	2.9	1.4	7.0
	%	60.4	—	1.5	—	10.5	7.1	3.4	17.1
Old Sneed Park	: ha	40.6	—	—	—	—	0.6	—	0.8
	%	96.7	—	—	—	—	1.4	—	1.9
Coombe Bridge Ave	: ha	48.7	—	0.4	3.0	—	5.8	—	1.1
	%	82.5	—	0.7	5.1	—	9.8	—	1.9

with the resident group of badgers, and foraged exclusively within a
range of 4.2 ha, composed entirely of gardens. Since this badger did
not belong to any of the social groups, details of its behaviour are
not included in the following analysis of habitat utilization.

Within the group foraging ranges, eight habitat types were recognized, namely private gardens; large "ornamental" gardens such as
are found in University Halls of Residence and Hotels; allotments;
sports fields and parks; horse paddocks; wooded areas, usually narrow
strips of trees on steep banks; blackberry scrub; waste land, mainly
areas of long rough grass and hawthorn scrub (Fig. 2). The distribution
of these habitat types within the seven group foraging ranges is shown
in Table V. As can be seen, no individual range contained all eight
habitat types, and the proportion of the various habitat types differed
markedly between the foraging ranges.

In rural Gloucestershire, Cheeseman (personal communication)
found that at all times of the year badgers actively selected pasture
land as their main foraging habitat, and were observed to spend
approximately twice as much time in this habitat as would be expected if they were foraging randomly and just using each habitat
in proportion to its availability. The same analysis was used here, but
in Bristol it was impossible to keep the badgers under visual observation for more than a few minutes at a time. Since it was easier to
observe badgers foraging in some habitats (such as playing fields)
than others (such as gardens) it would introduce a bias to calculate
the amount of time animals were observed to forage in a particular
habitat. So data obtained by direct observation were combined with
indirect data obtained when badgers were thought to be foraging,
this being determined by their behaviour and any noises heard. The
time spent foraging in each habitat was compared with the availability of that habitat (Fig. 6); allotments were omitted from this
analysis since they were so limited in occurrence. It can be seen that
badgers were using gardens in approximately the proportions they
were available at all times of the year, whereas usage of ornamental
gardens was most pronounced in May and June, when vegetables and
insects were available, and August and September, when windfall
fruit was abundant. Sports fields were only used to a limited extent,
and horse paddocks were utilized approximately in proportion to
their availability. Wooded strips appear to have been actively selected
as a foraging habitat. However, this was probably because 35% of
the setts were located in wooded strips, and these provided a relatively
quiet habitat in which the animals remained until human activity in
the surrounding streets was reduced. Blackberry scrub and waste
land were both utilized slightly more than might have been expected

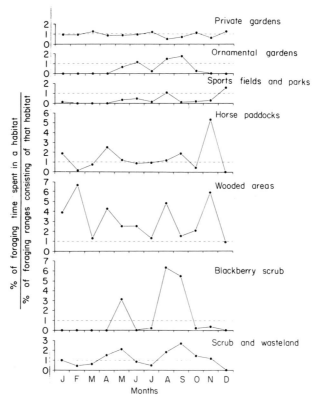

FIG. 6. Comparison of the time spent foraging in each habitat with the availability of that habitat on the group foraging ranges. If the badgers were using a particular habitat in proportion to its availability, a figure of one would be obtained. Figures above one suggest that the badgers were selecting that habitat, figures below one suggest they were avoiding it.

in May and August and September, firstly as a source of *Conopodium* tubers and later as a source of blackberries, but again both these habitats were extremely limited on the foraging ranges.

To determine whether the badgers in each social group were actively selecting any habitats within their group foraging ranges, a Shannon index was calculated for both the habitats available on each group's foraging range, and for the habitats used for foraging by the members of that social group. The diversity of habitats used for foraging increased in proportion to the diversity of habitats available; $y = 0.89x + 0.023$, $r = 0.904$, $P < 0.01$ (Fig. 7). If the badgers were showing no habitat selection at all, each point on the graph would lie on a line with slope $45°$ and variance zero. Comparing the slope of the regression line with that of the $45°$ line, $t = 0.171$, $0.5 < P < 0.9$.

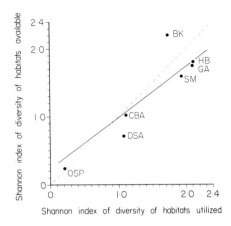

FIG. 7. Comparison of the diversity of habitats available on each foraging range with the diversity of habitats utilized by the members of each social group; $y = 0.89x + 0.023$, $r = 0.904$, $P < 0.01$. A line of slope $45°$ is marked with a dotted line; the slope of the regression line does not differ from that of this $45°$ line (see text). Letters denote the seven social groups as follows: BK, Bishops Knoll; CBA, Coombe Bridge Ave; DSA, Druid Stoke Ave; GA, Glen Avon; HB, Hiatt Baker; OSP, Old Sneed Park; SM, St Monicas.

Since there is no difference between the slopes of these lines, it is concluded that the animals within each social group were utilizing the habitats available on their foraging range in the proportions in which they were available, and were showing no habitat selection.

To examine for any similarities in diversity of habitats available, foraging range size and food selection by the seven social groups, four variables were selected for a principal components analysis (Table VI). The three significant eigenvectors are plotted in Fig. 8;

TABLE VI

Comparison of the seven group foraging ranges; the four variables were used in the principal components analysis (see Fig. 8)

	Foraging range size (ha)	% foraging range consisting of gardens and ornamental gardens	Shannon index of diversity of habitats available	Shannon index of diversity of food eaten
St Monicas	40	84.8	1.593	2.299
Hiatt Baker	81	79.6	1.809	2.320
Druid Stoke Ave	52	87.1	0.717	2.204
Old Sneed Park	42	96.7	0.240	2.192
Coombe Bridge Ave	59	82.6	1.016	2.241
Bishops Knoll	31	45.5	2.202	2.165
Glen Avon	41	60.4	1.744	2.074

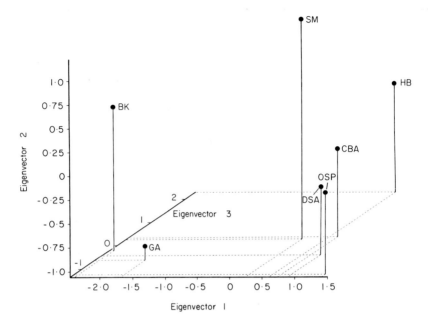

FIG. 8. Principal components analysis grouping the seven social groups according to the characteristics of their foraging ranges. Letters denote the seven social groups as follows: BK, Bishops Knoll; CBA, Coombe Bridge Ave; DSA, Druid Stoke Ave; GA, Glen Avon; HB, Hiatt Baker; OSP, Old Sneed Park; SM, St Monicas.

there were two distinct types of group foraging range on the study area. The Bishops Knoll and Glen Avon ranges were slightly smaller, with a lower percentage of the area devoted to gardens, with a higher diversity of habitat types available, and a lower diversity of food types eaten by the badgers. The other five foraging ranges were slightly larger, a higher proportion of the area consisted of gardens, there was a lower diversity of habitats available, and a higher diversity of food types eaten by the badgers.

SUMMARY

The pattern of food and habitat utilization by the badgers in Bristol is interesting. Most previous authors have agreed that earthworms are either the most or one of the most important food items in the diet of badgers (e.g. Andersen, 1954; Skoog, 1970; Bradbury, 1974; Kruuk & Parish, 1977; Neal, 1977). Kruuk & Parish (1977) and Neal (1977) suggested that the badger is a specialist earthworm eater, and that at least in some areas the availability of earthworms is the important factor determining badger population density. Kruuk *et al.* (1979) have also shown that badgers are most efficient in catching earthworms in short grass, and have suggested that the badger's need for short grass pasture could affect home range size in different areas.

In Bristol, by contrast, earthworms were only the major food item in one month (December). When the relative proportions of the various food items eaten each month were weighted according to how long the animals were active, then fruit was the most important food item for the Bristol badgers. Fruit was the main component of the diet during August to November, when badgers put on fat and increase their body weight by 20–30%, and occasionally up to 60% (Cheeseman, personal communication), prior to their period of reduced activity in December and January. When fruit did not predominate in the diet, the badgers in Bristol did not specialize in taking any one food type, but took a wide range of food items, and to obtain these the animals utilized the full range of habitats available to them. Eating a greater diversity of food types resulted in an increased rate of travel, with less time spent foraging in any one place.

It was surprising that badgers in Bristol were not utilizing earthworms as a food source to a greater extent. Short grassland was available in abundance, both in the form of garden lawns and as sports fields and parks. Visual counts of earthworms feeding on the surface in such habitats showed a mean of five to 10 worms per m^2 on suitable nights, with counts of 15 worms per m^2 and above not uncommon; these figures are comparable to Gloucestershire pastureland (Cheeseman, personal communication). There would appear to be no shortage of earthworms available to the Bristol badgers. Also, besides the grassland available within each foraging range, six of the seven foraging ranges studied adjoined extensive areas of short grass which were not used by any other groups of badgers for foraging. For instance, the foraging ranges of the St Monicas, Hiatt Baker and Bishops Knoll groups all bordered onto the Downs, a 179 ha area of short grassland with extensive earthworm populations (Fig. 2). Although foxes were frequently observed "worming" on this area, only twice did radio tracked badgers emerge onto this area to forage.

In fact urban areas might be expected to provide badgers with an opportunity to "worm" on more nights than in a rural area. Nights most suitable for earthworms to be active on the surface are when the grass air-temperature is above $2°C$, when the soil temperature does not exceed $10.5°C$, and when there has been some rain within the previous four days (Satchell, 1967). Kruuk (1978b) also found that wind was important in reducing earthworm availability. Hannell (1955) found that in Bristol, the mean number of rainy days each month was 14 or more with the exception of June, and mean rainfall per year was 35.8 inches. In towns generally, all the extreme climatic features likely to limit earthworm availability are reduced. There is 5–10% more precipitation in an urban area compared to the rural environs, annual mean wind speed is 20–30% less with 5–20% more calm periods (Landsberg, 1958), and the period between the last frost in spring and the first frost in the autumn may be three to four weeks longer in a city than in the surrounding countryside (Lowry, 1967). Also, during the summer months some garden lawns and sports fields are regularly watered. All these factors should combine to increase earthworm availability to the urban badger. Certainly earthworm chaetae were found in 88.1% of the Bristol dung and stomach samples (*cf.* 82% in Bradbury, 1974). Even in June to August, when earthworms were less frequent in rural dung samples, they were found in 86.5, 97.6 and 91.0% of the Bristol samples. But only rarely were chaetae common in the Bristol dung samples; they were normally present in very low numbers.

The earlier studies in Britain by Neal (1977), Kruuk (1978a,b) and Cheeseman & Mallinson (1980) have suggested that the badger is a territorial animal, living in well-defined social groups, and selectively foraging in habitats in which earthworms, particularly *Lumbricus terrestris*, could be caught on the surface in large numbers. In urban Bristol territorial boundaries were not well defined and foraging areas overlapped with those of adjacent social groups. Although earthworms and short grassland were extensively available, there was also a diversity of other food sources, and for most of the year the badgers did not specialize in eating any one food item, although in the autumn fruit did predominate. The present study suggests that the badger's social organization, foraging behaviour and patterns of habitat utilization may be more variable than the earlier studies have indicated.

ACKNOWLEDGEMENTS

The Nature Conservancy Council funded this work, and the Image Intensifiers were purchased with a Royal Society Scientific Investigations Grant. Part of the survey work was funded by the Manpower Services Commission. I am extremely grateful to these organizations. Messrs G. J. Crowther and N. P. Williams ably assisted with the fieldwork, and Mr W. G. D. Clark and Mrs J. Smith helped with the dung analysis. I am most grateful to Dr C. L. Cheeseman for his advice during the preparation of this manuscript. The maps used to illustrate this chapter have been reproduced by permission of the Geographers' A–Z Map Co. Ltd.

REFERENCES

Andersen, J. (1954). The food of the Danish badger (*Meles meles danicus* Degerbøl) with special reference to the summer months. *Dan. Rev. Game Biol.* 3(1): 1–75.

Asferg, T., Jeppesen, J. L. & Aaris-Sørensen, J. (1977). [The badger (*Meles meles*) in Denmark 1972/73]. *Danske Vildtunders.* 28: 1–56. [In Danish].

Bassindale, R. (1955). Fauna. In *Bristol and its adjoining counties*: 73–90. MacInnes, C. M. & Whittard, W. F. (Eds). Bristol: British Association for the Advancement of Science.

Bonnin-Laffargue, M. & Canivenc, R. (1961). Etude de l'activité du blaireau européen (*Meles meles* L.). *Mammalia* 25: 476–484.

Bradbury, K. (1974). The badger's diet. In *Badgers of Yorkshire and Humberside*: 113–125. Middleton, A. L. V. & Paget, R. J. (Eds). York: William Sessions.

Cheeseman, C. L. & Harris, S. (1982). Methods of marking badgers (*Meles meles*). *J. Zool., Lond.* 197.

Cheeseman, C. L. & Mallinson, P. J. (1980). Radio tracking in the study of bovine tuberculosis in badgers. In *A handbook on biotelemetry and radio tracking*: 649–656. Amlaner, C. J. & Macdonald, D. W. (Eds). Oxford: Pergamon Press.

Cowlin, R. A. D. (1972). The distribution of the badger in Essex. *Essex Nat.* 33: 1–8.

Gill, D. & Bonnett, P. (1973). *Nature in the urban landscape: A study of city ecosystems*. Baltimore: York Press.

Hannell, F. G. (1955). Climate. In *Bristol and its adjoining counties*: 47–65. MacInnes, C. M. & Whittard, W. F. (Eds). Bristol: British Association for the Advancement of Science.

Harris, S. (1977). Distribution, habitat utilization and age structure of a suburban fox (*Vulpes vulpes*) population. *Mammal Rev.* 7: 25–39.

Harris, S. (1980). Home ranges and patterns of distribution of foxes (*Vulpes vulpes*) in an urban area, as revealed by radio-tracking. In *A handbook on biotelemetry and radio tracking*: 685–690. Amlaner, C. J. & Macdonald, D. W. (Eds). Oxford: Pergamon Press.

Harris, S. (1981). An estimation of the number of foxes (*Vulpes vulpes*) in the city of Bristol, and some possible factors affecting their distribution. *J. appl. Ecol.* 18: 455–465.

Humphries, D. A. (1958). Badgers in the Cheltenham area. *Sch. Sci. Rev.* 39: 416–425.

Hunt, P. S. (1976). Anaesthesia of the European badger using ketamine hydrochloride. *Vet. Rec.* 98: 94.

Jayne, A. F. (1978). Avon mammal report, 1976. *Proc. Bristol Nat. Soc.* 36: 79–85.

Jayne, A. F. (1980). Avon mammal report, 1978. *Proc. Bristol Nat. Soc.* 38: x–xvi.

Jayne, A. F. (1981). Avon mammal report 1979. *Proc. Bristol Nat. Soc.* **39**: 65—72.

Kruuk, H. (1978a). Spatial organization and territorial behaviour of the European badger *Meles meles. J. Zool., Lond.* **184**: 1—19.

Kruuk, H. (1978b). Foraging and spatial organisation of the European badger, *Meles meles* L. *Behav. Ecol. Sociobiol.* **4**: 75—89.

Kruuk, H. & Parish, T. (1977). *Behaviour of badgers.* Cambridge: Institute of Terrestrial Ecology.

Kruuk, H., Parish, T., Brown, C. A. J. & Carrera, J. (1979). The use of pasture by the European badger (*Meles meles*). *J. appl. Ecol.* **16**: 453—459.

Landsberg, H. (1958). *Physical climatology* (2nd Edn). DuBois, Pennsylvania: Gray Printing Co.

Lowry, W. P. (1967). The climate of cities. *Scient. Am.* **217** (2): 15—23.

Mackintosh, C. G., MacArthur, J. A., Little, T. W. A. & Stuart, P. (1976). The immobilization of the badger (*Meles meles*). *Br. vet. J.* **132**: 609—614.

Middleton, A. L. V. & Paget, R. J. (1974). *Badgers of Yorkshire and Humberside.* York: William Sessions.

Ministry of Agriculture, Fisheries and Food (1976). *Bovine tuberculosis in badgers.* London: HMSO.

Ministry of Agriculture, Fisheries and Food (1979). *Bovine tuberculosis in badgers: Second report.* London: HMSO.

Ministry of agriculture, Fisheries and Food (1979). *Bovine tuberculosis in badgers: Third report.* London: HMSO.

Neal, E. (1972). The national badger survey. *Mammal Rev.* **2**: 55—64.

Neal, E. G. (1977). *Badgers.* Poole: Blandford Press.

Odum, E. P. (1971). *Fundamentals of ecology* (3rd Edn). Philadelphia: W. B. Saunders.

Satchell, J. E. (1967). Lumbricidae. In *Soil biology*: 259—322. Burges, A. & Raw, F. (Eds). London and New York: Academic Press.

Skoog, P. (1970). The food of the Swedish badger, *Meles meles* L. *Viltrevy* **7** (1): 1—120.

Stirling, E. A. & Harper, R. J. (1969). The distribution and habits of badgers on the southern outskirts of Durham city. *Bull. Mammal Soc.* **32**: 5—6.

Symes, R. G. (1972). Bristol mammal report, 1971. *Proc. Bristol Nat. Soc.* **32**: 141—148.

Symes, R. G. (1973). Bristol mammal report, 1972. *Proc. Bristol Nat. Soc.* **32**: 257—266.

Teagle, W. G. (1967). The fox in the London suburbs. *Lond. Nat.* **46**: 44—68.

Teagle, W. G. (1969). The badger in the London area. *Lond. Nat.* **48**: 48—75.

Tetley, H. (1940). Land mammals of the Bristol district. *Proc. Bristol Nat. Soc.* (4) **9**: 100—142.

DISCUSSION

J. Birks (Exeter University) — What interaction did you observe, if any, between badgers and domestic cats and dogs?

S. Harris — On the 5.5 km² study area there were 1225 pet cats, 1085 pet dogs and 47 adult foxes, potentially capable of producing 96 cubs. Most of the dogs were each confined to a single garden, but occasionally made contact and fought with foraging badgers. The cats, foxes and badgers freely roamed the area, and inter- and intra-specific contacts were frequent. All the interactions one would expect were observed, with no obvious pattern.

D. I. Chapman (Bury St Edmunds) — Is Bristol unusual amongst towns in having a very high density of badgers, and, if so, why?

Harris — The Bristol population density is high but some other towns have comparable populations. In Bristol the badger population is confined to certain limited areas of the city. Much of the surrounding urban area is suitable geologically for badgers and I do not know the reason why badgers are not present in these areas.

T. Parish (ITE) — I should like to comment on the amount of time your urban badgers spend resting each night compared to our rural badgers which do not rest at all.

Harris — I was surprised how inactive they could be. One can only assume that there was a diversity of food available to the badgers and that they did not have to expend too much effort in finding it.

Parish — Perhaps scavenged food is easily found as opposed to earthworms which require considerable effort to catch in quantity.

Harris — True, but since scavenged food is readily available, why is it not utilized more extensively? There are many anomalies in the behaviour of the Bristol badgers which I do not understand. For instance, apart from never seeming to scavenge systematically, neither do they ever spend long searching for earthworms, even when conditions are good. I am surprised the badgers do not forage for earthworms on the areas of grassland that are available to them.

D. W. Yalden (Manchester University) — Is it perhaps an historical accident that badgers are present in your study area in Bristol? Was the habitat there, before suburban development, more suitable than elsewhere?

Harris — Records are very poor and I have assumed that they were there before development took place, but one could equally assume that they should be present in other parts of Bristol which they are not.

D. I. Chapman — I would suggest that the historical explanation is the most likely, i.e. your population was there before development whereas those areas devoid of badgers before development do not have any now.

Symp. zool. Soc. Lond. (1982) No. 49, 325–340

Ranges and Food Habits of Lions in Rwenzori National Park, Uganda

KARL G. VAN ORSDOL

Uganda Institute of Ecology, P.O. Box 22, Lake Katwe, Uganda[*]

SYNOPSIS

A 32-month field study on the African lion (*Panthera leo* L.) in Rwenzori National Park, Uganda, concentrated on range utilization and feeding habits in two ecologically different areas. The first study area was at Mweya in the northern half of the Park where the biomass of potential prey species was 2800 kg km^{-2}. The second study area at Ishasha supported approximately five times the biomass of resident prey species (14,000 kg km^{-2}).

Data on nocturnal movements, hunting and feeding behaviour were collected from radio collared lions of three prides. Radio tracking data indicate that the Mweya pride occupied a range 1.8 times larger than either of the Ishasha prides. The density of lions was approximately five times higher in Ishasha (0.52 km^{-2}) than in Mweya (0.11 km^{-2}). Pride size in Mweya was half that of the Ishasha prides. Data from Ishasha suggest that in this area of high lion density, prides were composed of cohesive subgroups which infrequently interacted and which utilized different parts of the total pride range.

The nocturnal movement patterns and food intake of the three study prides are presented. Mean distances travelled per night varied little between prides, but significant differences did exist in regard to the pattern of movements. Food intake varied seasonally with dry season intake being lower than wet season intake for all three prides. During the dry season, the amount of food consumed by lions in the Mweya pride was approximately 20% below the estimated minimum physiological requirement. The influence of the dry season food shortage on both pride and group size and on cub mortality is examined.

INTRODUCTION

A 32-month field study of the African lion was carried out in Rwenzori National Park in south-west Uganda (Fig. 1). This paper represents preliminary findings from this work; further analyses can be found in Van Orsdol (1981). Study sites were selected in

---

[*] Present Address: Department of Applied Biology, University of Cambridge, Pembroke Street, Cambridge CB2 3DX, England.

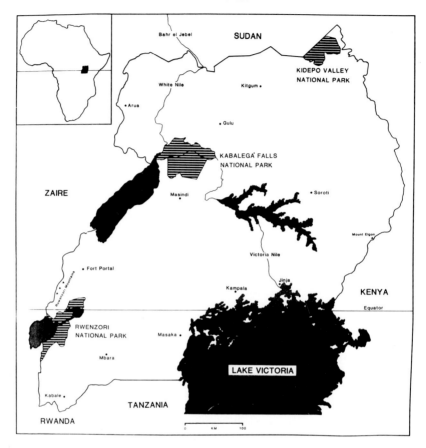

FIG. 1. The location of Rwenzori National Park in Uganda.

the Mweya and Ishasha regions of the Park (Fig. 2). The data were collected from three prides, one in Mweya and two in Ishasha. The latter two are referred to as the Southern Circuit and Northern Circuit prides. Of particular interest to this study was the hunting and feeding behaviour of the lions. As lions are largely nocturnal (Schaller, 1972; Rudnai, 1975; Van Orsdol 1981) it was necessary to use radio telemetering equipment in order to make observations at night. This chapter presents preliminary results from three aspects of the study which particularly benefited from radio telemetry: ranging behaviour, nocturnal movements and food habits.

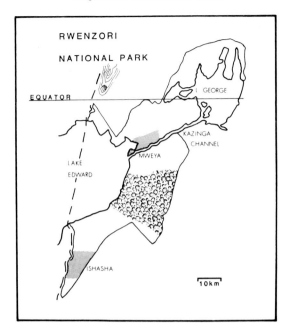

FIG. 2. Rwenzori National Park showing the Mweya and Ishasha study areas (stippled).

STUDY AREAS

The Mweya study area covers approximately 70 km² extending northward from the Kazinga Channel (Fig. 2). The vegetation of the area is dominated by *Sporobolus* and *Themeda* grassland dotted with *Capparis tomentosa* thickets. Grasses are generally short (less than 5 cm) owing to heavy grazing, particularly by the large numbers of hippopotamus (*Hippopotamus amphibius*) found along the shorelines of Lake Edward and the Kazinga Channel (Lock, 1967). *C. tomentosa* is the major source of cover for hunting lions. The major prey species in the area are the buffalo (*Syncerus caffer*), bushbuck (*Redunca redunca*), Uganda kob (*Adenota kob thomasi*), warthog (*Phacocherus aethiopicus*) and waterbuck (*Kobus defassa ugandae*). The prey species biomass is estimated at 2800 kg km⁻² (Table I).

The Ishasha study area covers 80 km² and lies in the southern end of the Park along the border with Zaire (Fig. 2). The area is bordered to the west and north by the Ishasha and Ntungwe Rivers. To the north of the study area lies thick *C. tomentosa* and *C. fasicularis* bushland, which merges with the swampy margin along Lake Edward.

TABLE I

Densities and biomass of prey species in the Mweya Study Area[†]

Species	Average weight[*] (kg)	Density (km^{-2})	(S.D.)	Biomass (km^{-2})
Buffalo	395	4.6	(2.2)	1817
Bushbuck	50	0.1	(0.1)	5
Kob	65	8.0	(8.6)	520
Warthog	50	0.8	(0.3)	40
Waterbuck	160	2.7	(1.8)	432
Total		16.2		2814

[*] Weights taken from Field & Laws (1970).
[†] Data from ground counts carried out by the author and described elsewhere (Van Orsdol, 1981). Seasonal differences in prey abundance are minimal with wet season biomass of 2860 kg km^{-2} and dry season biomass of 2760 kg km^{-2}.

To the east is the Kigezi Game Reserve, a narrow strip of land separating the Park from an area of intensive cultivation and settlement.

The vegetation of the Ishasha study area consists largely of open *Themeda* and *Hyparrhenia* grassland dotted with isolated *Ficus gnaphalocarpa* trees (Lock, 1977). Reduction of the elephant population in the region has resulted in substantial encroachment of open grassland by *Acacia sieberiana* woodland, which now covers roughly 50% of the total area (Yoaciel & Van Orsdol, 1981). The main prey species in the area are the buffalo, bushbuck, kob, topi (*Damaliscus korrigum jimela*), warthog and waterbuck. The biomass of these prey species is estimated at 14,000 kg km^{-2} (Table II).

TABLE II

Densities and biomass of prey species in the Ishasha Study Area[†]

Species	Average weight[*] (kg)	Density (km^{-2})	(S.D.)	Biomass (km^{-2})
Buffalo	395	13.6	(8.1)	5372
Kob	65	75.3	(18.7)	4895
Topi	100	38.1	(13.2)	3810
Warthog	50	0.2	(0.2)	10
Waterbuck	160	0.2	(0.2)	32
Total		127.4		14,119

[*] Weights taken from Field and Laws (1970).
[†] There is little difference between the wet season biomass (14,460 kg km^{-2}) and dry season biomass (13,340 kg km^{-2}).

As the prey species in both study areas are resident, there is little difference between the wet season (March to May and September to October) and dry season (December to February and June to August) biomass estimates (Fig. 3).

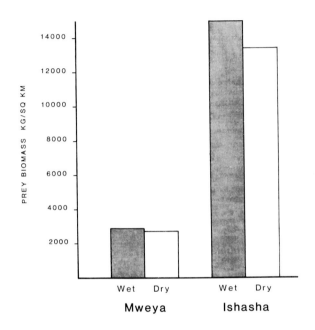

FIG. 3. The wet and dry season biomass estimates of prey species in the two study areas.

METHODS

Two weeks every month were spent in each study area with most observations being made on the three main study prides. All lions within the prides were known individually from scars, earcuts and whisker patterns (Pennycuick & Rudnai, 1970). For each lion sighting, the location was plotted to the nearest 100 m using a 1:50,000 map overlaid by a square kilometre grid system. Apart from the location, a number of environmental variables were recorded and behaviour monitored.

A radio collar was attached to one adult lioness from each pride. A total of nine immobilizations were carried out to fit or replace collars. All collars were prepared in the field. Each contained an AVM SB-2 type transmitter powered by either four 1.35 V mercury or two 1.35 V lithium cells. Dental acrylic was used to protect the transmitter/ battery package which was attached to the lion by 5 mm x 50 mm machine belting. A 30.5 cm whip aerial, covered by an additional layer of belting, ran from the transmitter along the outside of the collar. Two circuit wires were left disconnected and protruded into a small hollow in the acrylic. The collars were then filed and painted, the final electrical connection being made once the lion was immobilized.

Ketamine (CI-581) or Tilezole (CI-744) was used as the anaesthetic. Both drugs are manufactured by Parke-Davis Ltd. and were prepared from a powder form by dissolving in sterile water to a concentration of 200 mg ml^{-2}. The stock solutions were stored in 50 ml multi-dose vials, which were refrigerated except when taken into the field. Neither drug showed signs of precipitation or contamination.

Immobilizations were carried out in the early morning whenever possible to reduce the risk of hypothermia. The drugs were administered by a Cap-Chur dart system. All animals immobilized were treated with a topical antibiotic and injected with a 2.0 g dose of ampicillin.

To compare range use by different subgroups of lions within a pride, the centres of the ranges of each subgroup were obtained by determining the mean X and Y values from the grid co-ordinate data. The 95% confidence limits around the mean were calculated based on the bivariate normal distribution (e.g. Jennrich & Turner, 1969; Koeppl, Slade & Hoffmann, 1975).

The amount of edible meat available from each lion kill was estimated as 75% of carcass liveweight. The age of the victim was determined from tooth wear and the liveweight estimated from growth curves of captive animals or from records of weights of animals shot in the Park. An adult female lion was estimated to require approximately 5 kg of meat per day (Schaller, 1972) to meet physiological requirements.

RESULTS

Ranges

Ranges were estimated using the convex polygon method, which was slightly modified by the exclusion of areas unsuitable for lions. This exclusion was particularly necessary in constructing a polygon for the Mweya pride as the minimum range would otherwise include portions of Lake Edward and the Kazinga Channel. With these modifications, the total estimated range of the Mweya pride was 68 km² (Fig. 4).

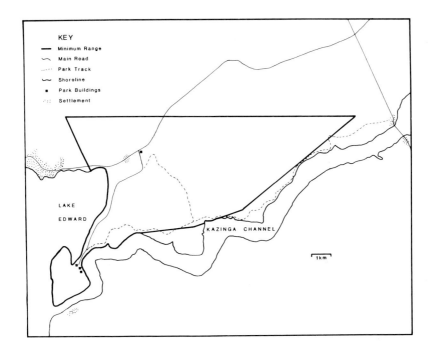

FIG. 4. The minimum range of the Mweya pride.

The minimum range of the Southern Circuit pride was estimated at 31 km² (Fig. 5). This estimate was conservative for it was likely that the actual range extended from the Ishasha River eastward to the escarpment in the Kigezi Game Reserve. If this additional area were included, the estimated range of the pride would be 38 km².

The Northern Circuit pride ranged over an observed 32 km² (Fig. 5) although it was likely that the range covered a minimum of 37 km².

All three pride ranges were partially constrained by the presence of physical or vegetation barriers. In Mweya, the Kazinga Channel in the south and human habitation to the east restricted movements while in Ishasha, lion ranges were limited by the swamp and thicket vegetation to the north and the escarpment and human habitation to the east.

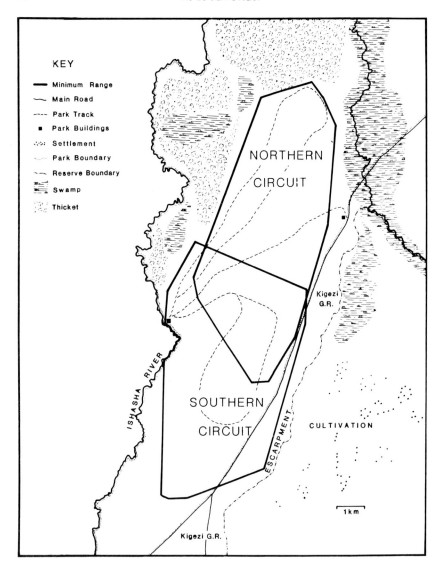

FIG. 5. The minimum range of the Southern Circuit and Northern Circuit prides in Ishasha.

Pride Size and Biomass

Mean pride size in Mweya was 7.7 with a pride biomass of 6.2 female equivalents (Fig. 6). A female equivalent was a unit of biomass based on the weight of an adult female as used by Bertram (1973). Pride

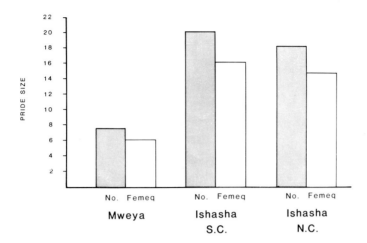

FIG. 6. The size of the three study prides in terms of total numbers of animals (stippled histogram) and biomass in female equivalents (femeq) (open histogram).

size in Ishasha was substantially larger, with the Southern Circuit pride containing a mean of 20.2 lions and a biomass of 16.2 female equivalents. The Northern Circuit pride consisted of a mean of 18.3 lions with a biomass of 14.7 female equivalents (Fig. 6). Using the minimum range estimated above, lion density in Mweya was 0.11 km^{-2} and 0.52 km^{-2} in Ishasha. The corresponding biomass was 0.09 female equivalents km^{-2} and 0.42 female equivalents km^{-2} in Mweya and Ishasha respectively.

Pride Association and Ranging Patterns

Data from both direct observation and radio tracking strongly suggest that prides in Ishasha fragment into semi-permanent subgroups. Such fragmentation may well be the result of high lion density in the area leading to increased aggression at feeding sites, especially when large prey items are rarely killed. Pride fragmentation is particularly advanced in the Southern Circuit.

A dendrogram of the association values (a_n) of the Southern Circuit pride in 1979 (Fig. 7) illustrates the extent of subgroup formation. The pride consisted of four main groups which seldom interacted; the mean intragroup association value was 0.90 as compared to the mean intergroup value of 0.12. Of these four subgroups, two were composed of adult females with cubs, one of three adult females and a sub-adult male and one of two adult males which associated completely with each other and frequently with an adult female and her sub-adult offspring.

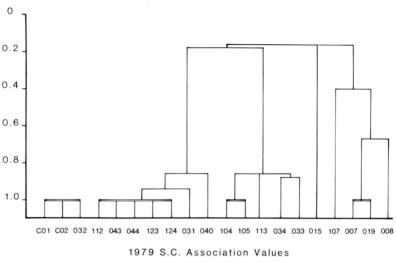

1979 S.C. Association Values

```
000-099 : FEMALE
100-199 : MALE
      C : UNSEXED CUB
```

FIG. 7. Dendrogram of association patterns of the Southern Circuit pride in 1979. The association values (a_n) were calculated as: $a_n = (2N/n_1 + n_2)$ where N is the number of times two animals were observed together and n_1 and n_2 the total number of observations of each animal. Two lions were considered associated if they were within sight of each other, or less than 100 m apart. Two animals consistently observed together would have an association value of 1.0 while animals never seen together would be assigned a value of 0.0.

Did these separate subgroups use different portions of the pride range? To assess differences in range use, the "centres" of the ranges of the four groups were compared (Fig. 8). These centres vary little in the mean X value (east—west) although considerable differences did exist in the Y values (north—south). The group which consisted of adult females Nos 007, 008 and 019 and the sub-adult male No. 107 was found consistently north of the other groups while females Nos 031 and 040 and their cubs Nos 043, 044, 112, 123 and 124 were found furthest south. The ranging centres of the other two groups were intermediate of these extremes.

The underlying assumption of bivariate normality in animal movements was unlikely to be met given the pattern of resource partitioning which influences these movements (Macdonald, Ball & Hough, 1980). The co-ordinate data were tested for normality by the Shapiro-Wilk W and Kolomogorov-Smirnov tests. Nine out of ten X and Y co-ordinate sets departed significantly from normality at the 0.01 level. In view of the departure from normality of the data, further

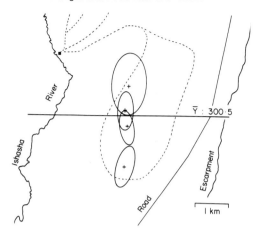

FIG. 8. The range centres, with 95% confidence ellipses of the mean, of the four subgroups of the Southern Circuit pride.

analyses of ranging patterns were carried out using non-parametric techniques.

The mean Y co-ordinate value of the Southern Circuit pride was calculated as 300.5 (Fig. 8) on the grid system. The number of observations of each subgroup north and south of the pride mean were tabulated and compared with the expected number by a chi-square test. These tests strongly suggest that the subgroups used different portions of the pride range with the adult females and cubs found significantly more often south of the mean ($P < 0.001$) and the sub-adult females significantly more often north of the overall mean ($P < 0.001$).

Nocturnal Movements

The Mweya pride travelled a mean of 2.7 km per night (s.d. = 2.1). There was no significant difference between the distances travelled at night during the wet and dry season ($P > 0.05$). In Ishasha, the Southern Circuit pride travelled a mean of 2.4 km (s.d. = 1.4) at night with no significant differences between seasons ($P > 0.05$). The Northern Circuit pride travelled on average slightly less per night (mean = 2.2 km, s.d. = 0.76) than the other two prides.

The pattern of nocturnal movements varied significantly between prides. The distances travelled by each pride were lumped into units of one kilometre and compared by a chi-square test. The differences

between the Southern and Northern Circuit prides were not significant $(P > 0.05)$ and the data were pooled for comparison with the
Mweya pride. This difference was highly significant $(P < 0.001)$. The
largest difference between the study areas was in the disproportion-
ally large number of nights on which the Mweya pride travelled less
than 1.0 km $(P < 0.02)$.

Food Intake

To estimate the level of food intake of lions, it was assumed that the
amount of edible meat on a carcass was 75% of the liveweight of
animals up to 100 kg, 65% for animals between 101 and 250 kg and
60% for animals greater than 250 kg. The rate of food intake of lions
was calculated as:

$$\frac{\text{weight of meat available per carcass}}{\text{lion biomass (in female equivalents) feeding}} \times \frac{1}{\text{interval between kills (days)}}$$

These results suggest that the study prides experienced a lower
rate of food intake during the dry season than in the wet (Fig. 9).
In Mweya, the pride was observed to kill once every 4.0 days, with
a mean biomass of lions feeding at these kills of 3.0 female equivalents in the wet season and 3.6 in the dry season. The mean amount
of edible meat per feeding bout was higher in the wet season (81 kg)

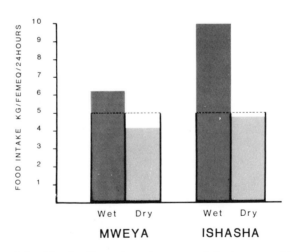

FIG. 9. Seasonal food intake. The estimated amount of food consumed by lions in the
two study areas in the wet and dry season (stippled) in terms of kg of meat female
equivalent^{-1} 24 h^{-1}. The 5 kg female equivalent^{-1} 24 h^{-1} histogram represents the estimated physiological minimum requirements of an adult lioness.

than in the dry season (50 kg) although this difference was not significant ($P > 0.05$). The resulting estimates of food intake were 5.4 kg of meat per female (or female equivalent) per day in the wet season and 4.2 kg meat per female equivalent per day in the dry season.

Food intake of lions in Ishasha was considerably higher than of those in Mweya. This difference is due to more frequent kills being made (once every 3.1 days) and a larger amount of meat available per kill (94 kg). As a result, food intake during the wet season was estimated at 10.0 kg per female equivalent per day and 4.8 kg during the dry season (Fig. 9).

DISCUSSION

The data presented above suggest significant variation in lion ranging patterns, social relationships and feeding ecology between two areas of strikingly different prey abundance. To attribute the differences in ranges or pride size to prey density alone would ignore the variety of factors which influence the "catchability" (Bertram, 1973) of prey and thus their potential availability as food items. Results not presented here suggest, for example, that vegetation and prey group size significantly influence hunting success of lions.

The dichotomy in both lion and prey densities between study areas presents an opportunity to examine some of the ecological and social factors which influence grouping patterns. Group size varied little between prides and there was no correlation between pride size and group size ($r_s = -0.06$, $P > 0.05$). The size of prey items has been suggested as an important determinant of lion group size (Kleiman & Eisenberg, 1973; Caraco & Wolf, 1975). Aggression at feeding sites would accelerate pride fragmentation, particularly if large carcasses were not readily available. The similarity in observed mean group size between the three prides may well be the result of characteristics of carcass weights. The modal carcass weight of kills in the Mweya pride was between 40 and 50 kg. Lions feeding largely on animals of this size and killing at the rate of once every four days, as observed in Mweya, could only survive in groups of less than 1.9 female equivalents. The observed feeding group size in the Mweya pride of 3.1 female equivalents was only energetically feasible because the pride occasionally killed larger animals such as waterbuck which weigh up to 240 kg.

In Ishasha, the upper limits of group size also appeared to be set by carcass weight. The two main prey species in this area are the kob and topi which have mean liveweights of 65 and 100 kg respectively. Assuming a kill rate of once every 3.1 days, lion groups feeding solely on kob could be as large as 3.1 female equivalents and still meet the minimum physiological requirements. Groups feeding on topi, on the other hand, could be as large as 4.8 female equivalents. The observed group sizes of lions feeding on kills closely approximate those predicted from carcass weight distribution and kill frequency. The Southern Circuit pride, which feeds largely on kob, had an observed group size of 3.0 while the Northern Circuit pride feeding on topi had an observed group size of 4.1. This also suggests that the high prey densities could support more subgroups within an area, and thus allow for larger prides and higher lion density, but not necessarily larger groups.

Food intake was lowest for prides in both study areas during the dry season. The Mweya pride consumed about 4.2 kg female equivalent^{-2} 24 h^{-1} during the dry season, a value approximately 20% below the estimated minimum requirements. As the relatively higher level of energetic requirements of cubs and sub-adults was not accounted for in the above calculations, the actual level of food shortage experienced by the pride was greater than that suggested by the data. Furthermore, unlike the prides in Ishasha, the Mweya pride did not experience a surplus of food during the wet season (Fig. 9).

The effect of the differences in food intake was manifested in the levels of cub mortality. The number of cubs dying before 18 months of age was significantly higher in Mweya than in Ishasha (Fisher Exact Prob. Test, $P < 0.05$) (Fig. 10). While sample size is small, the data

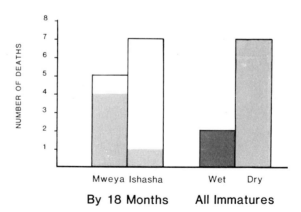

FIG. 10. Cub mortality. Left: The number of cubs born (histogram) and dying (shading within histogram) before 18 months of age in the two study areas. Right: The number of immatures dying in the wet and dry seasons (data from both study areas combined).

are indicative of the more adverse conditions in Mweya. A comparison of the time of death of all immature animals suggests higher mortality in the dry season (Fig. 10). This difference in seasonal mortality is not significant (Fisher Exact Prob. Test, $P > 0.05$), but the periods of low food intake in both study areas coincide with the period of higher cub mortality.

The use of radio tracking equipment has been particularly useful in observing nocturnal movements. Why did the Mweya pride frequently travel distances of less than one kilometre per night when the prey density, and their level of food intake, was low? It is likely that in this area of low prey density, travelling longer distances would not be energetically rewarding. The main prey species of the Mweya pride was the warthog, a diurnal species generally not available to lions at night. The rate of locating suitable prey at night is probably not substantially increased by moving over large distances. Environmental factors may also decrease the chances of lions successfully hunting prey. During this study lion hunts were more successful at night when the moon was below the horizon than when it was above ($P < 0.01$), and the distances lions travelled at night correlated with cloud cover ($r_s = 0.48, P < 0.001$). This suggests that the moon and stars provide sufficient light for prey to detect hunting lions.

CONCLUSIONS

Observations of radio collared lions indicate substantial variation in range size and pride size between areas with large differences in prey abundance. Under conditions of high prey and lion densities, prides may fragment into cohesive subgroups, which utilize different portions of the total range. The subgroups are composed of animals of similar age and sex. The grouping patterns of prides also appear to be influenced by prey size with lion group size limited in part by prey carcass weight. The three study prides experience shortages of food in the dry season. This shortage is particularly severe in the Mweya pride, which is unable to maintain its physiological requirements during that time of year. The shortage of food in Mweya during the dry season results in higher cub mortality than in Ishasha. The data also suggest that dry season cub mortality is higher than wet season mortality in both study areas.

ACKNOWLEDGEMENTS

The cooperation and support of the Uganda Institute of Ecology, particularly its director Dr E. Edroma, the National Research Council, Uganda National Parks and the staff of Rwenzori National Park are gratefully acknowledged. Messrs Parke-Davis and Co. Ltd. kindly supplied the Ketamine and Tilezole for use in the immobilizations. Drs T. T. Struhsaker and S. K. Eltringham provided valuable advice throughout the course of this study, for which I am most grateful. Dr J. P. Rood generously gave me his file of lion identification cards. Dr P. Moehlman and D. Ray gave helpful comments and criticisms during the preparation of this chapter.

REFERENCES

Bertram, B. C. (1973). Lion population regulation. *E. Afr. Wildl. J.* 11: 215—225.

Caraco, T. & Wolf, L. (1975). Ecological determinants of group sizes in foraging lions. *Am. Nat.* 109: 343—352.

Field, C. R. & Laws, R. M. (1970). The distribution of larger herbivores in Queen Elizabeth National Park Uganda. *J. appl. Ecol.* 7: 273—294.

Jennrich, R. I. & Turner, F. B. (1969). Measurement of non-circular home range. *J. Theoret. Biol.* 22: 227—237.

Kleiman, D. G. & Eisenberg, J. F. (1973). Comparisons of canid and felid social systems from an evolutionary perspective. *Anim. Behav.* 21: 637—659.

Koeppl. J. W., Slade, N. A. & Hoffmann, R. S. (1975). A bivariate home range model with possible application to ethological data analysis. *J. Mammal.* 56: 81—90.

Lock, J. M. (1967). *Vegetation in relation to grazing and soils in Queen Elizabeth National Park Uganda.* Ph.D. Thesis: Unversity of Cambridge.

Lock, J. M. (1977). The vegetation of Rwenzori National Park, Uganda. *Bot. Jb.* (Syst.) 98: 372—448.

Macdonald, D. W., Ball, F. G. & Hough, N. G. (1980). The evaluation of home range size and configuration using radio tracking data. In *A handbook on biotelemetry and radio tracking*: 405—424. Amlaner, C. J. & Macdonald, D. W. (Eds). Oxford: Pergamon Press.

Pennycuick, C. & Rudnai, J. (1970). A method of identifying individual lions, *Panthera leo*, with an analysis of the reliability of identification. *J. Zool., Lond.* 160: 497—508.

Rudnai, J. A. (1975). *The social life of the lion.* Lancaster: Medical and Technical Publ. Co.

Schaller, G. B. (1972). *The Serengeti lion.* Chicago: University Press.

Van Orsdol, K. G. (1981). *Predator-prey relationships in Rwenzori National Park, Uganda.* Ph.D. Thesis: Unversity of Cambridge.

Yoaciel, S. M. & Van Orsdol, K. G. (1981). The influence of environmental changes on an isolated topi (*Damaliscus lunatus jimela* Matschie) population in the Ishasha Sector of Rwenzori National Park, Uganda. *Afr. J. Ecol.* 18: 167—174.

Symp. zool. Soc. Lond. (1982) No. 49, 341—352

Leopard Ecology as Studied by Radio Tracking

B. C. R. BERTRAM

The Zoological Society of London, Regent's Park, London NW1, England

SYNOPSIS

Three leopards were radio tracked in the northern Serengeti over a total of 12 leopard months. Great care was taken to get and keep the animals easily observable. Comparison was possible with radio tracked lions in precisely the same area at the same time. An adult female leopard occupied a range (probably a territory) of 15.9 km², 57% as big as the territory of the largely co-resident lion pride of four females. Leopards fed on a wider range of prey species than lions, with little overlap; separation was mainly on the basis of size.

INTRODUCTION

The leopard (*Panthera pardus*) is still a relatively unstudied animal in the wild. Its secretive and usually timid nature and generally nocturnal habits make it a difficult animal both to find and to observe. The species has been the main subject of periods of study in Sri Lanka (Muckenhirn & Eisenberg, 1973), in Kenya (Hamilton, 1976), and in South Africa. In addition, incidental observations of leopards have been published by observers of other predator species (Schaller, 1972; Kruuk, 1972, 1975). From these reports, it is clear that in some respects leopard ecology and behaviour are broadly similar to those of some of the other large predator species. The leopard's closest relative as well as one of its closest competitors is the lion (*Panthera leo*), weighing 120—180 kg (Bertram, 1975a) compared with the leopard's 35—55 kg (King, Bertram & Hamilton, 1977). Lions are found over a large part of the leopard's geographical range in Africa (Myers, 1976). They often live in the same habitat. It is known that both lions and leopards feed on a range of species of mammals, but the extent of competition between the two species has been unknown. Studies of lions in one region and of leopards in another can tell one little about how different or how similar are the ecological niches of the two species. It is possible that the niche of the leopard can be best described by comparing it with that of the much more extensively studied lion (Schaller, 1972; Bertram, 1973, 1975a, b, 1979; Rudnai, 1974).

This chapter therefore reports the results of a study of leopards which was undertaken concurrently with a study of lions. For the first time, the inter-species comparisons made are between animals living in exactly the same habitat, selecting food from among the same spectrum of potential prey species, and being observed at the same time and by the same methods.

The four-year study of lions in the Serengeti National Park in Tanzania was conducted from 1969 to 1973 (Bertram, 1973, 1975b, 1976a, 1980; Bygott, Bertram & Hanby, 1979). Most of the information reported in this chapter was collected within a woodland study area at Lobo at the north-east edge of the National Park in the second halves of 1972 and 1973. Descriptions of the environment have been given by Schaller (1972) and Sinclair (1977, 1979).

METHODS

The effective study of leopards in the northern Serengeti depended totally upon radio tracking. The equipment requirements for the radio tracking programme in the Serengeti have been described elsewhere (Bertram, 1980); the programme made use almost entirely of the SM2 transmitters and LA12 receiver produced by the AVM Instrument Company of Champaign, Illinois. For use on leopards, the transmitters were sealed inside a dental acrylic block which hung at the bottom of a machine-belting collar; the details of the construction have been described elsewhere (Bertram, 1976b).

The leopard to be collared, when eventually discovered, was darted at a range of up to 25 m. Approaching to within this range was impossible with some leopards, and where it was possible it took a great deal of time and patience. It was essential to avoid alarming or disturbing the animal. The leopard was anaesthetised with C1-744 (Bertram & King, 1976; King, Bertram & Hamilton, 1977). While it was sedated, the collar was fitted and tested and the animal was weighed and measured (Bertram, 1975b).

Tracking, with a three-element Yagi aerial, was done almost entirely from the ground, making use where possible of high vantage points such as rock outcrops and the roof of the vehicle. Signals were first detected at ranges ranging usually between 0.5 and 10 km, depending on the activity and particularly on the height of the leopard above the ground.

When the loudness of the signal indicated that a leopard was nearby, it was imperative not to frighten it by noisily taking the vehicle too close to its resting or hiding place. When eventually located visually, the animal was watched at distances of 20—200 m. Data were collected on its location, the vegetation type it was in, its companions, if any, any prey or signs of feeding, and behaviour. Most observations were made in the early morning. At night, lights were used if necessary to find the leopard or to identify its prey. Observation periods during the day lasted between 0.25 and 4 h, depending on the activity of the animal and on how many other leopards and lions needed to be located that day.

There were four main advantages resulting from the use of radio tracking in studying leopards. First, there was an enormous increase (of the order of thirty-fold) in the rate of sightings of leopards — from around one per two to three weeks on average, to one or two per day. Second, because the same individual was found in each case, continuous data could be obtained which made it possible, for example, to plot range size and use pattern, to study food preferences, and to examine changes in social behaviour with age. Third, it became possible to watch the animal at close range, because it quickly became habituated to the presence of the vehicle. Provided that disturbance was minimized when darting and when tracking, the leopard ceased to avoid the vehicle after a very few observations. It could therefore be observed easily while it was behaving in a normal manner nearby. Fourth, and conversely, radio tracking also made possible observations from further away. Because the animal could always be re-located, it could be allowed to disappear from sight for considerable periods, for example in the course of its hunting. This is particularly necessary for an animal such as the leopard which relies for its success in catching its food on being able to ambush or to stalk up to unalerted prey animals.

Three leopards were radio collared and observed subsequently: a 2¾ year-old male during June to October 1972, a 1¾ year-old female during May to October 1973, and the latter's mother, an adult female, during July to October 1973.

RESULTS

Range

Figure 1 shows the area used by the adult female leopard and her female cub during a five-month period in 1973. It can be seen that both animals spent their time, in so far as could be determined, within an area of 15.9 km².

A small proportion of the points shown in Fig. 1 are where the

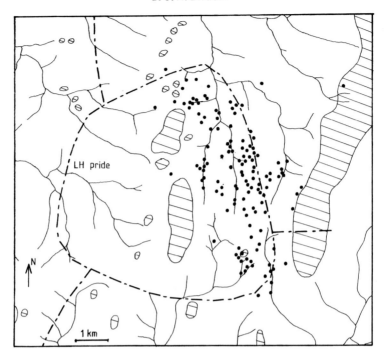

FIG. 1. Range of an adult female leopard and her full-grown female cub over five months.
● indicates the location of the first sighting of one or both individuals each day. — — — indicates the territory boundaries of resident lion prides. ▢ indicates hills or rocks.

two animals were together, usually at a kill. Most are of one individual alone. One point is well outside the edge of the range indicated in
Fig. 1; on that morning the adult female was coming into season and
mating with an unknown adult male; on the subsequent four days
she mated with him inside her normal range. The points plotted are
those at which the individual was first located on any day. No further points were plotted for that animal on that day, regardless of
how far she moved. Thus the points gave a very rough indication of
the intensity of usage of different parts of the range. It can be seen
that the area was in fact used fairly evenly. The edges of the range
were slightly less frequented, particularly by the juvenile female. The
streams and drainage lines indicated in Fig. 1 are slight, and were all
dry during the period of study. They could not in any way restrict
the animals' movements, and it can be seen that they did not act as
boundaries. On the other hand, they were frequented for a dispro-
portionate amount of time, because they did tend to provide shady
refuges in which the leopards often spent the middle of the day, and

large trees suitable for leopards to rest in were more common near the streams. The steep hillside along the south-east corner of the range apparently did form a boundary, despite the fact that leopards could walk up it without difficulty.

The young male radio tracked in the previous year kept to an area of 17.8 km² ; this was probably the size of the range of the adult female presumed to be his mother, whom he fairly often accompanied (19% of observations).

From night observations, there was no indication that the leopards were going further away from the centre of their ranges at night; thus the areas plotted in Fig. 1, although made up predominantly of day-time observations, probably do represent the full extent of those individuals' ranges.

It was not possible to radio-track two adjacent adult females con-currently, so the extent of range overlap or separation is unknown. Almost nothing was known about neighbouring leopards. No other adult female was observed inside the radio collared female's range during the four months she was being tracked, but sightings of leopards without collars were so occasional (four during those four months, all outside her range) that little can be deduced about territoriality. Indications from elsewhere (Schaller, 1972; Hamilton, 1976; Bertram, unpublished observations) suggest that the ranges of adult females scarcely overlap one another.

Comparisons with Lion Ranges

Figure 1 shows also the edges of the range occupied by the four females of a pride of lions (the LH pride) over a period of 2¾ years. This period included the four months during which the female leo-pard was being tracked, and during those four months the lions used virtually the whole of their range shown in Fig. 1. Parts of the terri-torial boundaries of three other adjacent prides of lions are also shown in Fig. 1.

It can be seen that the two large cat species were occupying and using the same area, despite the fact that the LH lion pride was ex-cluding the other lion prides from its range and the leopard was probably excluding other female leopards from hers. The leopard's range included an intersection of the boundaries of three lion prides' territories. There is no relation between the boundaries of the lion territories and of the leopard's range.

The LH lion pride's territory was considerably larger than that of the leopard (27.9 versus 15.9 km²). However, because there were four adult females in the lion pride, the area per adult was smaller for the lions than for the leopard.

The comparison so far has been between only one social group of each species, because these were the individuals whose ranges were superimposed on top of one another. There was only scanty information (above) on the range sizes of other leopards nearby. However, abundant information was collected on the areas used by the three neighbouring lion prides. As shown in Table I, these prides all contained more adult females and occupied larger territories. Bigger prides tended to occupy bigger territories. None the less, the area available per adult female did not vary greatly, and was again considerably smaller than for the leopards.

TABLE I

The areas occupied by leopards and lions

	Leopards		Lions			
Identity	♀ + daughter	Young ♂ (+ mother)	LH	LS	GN	LN
Area used (km²)	15.9	17.8	27.9	48.7	42.4	66.9
Mean number of adult ♀♀	1	1	3.4	5.8	5.3	9.9
Area per adult ♀ (km²)	15.9	17.8	8.2	8.4	8.0	6.8

Food Taken

Published lists (for example Kruuk & Turner, 1967; Pienaar, 1969; Makacha & Schaller, 1969; Schaller, 1972; Rudnai, 1974) of the prey species taken by leopards and by lions overlap greatly — most prey species are somewhere or at some time eaten by each of the two predator species. However, both species are known to be adaptable and variable in their feeding. They can only take what is available, and this varies from place to place, and often from one season to another. There has been no information on the extent to which at the same place and at the same season the two large cats differ in the prey they take. It has now proved possible to determine this by radio tracking the lions and leopards as described above.

Figure 2 shows the distribution of prey species taken by the three radio tracked leopards and by the lions of the LH pride during the period that the leopards were being radio tracked. The range of prey species available to the two predator species was the same, because the ranges the predators occupied were superimposed. It can be seen that at the same place and at the same time the overlap in the diet of the two species was slight — only two of the 20 prey species were taken by both predators, and in neither case in large numbers by both.

The main reason for the divergence in diet is shown in Fig. 2

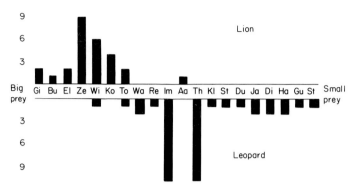

FIG. 2. Distribution of kills of radio collared leopards and of a co-resident lion pride, Columns indicate the number of kills of each species of prey. The prey species are arranged in decreasing order of size, vis. giraffe, buffalo, eland, zebra, wildebeest, kongoni, topi, warthog, reedbuck, impala, aardvark, Thomson's gazelle, klipspringer, steinbock, duiker, jackal, dikdik, hare, guinea fowl, starling.

where the prey species taken are arranged in roughly descending order of adult size. It can be seen that lions were on the whole taking the larger prey species. The single wildebeest (*Connochaetes taurinus*) and topi (*Damaliscus korrigum*) taken by leopards were both young animals, about four months old and almost newborn respectively. By contrast, most (75%) of the lion prey were adult. Adults of these larger prey species appear to be largely invulnerable to the much smaller leopard, which in many cases made no attempt to stalk them in circumstances where lion groups would have done so.

There are indications in Fig. 2 that leopards fed on a wider range of different prey species than lions did. This is reinforced by kill data collected less systematically over the whole of the Serengeti over several years by a number of different observers and gleaned from Serengeti Research Institute records. The sample of only 137 leopard kills included 31 different prey species, whereas Schaller's (1972) sample, for example, of 1180 lion kills included only 22 species.

Observations of leopards hunting could rarely be made except in the daytime. Almost invariably the leopard hunted alone. It was apparent to the observer that the major problem the leopard had to overcome was a considerable one — namely to get within a few yards of an unalerted prey animal. Overall, only three (5%) of 64 daytime hunts by radio collared leopards were successful. A "hunt" was defined as an orientation and an approach towards a particular prey animal. Thus many hunts ended because the potential prey wandered away, totally oblivious of the leopard. It is likely that the hunting success rate of fully adult animals only, and at night when most hunting occurs, is somewhat higher.

Leopards were observed to use three methods for getting to within range of their prey. By far the most common was stalking. Having detected potential prey, the leopard approached it slowly and stealthily, body low to the ground, making use of cover, and advancing only when the prey animal was not looking. Sometimes, the leopard was resting inactive when it first detected its prey, in which case, if it was concealed at a suitable place such as near a stream crossing, it remained there and in effect used the second hunting method, namely ambushing. Ambushing was rare, however, probably because there were few places in the environment where prey animals would come predictably. The third hunting method was a little more common and largely opportunistic. In the course of its travels the leopard investigated clumps of vegetation and dashed after any small animal (particularly hares (*Lepus capensis*)) flushed out of them.

DISCUSSION

As a result of radio tracking, the following conclusions can be drawn about leopards in the Serengeti region. The species is essentially solitary, occupying a relatively small range from which it probably excludes other adults of the same sex. Hamilton's (1976) data, mainly on male leopards in Tsavo West National Park in Kenya, agree with these findings.

Leopards feed on a variety of small prey species which they catch by hunting alone. The work reported here showed that at the same place and at the same time there was little overlap with lions in the prey species taken. However, these observations were made during the dry seasons when the study area contained large numbers of migratory prey species, particularly the wildebeest and zebra (*Equus burchelli*), which together comprised more than 50% of the lions' diet. For about three months each year, in the wet season, the study area was empty of migratory species, and the total prey biomass was down to about 30% of that during the dry season. During the wet season the lions were feeding more on impala (*Aepyceros melampus*) and warthog (*Phacochoerus aethiopicus*). Competition between lions and leopards would therefore have been greater during the wet season than was observed during the dry season. On the other hand, it is possible that conditions in the wet season made hunting easier for leopards.

The range sizes measured for leopards were considerably smaller than those of 40–60 km² quoted by Schaller (1972) in the Seronera region of the Serengeti. The latter region consisted of a few rock outcrops and streams and their tributaries, which formed good leopard habitat, separated by large areas of open grassland which was scarcely suitable leopard habitat. Joining the outermost points on each stream at which each leopard was seen encloses an area containing a large proportion of unsuitable habitat. By contrast, in the woodland study area, the radio tracked leopards made use of the whole of their range as described.

As was shown in Table I, leopards in the woodland study area used smaller ranges than lions, but because they were solitary animals, they occurred at lower density than did lions. This is at first sight surprising — one might expect that the smaller predator would occur at higher density, particularly as it was feeding on a wider variety of smaller prey species so its food would likewise be expected to be more abundant. The ecosystem may be unusual in having only a small proportion of its herbivores neither very large nor very small (Sinclair, 1975). Of the ungulate herbivores in the study area in the dry season, an estimated 80% of individuals were of a size too large for leopards to take. Lions by contrast could prey upon them, partly because of the lion's much greater body size and partly because of the lion's social hunting methods which result in an improved hunting success rate for larger ungulates (Schaller, 1972; Caraco & Wolf, 1975). The biomass of rodents and lagomorphs is relatively small, whereas that of the invertebrate herbivores is very high (Sinclair, 1975). Considering the productivity of the area, there is a relatively meagre supply of food between 2 and 20 kg in weight, which is the most practicable size range for leopards.

A surprisingly high proportion (9%) of leopard kills were of species which were themselves predators, such as the whitetailed mongoose (*Ichneumia albicauda*), genet (*Genetta* sp.), and jackals (*Canis mesomelas*), and the offspring of lions, spotted hyaenas (*Crocuta crocuta*), cheetahs (*Acinonyx jubatus*) and wild dogs (*Lycaon pictus*). It is not clear why these carnivores should themselves apparently be so vulnerable to predation.

The interactions between leopards and the other large predator species are complicated. Radio collared leopards were seen chasing, but also being chased by, cheetahs and hyaenas. Cheetahs and wild dogs were both so scarce as to be unimportant in the study area. Spotted hyaenas deprived leopards of parts of some of their larger kills or scavenged carcasses, and are probably one reason why leopards kill small prey and why they carry that prey up into trees if possible. The presence of hyaenas, with their adaptions as highly efficient scavengers (Kruuk, 1972), probably helps to prevent leopards from getting a significant part of their food by scavenging.

Lions were observed to chase leopards whenever they saw them. In these cases the leopard invariably escaped safely into a tree, although it may not always succeed in doing so (Schaller, 1972). The presence of trees or rocks as retreats is presumably what makes it possible for leopards to co-exist with their larger rivals. Lions are able to keep their territories empty of other lions, but they cannot exclude the more agile and more arboreal leopards. As we have seen, ecological competition between the two species is relatively slight, so there would be little real advantage in their trying to do so. The presence of lions, killing larger prey, probably helps to set the boundaries of the leopard's ecological niche, as the tiger (*Panthera tigris*) apparently does in Asia (Seidensticker, 1976). The puma (*Felis concolor*) in North America, without a sympatric larger cat species, takes disproportionately larger prey than leopards do (Hornocker, 1970). In other respects, the leopard's territorial system apparently resembles that of the puma (Seidensticker *et al.*, 1973).

ACKNOWLEDGEMENTS

I am grateful to the Directors and staff of Tanzania National Parks and the Serengeti Research Institute for permission, assistance and facilities for my research in the Serengeti; to the African Wildlife Leadership Foundation, the Royal Society, and the Natural Environment Research Council for financial support; and to Mr Royal Little and the New York Zoological Society for a grant for radio tracking equipment.

REFERENCES

Bertram, B. C. R. (1973). Lion population regulation. *E. Afr. Wild. J.* 11: 215–225.
Bertram, B. C. R. (1975a). Weights and measures of lions. *E. Afr. Wildl. J.* 13: 141–143.
Bertram, B. C. R. (1975b). Social factors influencing reproduction in wild lions. *J. Zool., Lond.* 177: 463–482.
Bertram, B. C. R. (1976a). Kin selection in lions and in evolution. In *Growing points in ethology*: 281–301. Bateson, P. P. G. & Hinde, R. A. (Eds). Cambridge: Cambridge University Press.
Bertram, B. C. R. (1976b). *Studying predators*. Handbk. 3. Nairobi: African Wildlife Leadership Foundation.

Bertram, B. C. R. (1979). Serengeti predators and their social systems. In *Serengeti: dynamics of an ecosystem*: 221—248. Sinclair, A. R. E. & Norton-Griffiths, M. (Eds). Chicago: Chicago University Press.

Bertram, B. C. R. (1980). The Serengeti radio tracking program, 1971—73. In *A handbook on biotelemetry and radio tracking*: 625—631. Amlaner, C. J. & Macdonald, D. W. (Eds). Oxford: Pergamon Press.

Bertram, B. C. R. & King, J. M. (1976). Lion and leopard immobilization using C1-744. *E. Afr. Wildl. J.* 14: 237—239.

Bygott, J. D., Bertram, B. C. R. & Hanby, J. P. (1979). Male lions in large coalitions gain reproductive advantages. *Nature, Lond.* 282: 839—841.

Caraco, T. & Wolf, L. L. (1975). Ecological determinants of group sizes of foraging lions. *Am. Nat.* 109: 343—352.

Hamilton, P. H. (1976). *The movements of leopards in Tsavo National Park, Kenya, as determined by radio-tracking.* M.Sc. Thesis: Nairobi University.

Hornocker, M. (1970). An analysis of mountain lion predation upon mule deer and elk in the Idaho Primitive Area. *Wildl. Monogr.* 21: 1—39.

King, J. M., Bertram, B. C. R. & Hamilton, P. H. (1977). Tiletamine and Zolazepam for immobilization of wild lions and leopards. *J. Am. Vet. Med. Assoc.* 171: 894—898.

Kruuk, H. (1972). *The spotted hyena.* Chicago: Chicago University Press.

Kruuk, H. (1975). Functional aspects of social hunting by carnivores. In *Function and evolution in behaviour*: 119—141. Baerends, G., Beer, C. & Manning, A. (Eds). Oxford: Oxford University Press.

Kruuk, H. & Turner, M. (1967). Comparative notes on predation by lion, leopard, cheetah and wild dog in the Serengeti area, East Africa. *Mammalia* 31: 1—27.

Makacha, S. & Schaller, G. B. (1969). Observations on lions in the Lake Manyara National Park, Tanzania. *E. Afr. Wildl. J.* 7: 99—103.

Muckenhirn, N. A. & Eisenberg, J. F. (1973). Home ranges and predation of the Ceylon leopard (*Panthera pardus fusca*). In *The world's cats* 1: 142—175. Eaton, R. L. (Ed.). Winston, Oregon: World Wildlife Safari.

Myers, N. (1976). *The leopard* Panthera pardus *in Africa.* [I.U.C.N. Monograph 5.] Morges, Switzerland: International Union for the Conservation of Nature and Natural Resources.

Pienaar, U. de V. (1969). Predator prey relations amongst the larger mammals of the Kruger National Park. *Koedoe* 12: 108—176.

Rudnai. J. (1974). The pattern of lion predation in Nairobi Park. *E. Afr. Wildl. J.* 12: 213—225.

Schaller, G. B. (1972). *The Serengeti lion.* Chicago: Chicago University Press.

Seidensticker, J. (1976). On the ecological separation between tigers and leopards. *Biotropica* 8: 225—234.

Seidensticker, J. C., Hornocker, M. G., Wiles, W. V. & Messick, J. P. (1973). Mountain lion social organisation in the Idaho Primitive Area. *Wildl. Monogr.* 35: 1—60.

Sinclair, A. R. E. (1975). The resource limitation of trophic levels in tropical grassland ecosystems. *J. Anim. Ecol.* 44: 497—520.

Sinclair, A. R. E. (1977). *The African buffalo.* Chicago: Chicago University Press.

Sinclair, A. R. E. (1979). The Serengeti environment. In *Serengeti: dynamics of an ecosystem*: 31—45. Sinclair, A. R. E. & Norton-Griffiths, M. (Eds). Chicago: University Press.

DISCUSSION

M. J. Delany (Bradford) — We have seen a separation of food size with regard to the lion and the leopard. Could you tell us where the cheetah fits into this picture?

B. C. R. Bertram — There is certainly a habitat difference, with cheetahs inhabiting more open country. There is also a difference in the mode of hunting. Cheetahs fed largely on Thomson's gazelles and hares.

Author Index

Numbers in italics refer to pages in the References at the end of each article

A

Aaris-Sørensen, J., 302, *321*
Ables, E. D., 266, *284*
Akande, M., 232, *255*
Allen, D. L., 216, *228*
Amadon, D., 156, *159*
Amlaner, C. J., 130, 134, *136*, *137*,
    141, *149*, 172, *172*, 177, 181, *193*,
    268, 269, 270, *284*, *287*
Andersen, H. T., 116, *124*
Andersen, J., 318, *321*
Anderson, F., 212, *228*
Anderson, R. M., 262, *284*
Andres, R. D., 266, *288*
Arnold, G. P., 75, 80, 81, 83, 85, *91*
Asferg, T., 302, *321*

B

Bailey, E. D., 255, *256*
Ball, F. G., 202, *205*, 216, 224, *229*,
    334, *340*
Banks, E. M., 198, *204*
Barr, N. L., 129, *136*
Bassindale, R., 304, *321*
Bauer, K., 156, *159*
Baumgartener, L. L., 195, *195*
Bayley, D., 63, 65, *69*
Beale, D. M., 182, *193*
Begon, M., 215, *228*
Bellrose, F. C., 119, *126*, 130, *137*
Beneš, J., 259, *285*
Berger, M., 120, 122, 124, *124*
Bernstein, M. H., 121, *124*
Berthoud, G., 217, 222, 225, *228*
Bertram, B. C. R., 332, 337, 340, 341,
    342, *350*, *351*
Bezzel, E., 156, *159*
Birks, J. D. S., 231, 232, 233, 238,
    239, 240, 241, 242, 244, 245, 248,
    250, 251, 253, 255, *255*, *256*
Bishop, R. A., 266, *288*
Black, C. P., 115, *124*

Bohus, B., 181, *193*
Boitzov, L. V., 261, 262, 267, *285*
Bond, C. F., 115, *124*
Bonnett, P., 302, *321*
Bonnin-Laffargue, M., 310, *321*
Bourne, W. R. P., 232, *255*
Bowers, C. A., 14, *18*
Bradbury, K., 318, 320, *321*
Braestrup, F. W., 262, 265, *285*
Brander, R. B., 164, *172*
Brockie, R. E., 210, *228*
Brooks, A., 114, *124*
Brooks, R. J., 198, 204, *204*, *205*
Brown, C. A. J., 319, *322*
Brown, J. L., 282, *285*
Brown, J. M., 254, *255*
Brown, L., 156, *159*
Brüll, H., 156, *159*
Bryan, R. M., 109, *126*
Burne, R. H., 108, *124*
Burt, W. H., 215, 217, *228*
Butler, P. J., 107, 109, 110, 111, 112,
    113, 114, 115, 116, 119, 121, 122,
    123, 124, *124*, *125*, *127*, 130, *136*
Bygott, J. D., 342, *351*

C

Calder, W. A., 115, 122, *125*, *126*
Campbell, P. A., 210, 217, 222, 225,
    228
Campbell, W. B., 118, *126*
Canivenc, R., 310, *321*
Caraco, T., 337, *340*, 349, *351*
Carey, F. G., 90, *92*
Carrera, J., 319, *322*
Carroll, D. S., 90, *91*
Casey, T. M., 261, *286*
Castellini, M. A., 118, *126*
Cederlund, G., 154, 155, *159*
Chanin, P. R. F., 232, 233, 239, 249,
    250, 254, *255*, *256*
Cheeseman, C. L., 292, *298*, 303, 320,
    321

Chesemore, D. L., 266, 267, *285*
Chitty, D., 267, *285*
Chitty, H., 267, *285*
Christian, J. J., 129, *137*
Church, D. W., 89, *92*
Church, K. E., 6, *17*
Chute, F. S., 180, *193*, 198, *204*
Clark, B., 109, *126*
Clutton-Brock, J., 261, *285*
Coah, R. S., 182, *193*
Cochran, W. W., 13, *18*, 119, *126*,
   130, *136*, *137*, 141, *149*, 164, *172*,
   177, *193*
Colacino, J. M., 115, *125*
Cole, I. J., 260, 261, *285*
Collett, R., 265, 266, *285*
Cook, J. C., 76, *92*
Corbet, G. B., 261, *285*
Coutant, C. C., 90, *91*
Cowardin, L. M., 15, *18*
Cowlin, R. A. D., 302, *321*
Craig, J. W., 63, *69*
Craighead, F. C., 130, *136*
Craighead, J. J., 130, *136*
Creasy, R. K., 109, *127*
Cupal, S. S., 9, *18*
Cushing, D. H., 76, 89, *91*
Cuthbert, J., 232, *255*

D

Davies, R. S., 130, *136*
Davis, R. W., 118, *126*
Davis, S., 262, *285*
Day, M. G., 232, *255*
Deane, C. D., 232, *256*
Deat, A., 221, *228*, 297, *298*
Delany, M. J., 210, *228*
DeMaster, D. P., 11, *18*
DeMoor, P. P., 212, *228*
Dewar, J. M., 111, 114, 116, *125*
Dickson, W., 82, *91*
Dimelow, E. J., 219, *228*
Doell, A. C., 180, *193*
Doty, H. A., 250, *256*
Drewry, W. F., 109, *126*
Dreyfert, T., 154, 155, *159*
Drummond, P. C., 109, *125*
Dubrovskii, N., 267, *285*
Duke, G. E., 130, *136*
Dunstan, T., 154, *159*
Duval, R. L., 15, *18*

E

Eagle, T., 16, *18*
Eberhardt, L. E., 266, *285*
Eisenberg, J. F., 337, *340*, 341, *351*
Eliassen, E., 116, 120, *125*, 130, *136*
Elsner, R., 109, *125*
Elton, C., 262, 267, *285*
Englund, J., 266, *285*
Eriksson, L.-O., 90, *91*
Erlinge, S., 210, 212, 217, 221, 222,
   228, 249, 253, 254, *256*
Errington, P. L., 250, *256*
Erzepky, R., 156, *159*
Evens, W. E., 14, *18*
Ewer, R. F., 219, *228*

F

Fairley, J. S., 232, *256*
Fallek, H. G., 11, *18*
Falls, J. B., 181, *193*, 204, *204*
Fenton, M. B., 163, *172*
Fetherston, K., 262, 263, *285*
Field, C. R., 328, *340*
Fischer, A. H., 264, *285*
Fitchen, F., 49, *59*
Fleming, T. H., 163, *172*
Folkow, B., 109, *125*
Foote, K. G., 88, *91*
Forbes, J. E., 135, *136*
Fox, M. W., 265, 267, *285*
Fromm, E., 261, *286*
Fryer, T. B., 111, *125*
Fuller, M. R., 130, *136*
Fuller, W. A., 198, *204*

G

Gabrielsen, G., 116, *126*
Gerell, R., 232, 235, 239, 240, 247,
   249, 251, 252, 253, 254, *256*
Gilbert, P. W., 115, *124*
Gill, D., 302, *321*
Gilmer, D. S., 15, *18*
Glazener, W. C., 182, *193*
Glutz, V., 156, *159*
Godfrey, G. A., 141, *149*
Godfrey, G. K., 198, *204*
Goldberg, J. S., 182, *193*

Gorman, M., 291, 293, *298*
Greenewalt, C. H., 121, *125*
Greer Walker, M., 80, 81, 83, 85, 90, *91*, *93*
Grimaldi, E., 265, *286*
Grinnell, S. W., 109, *126*, *127*
Grue, H., 279, *285*

H

Haas, W., 182, *193*
Hackett, D. F., 216, *229*
Hagmeier, E. M., 264, *285*
Häkkinen, U., 198, *205*
Hamilton, D., 49, *59*
Hamilton, P. H., 182, *193*, 341, 342, 345, 348, *351*
Hamilton, P. M., 12, *18*
Hamley, J. M., 181, *193*, 204, *204*
Hanby, J. P., 342, *351*
Hannell, F. G., 319, *321*
Hanson, W. C., 266, *285*
Harden Jones, F. R., 75, 76, 80, 81, 83, 85, 86, 89, 90, *91*
Harding, P. J. R., 180, *193*, 198, *204*
Harper, C. A., *59*
Harper, R. J., 302, *322*
Harris, S., 266, *285*, 301, 302, 303, *321*
Harrison, C. W., 61, *69*
Hart, J. S., 120, 122, 123, 124, *124*, *125*, *126*, *127*
Hatch, D. R. M., 266, *288*
Haug, J. C., 11, *18*
Haugen, A. O., 216, *228*
Haukioja, E., 264, *286*
Hawkins, A. D., 89, *92*
Hector, D. H., 115, *125*
Heezen, K. L., 201, *204*
Heithaus, E. R., 163, *172*
Henshaw, R. E., 261, *286*
Herman, T. B., 198, *204*, *205*
Hersteinsson, P., 259, 267, 275, *286*
Herter, K., 210, *228*
Hewlett Packard, 43, *45*
Hilborn, R., 215, *228*
Hildebrand, M., 261, *286*
Hill, G., 253, *256*
Hills, M., 261, *285*
Hirons, G. J. M., 129, 130, 135, 139, 140, 141, *149*

Hochachka, P. W., 109, *127*
Hock, R., 261, *287*
Hoffmann, R. S., 330, *340*
Holley, M. L., 76, *92*
Holliday, F. G. T., 89, 90, *93*
Holmes, P., *59*
Hornocker, M. G., 11, *18*, 350, *351*
Horrall, R. M., 90, *92*
Hough, N. C., 202, *205*, 216, 224, *229*, 334, *340*
Humphries, D. A., 302, *321*
Hunt, P. S., 303, *321*

I

Ireland, L. C., 75, *92*
Irving, L., 108, 109, *126*, *127*, 261, *286*, *287*

J

Järvinen, O., 262, *286*
Jayne, A. F., 302, *321*, *322*
Jennrich, R. I., 216, *228*, 330, *340*
Jensen, B., 279, *285*, *286*
Jeppesen, J. L., 302, *321*
Jewell, P. A., 215, *228*
Johansen, K., 109, 117, *126*
Johnson, F., 261, *287*
Johnston, D. W., 119, *126*
Jones, D. R., 109, 110, 121, 123, *125*, *126*
Jordan, E. C., 70, *71*

K

Kaikusalo, A., 266, *286*
Kanwisher, J. W., 75, 90, *92*, 116, 120, *126*
Kanwisher, N., 116, *126*
Kaufmann, J. H., 244, *256*
Kauri, H., 264, *286*
Kawashiro, T., 115, *127*
Keeler, C., 261, *286*
Kenward, R. E., 129, 130, 131, 133, 135, *137*, 141, 144, 153, 154, 155, 156, 157, 158, *159*, 175, 194, *195*
Kern, W., *59*
Kimmich, H. P., 129, *137*

King, C. M., 249, *256*
King, J. M., 341, 342, *351*
King, R. W. P., 61, *69*
Kleiman, D. G., 337, *340*
Knight, A. E., 99, *104*
Knowlton, F. F., 11, *18*
Ko, W. H., 134, *137*, 269, *286*
Koeppl, J. W., 330, *340*
Kolz, A. C., 11, *18*
Kooyman, G. L., 118, *126*
Koponen, S., 264, *286*
Krebs, C. J., 215, *228*
Kristiansson, H., 210, 212, 217, 221, 222, *228*
Kristoffersson, R., 222, *229*
Krog, H., 261, *286*
Krog, J., 109, *126*
Kruuk, H., 274, 282, *286*, 291, 292, 293, 294, *298*, 303, 318, 319, 320, *322*, 341, 346, 350, *351*
Kuechle, V. B., 1, 11, 12, 13, 14, 15, 16, *18*
Kurtén, B., 259, *286*

L

Lance, A. N., 170, *172*
Landsberg, H., 320, *322*
Larochelle, J., 121, *127*
Lasiewski, R. C., 115, *126*, 254, *255*
Lavrov, N. P., 265, 266, *286*
Laws, R. M., 328, *340*
Lawson, K. D., 90, *92*, 120, *126*
Layzer, J. B., 99, *104*
Leinati, L., 265, *286*
Lemnell, P.-A., 154, 155, *159*
Le Munyan, C. D., 129, *137*
Lentfer, J. W., 11, *18*
Leuze, C. C. K., 181, *193*, 198, *205*
Lever, C., 232, *256*
Leyhausen, P., 217, *229*
Liggins, G. C., 109, *127*
Likens, G. E., 262, 264, *288*
Lindemann, W., 219, *229*
Lindquist, B., 154, *159*
Linn, I. J., 197, 231, 232, 238, 239, 240, 241, 242, 244, 245, 248, 251, 253, *255*, *256*
Lisander, B., 109, *125*
Little, T. W. A., 303, *322*
Lloyd, H. G., 31, *42*, 212, *229*, 234, *257*, 266, *286*, 292, *298*

Loasby, R., *59*
Lock, J. M., 327, 328, *340*
Lockie, J. D., 249, *256*
Long, F. M., 9, *18*
Longstaff, T. G., 266, *286*
Lord, R. D., 109, 119, *126*, 130, *136*, *137*, 141, *149*
Lowry, W. P., 320, *322*
Luke, D. McG., 89, 90, *92*
Lund, Hj. M. K., 260, *286*
Lynn, C., 49, *59*

M

MacArthur, A. H., 264, *286*
MacArthur, J. A., 303, *322*
Macdonald, D. W., 130, 134, *137*, 164, 172, *172*, 187, *193*, 202, *205*, 216, 224, *229*, 249, *256*, 259, 265, 266, 267, 268, 269, 271, 272, 273, 274, 282, *284*, *286*, *287*, 334, *340*
MacFarlane, A., 90, *93*
Mackintosh, C. G., 303, *322*
MacLennan, R. R., 255, *256*
Madison, D. M., 198, *205*
Makacha, S., 346, *351*
Maksimov, A. A., 267, *287*
Mallinson, P. J., 292, *298*, 303, 320, *321*
Mandelli, G., 265, *286*
Marancik, G., 99, *104*
Margetts, A. R., 80, 81, *91*
Marshall, W. H., 254, *256*
Marten, H., 14, *18*
Martin, D. E., 11, *18*
Mauget, C., 221, *228*, 297, *298*
Mauget, R., 221, 224, *228*, *229*, 297, *298*
Maurel, D., 221, *228*, 297, *298*
Maxfield, L., 130, *136*
McCleave, J. D., 101, 102, *104*
McCleery, R., 141, *149*, 181, *193*
McFarlane, R. W., 119, *126*
Mechlin, L. M., 15, *18*
Mellinger, T., 261, *286*
Melquist, W. E., 11, *18*
Messick, J. P., 350, *351*
Meyer, C. S., 49, *59*
Middleton, A. L. V., 302, *322*
Millard, R. W., 117, *126*
Millen, J. E., 109, *126*
Milsom, W. K., 117, *126*

Mineau, P., 198, *205*
Mitson, R. B., 56, *59*, 75, 76, 79, 86, 90, *92*, *93*
Monson, M., 261, *286*
Morris, P. A., 207, 210, 212, 217, 222, *229*
Morrison, D. W., 163, *172*
Muckenhirn, N. A., 341, *351*
Murdaugh, H. V., 109, *126*
Murray, T. L., 63, 65, *69*
Myers, N., 341, *351*
Myllymäki, A., 198, *205*

N

Neal, E. G., 302, 303, 306, 307, 318, 320, *322*
Nelson, D. R., 75, 90, *92*
Newdick, M. T., 266, *287*
Nielsen, B. L., 279, *286*
Niewold, F. J. J., 266, 267, *287*
Novikov, G. A., 266, *287*
Nybert, E., 129, *137*

O

Öberg, B., 109, *125*
Odum, E. P., 312, *322*
O'Gorman, F., 232, *256*
Osgood, D. W., 189, *193*
Osgood, W. H., 265, *287*
Østbye, E., 264, *286*
Oswald, R. L., 90, *92*
Owen, R. B., Jr, 130, 135, 139

P

Paasikallio, A., 198, *205*
Paget, R. J., 302, *322*
Parish, T., 291, 292, 293, *298*, 303, 318, 319, *322*
Parker, G. H., 265, *287*
Parkes, J., 217, *229*
Påsche, A., 109, *126*
Pearson, N. D., 86, *92*
Pennycuick, C. J., 121, *126*, 162, *172*, 329, *340*
Phillips, R. L., 266, *288*

Pienaar, U. de V., 346, *351*
Pincock, D. G., 75, 89, 90, *92*, *93*, 96, 98, *104*
Power, J. H., 101, 102, *104*
Prange, H. D., 116, *126*
Pratt, A. R., 76, *92*
Preble, E. A., 265, *287*
Priede, I. G., 90, *92*, 98, *104*
Proby, C. E., 189, *193*
Pulliainen, E., 260, 266, *287*
Purves, M. J., 110, *126*

Q

Qvist, J., 109, *127*

R

Ralphs, J. D., 63, 65, *69*
Rausch, R. L., 260, *287*
Rausch, V. R., 260, *287*
Rayner, J. M. V., 121, *126*
Redfield, J. A., 215, *228*
Reeve, N. J., 207, 212, *229*
Reichle, R. A., 14, *18*
Richet, C., 108, 116, *126*
Robin, E. D., 109, *126*
Robin, H. K., 63, 65, *69*
Rommel, S. A., 90, *93*, 102, *104*
Ross, M. J., 14, 16, *18*
Rowe, J. J., 194, *195*
Roy, O. Z., 120, 122, 123, 124, *124*, *125*, *126*, *127*
Rudd, R. L., 153, 158, *159*
Rudnai, J. A., 326, 329, *340*, 341, 346, *351*

S

Saemundsson, B., 275, *287*
Sanderson, G. C., 202, *205*, 216, *229*, 248, *256*
Sargeant, A. B., 11, *18*, 177, 192, *194*, 250, *256*, 266, *287*
Satchell, J. E., 319, *322*
Schaller, G. B., 326, 330, *340*, 341, 342, 345, 346, 347, 349, 350, *351*
Scheid, P., 115, *127*
Scheipers, G., 115, *127*

Schmidt-Nielsen, K., 107, 115, 116, 121, *124*, *125*, *126*, *127*
Schneider, R. C., 109, *127*
Schnell, J., 198, *204*
Scholander, P. F., 108, 109, *126*, *127*, 261, *287*
Scholes, P., 85, *91*
Schwartz, S. S., 266, *287*
Sciarrotta, T. C., 90, *92*
Scott, T. G., 265, *287*
Seidensticker, J. C., 350, *351*
Sempéré, A., 221, *228*, 297, *298*
Sequiera, D., 265, 287
Shackelford, R. M., 260, 261, *285*, 288
Shaiffer, C. W., 15, *18*
Sharp, W. M., 195, *195*
Shilyaeva, L. M., 266, *288*
Shorten, M., 139, 140, *149*
Siegel, S., 217, *229*
Silby, R., 141, *149*, 181, *193*
Sinclair, A. R. E., 342, 349, *351*
Siniff, D. B., 11, 12, *18*, 266, *288*
Sinnett, E. E., 118, *126*
Skiffins, R. M., 19, 176, 177
Skolnik, M. I., 67, *69*
Skoog, P., 318, *322*
Skrobov, V., 266, *288*
Slade, N. A., 330, *340*
Slagsvold, P., 262, *288*
Slama, H., 115, *127*
Smith, A. D., 182, *193*
Smith, G. W., 89, *92*
Smith, H. R., 181, 185, *194*
Smith, R. M., 198, *205*
Snider, M. T., 109, *127*
Soper, J. D., 267, *288*
Southwick, E. E., 120, *127*
Southwood, T. R. E., 215, *229*
Speller, S. W., 266, *288*
Standora, E. A., 90, *92*
Stasko, A. B., 16, *18*, 75, 89, 90, *92*, *93*, 96, 98, *104*
Stebbings, R. E., 161, 162, *172*
Stevenson, J. H. F., 232, *256*
Stickel, L. F., 216, *229*
Stirling, E. A., 302, *322*
Stoddart, L. C., 135, *137*
Storeton-West, T. J., 31, 79, 86, 90, *92*, *93*, 102
Storm, G. L., 266, *288*
Strandgaard, H., 215, *229*

Stuart, P., 303, *322*
Stults, C. D., 264, *285*
Sundnes, G., 90, *92*
Swan, L. W., 119, *127*
Swanson, G. A., 11, *18*, 250, *256*
Symes, R. G., 302, *322*

T

Tarboton, W., 153, *159*
Taylor, E. W., 110, *125*
Taylor, K. D., 31, *42*, 212, *229*, 234, *257*, 292, *298*
Teagle, W. G., 302, *322*
Teal, J. M., 120, *126*
Tembrock, G., 267, *288*
Tenney, S. M., 115, *124*
Tester, J. R., 12, 13, *18*, 181, *194*, 201, *204*, 266, *288*
Tetley, H., 304, *322*
Thomas, D. W., 163, *172*
Thomas, J., 16, *18*
Thomas, S. P., 121, *124*
Thompson, D. C., 187, *194*
Thompson, H. V., 232, *257*
Tinline, R. R., 202, *205*, 216, 221, 223, 224, *229*
Torre-Bueno, J. R., 115, 120, 121, *127*
Trainer, D. O., 182, *193*
Trevor-Deutsch, B., 198, *205*, 216, *229*
Tucker, V. A., 121, *127*
Turner, F. B., 216, *228*, 330, *340*
Turner, M., 346, *351*
Twigg, G. I., 210, *229*
Tytler, P., 89, 90, *93*

U

Ulveland, S., 90, *91*
Underwood, L. S., 261, *286*
Urquhart, G. G., 89, *92*

V

Väisänen, R. A., 262, *286*
Van Orsdol, K. G., 325, 326, 328, *340*
Verts, B. J., 130, *137*

Videsott, R., 265, *286*
Voegeli, F. A., 90, *93*
Voigt, D. R., 202, *205*, 216, 221, 223, 224, *229*
Vossen, J. L., *59*

W

Wade, L., 261, *286*
Wahrenbrock, E. A., 118, *126*
Wakeley, J., 153, *159*
Walters, V., 261, *287*
Warner, D. W., 13, *18*, 135, *136*
Warner, J. S., 153, 158, *159*
Watson, A., 170, *172*
Watson, D. W., 67, *69*
Webster, A. B., 204, *205*
Weeks, R. W., 9, *18*
Weiss, E., 109, *126*
West, N. H., 109, 121, 123, *125*, *126*
White, M., 182, *193*
White, W., 129, *137*
Whittaker, R. H., 262, 264, *288*
Widén, P., 135, 153, 155, 157, *159*
Wielgolaski, F. E., 264, *288*
Wiles, W. V., 350, *351*
Wilkie, D. R., 121, *127*

Williams, J. M., 162, *172*
Williams, P., 63, *69*
Williams, T. C., 120, *126*, 162, *172*
Willmer, H., 115, *127*
Winn, R. T. E., 67, *69*
Winter, J. D., 12, *18*
Wipf, L., 260, *288*
Wise, M. H., 232, 233, 252, 253, *257*
Woakes, A. J., 107, 110, 111, 112, 113, 114, 119, 122, 123, 124, *125*, *127*
Wolf, L., 337, *340*, 349, *351*
Wood, D. A., 177, 185, 189, *194*
Wright, H. E., 67, *69*
Wrigley, R. E., 266, *288*

Y

Yoaciel, S. M., 328, *340*
Young, A. H., 56, *59*, 89, 90, *92*, *93*

Z

Zapol, W. M., 109, *127*
Ziesemer, F., 129, 141
Zinnel, K. E., 14, *18*

Subject Index

Numbers in italics refer to figures

A

Aardvark, *347*
Acacia sieberiana, 328
Accipiter gentilis, 131, 153—160
Acer pseudoplatanus, 194, 241
Acinonyx jubatus, 349
Activity monitoring
   woodcock, 142, *146*, 149, *150*
Activity patterns
   badgers, 294, 301—323, *307—309*
   free-living birds, 129—137
   goshawks, 153—160, *156, 157*
   mink, 247, 253
Acoustic telemetry, marine fisheries, 75—93
Acoustic transponding
   compass tag, acoustic telemetry, 85 *87*
Adenota kob thomasi, 327
Aepyceros melampus, 348
Aerial calibration, radio tagging, voles, 199
Aerials, tag systems, 37, *37*
African lion, 325—340
Alces alces, 7
Alcids, 265
Alnus glutinosa, 233
Alopex lagopus, 259—289
America, North, 1—18
Anas platyrhynchos, 109, 119
Anas rubripes, 120, 123
Anser, 265
Antennas, receivers, state of the art, 15
Antilocapra americana, 182
Apodemus sylvaticus, 203, 279
Apples, 274, 312
Arctic foxes, 259—289
Artibeus jamaicensis, 163
Artibeus hirsutus, 163
Arvicola terrestris, 181
Attachment design, transmitters, state of the art, 11
Automatic activity monitoring
   badgers, 293

Automatic activity monitoring *cont.*
   free-living birds, 135
   woodcock, *150*
Automatic data recording, state of the art, 13
Automation, 13, 135, *150*, 197—205, 293
Aythya australis, 110
Aythya ferina, 111
Aythya fuligula, 111, *112, 113*

B

Badgers, 31, 291—299
Bait colour-marking, badgers, 293, *294*
Batteries, transmitters, state of the art, 5, *6*
Battery life, transmitters, greater horse-shoe bats, *164*
Behaviour
   diving and flying birds, 107—128
   free-living birds, 129—137
   migratory fish, 75—93
   red and Arctic foxes, 267
Biomass, lions, 332
Biotelemetry, state of the art, 1—18
Birds, 11, 61, 107—128, 129—137, 274, 281
Blackberries, 312, 315
Boars Hill, Oxford, 272
Breeding
   grey squirrels, *186*
   woodcock, 144, *147*
Bradycardia, diving and flying, birds, 107—128, *112, 113, 117, 118, 122*
Bramble, 306
Branta leucopsis, 121
Branta spp., 265
Bristol, 301—323
Buffalo, 328, *347*
Bushbucks, 328

C

Canis mesomelas, 349
Capparis fasicularis, 327
Capparis tomentosa, 327
Capture, hedgehogs, 208
Carollia perspicillata, 163
Carrion, 265, 279, *280*, 281
Cats, 217
Charadriidae, 279
Cheetahs, 352
Cherries, 274
Circadian rhythm, woodcock, *147*
Clethrionomys glareolus, 203
Clione (research vessel), 76
Cod, 78, 79
Coelopa frigida, 279
Collar irritation, mink, 235
Collars, grey squirrels, 177, *178*
Compass tag, acoustic transponding, 85
Computers, 13, 197—205, 207—230
    radio tracking, hedgehogs, 207—230, *220*
    radio tracking, voles, 197—205
Connochaetes taurinus, 347
Conopodium, 316
    majus, 294
Corella (research vessel), 80
Corvus ossifragus, 121
Coturnix coturnix, 123
Crocuta crocuta, 349
Cruelty to Animals Act, (1876), 181
Cub mortality, lions, *338*
Cyclopterus lumpus, 279
Cystophora cristata, 109

D

Damaliscus korrigum, 347
Damaliscus korrigum jimela, 328
Data analysis, radio tagging, voles, 202
Data collection, grey squirrels, 186
Data recording, receivers, radio tagging, voles, 199
Daytime, radio location, hedgehogs, 214
Den types, mink, 240, *241*
Den use, patterns, mink, 241, *242*
Denning behaviour, mink, 240, 250
Design considerations, tag system, 31—45

Design, transmitters, state of the art, 3
Development licences, regulatory control, 23
Dicrostonyx, 265
Dikdik, *347*
Diet
    radio-tagged mink, 236
    red and Arctic foxes, 262, *264*, 274, 279
Distance travelled, home range, hedgehogs, 219, 225
Diving birds, 107—128
Dog-foxes, 277
Dolphins, 11
Domestic ducks, 107—128
Ducks, 4, 107—128
Duiker, *347*

E

Earthworms, 147, *148*, 282, 294, 310, *311*, 318—320
Ecology
    leopards, 341—352
    societies, red and Arctic foxes, 282
Eider duck, 283
Efficiency, otter trawl, acoustic telemetry, 80, *81*
Eland, *347*
Elephants, 328
Elks, 9
Empetrum nigrum, 279
Encapsulation design, state of the art, 5
Epomophorus gambianus, 163
Equipment performance specification, tag system, United Kingdom, 38
Equipment, radio tracking
    fish, 102
    greater horseshoe bats, 163
    grey squirrels, 175
    hedgehogs, 212
    red and Arctic foxes, 268
Equipment standards, regulatory control, 22
Equus burchelli, 348
Erinaceus europaeus, 207—230
Exclusive bands, biomedical telemetry, 20

F

Faeces, badgers, *296*, 297
Fagus sylvatica, 194
Feeding
    badgers, 294
    free-living birds, 135
    goshawks, 155, 156
    greater horseshoe bats, 161–173, *171*
    woodcock, 145, *148*
Felis concolor, 350
Feral mink, 231–257
Ficus gnaphalocarpa, 328
Ficus sp., 163
Field voles, 197–205
Fish, 15, 31–45, 61, 75–93, 95–105, 233, 252, 265
Fish, marine, acoustic telemetry, 75–93
Fish tags, state of the art, 15
Fish tracking, radio tags, 95–105
Flight monitoring packages, woodcock, *132*, *133*
Flight patterns, greater horseshoe bats, 161–173
Flying birds, 107–128
Flying, goshawks, 154, 156
FM transmitter, diving and flying bird physiology, 107–128
Food
    badgers, 310, *311*, *312*
    leopards, 346, *347*
    lions, 346
Food habits, lions, 325–340
Food intake, lions, 336, *336*, *338*
Foraging
    Arctic fox, 279, *280*
    badgers, *316–318*
    grey squirrels, *189*
    red fox, 274
Foxes, 31
Fraxinus excelsior, 194
Free-flying birds, 121
Free-living birds, 129–137
Frequency allocation, regulatory control, 21
Frequency assignment, regulatory control, 21
Frequency, transmitter tag, design and performance, 39
Fruit, *311*

G

Gazelles, 352
Geese, 111, 121, 122, 124
Genetta sp., 349
Giraffe, *347*
Goshawks, 129–137, *132*, 153–160
Gossamer (research vessel), 76
Greater horseshoe bats, 161–173
Grey squirrels, 175–196
    effect of collars, 182
Grid method, home range area, hedgehogs, 223, *224*
Group foraging range, badgers, 313, *314–318*
Group range size, badgers, *295*
Grouse, 6
Guinea fowl, *347*

H

Habitat, red and Arctic foxes, 262, *263*
Habitat utilization, badgers, 301–323, *314*
H-Adcock array, aerial tag system, 38
Haddock, 78
Halichoerus grypus, 109
Handling, grey squirrels, 194
Hares, 252, *347*
Hawthorn, 315
Heart rate, diving and flying birds, 107–128, *112*, *113*, *117*, *118*, *122*
Hedgehogs, 207–230
Hesperiphona vespertina, 120
High resolution sector scanning sonar, acoustic telemetry, 76
Hippopotamus amphibius, 327
Home Office, Radio Regulatory Department, 19–30
Home range
    badgers, *295*, 313
    hedgehogs, 207–230
    lions, 325–340
    mink, 231–257, *238–240*, *246*
    voles, 202
    woodcock, 144
Hunger, goshawks, 155, 156
Hyaena hyaena, 265
Hybrid/microcircuit technology, 53
Hyparrhenia, 328

I

Ichneumia albicauda, 349
Impala, *347*
Implantable transmitters, diving and flying birds, 107—128
Incubation, woodcock, 144
Inter-den movements, home range, mink, 244
Invertebrates, 265, 279, 281, *311*, 349
Irish mink, 232
Isotope recovery, faeces, badgers, 293, *296*

J

Jackal, *347*

K

Kinkajou, 219
Klipspringer, *347*
Kob, 328
Kobus defassa ugandae, 327
Kongoni, *347*

L

Lagomorphs, 349
Lagopus mutus, 265, 279
Larus argentatus, 120
Larus atricilla, 121
Larus delawarensis, 120
Larus marinus, 120
Leadless chip carriers, *58*
Leg beat cycle, diving birds, *114*
Lemmings, 266
Lemmus, 265
Leopards, 341—352
Leptonychotes weddelli, 109
Lepus capensis, 348
Licence application, low power telemetry or telecontrol system, 29
Licensing, regulatory control, 23
Lions, 325—340, 341—352
Litter absence, grey squirrels, *188*
Litter frequency, grey squirrels, 185
Lumbricus terrestris, 265, 274, 320
Lycaon pictus, 349

M

Maize, 182
Male sexual status, grey squirrels, 182
Mallard, 109, 115, 116
Mammals, 31—45, 98, 103, 108, 109, 129, 181, 197, 208, 233
effects of radio packages, 181
Manatees, 7
Marine fish, acoustic telemetry, 75—93
Marine fisheries, 75—93
Marking, hedgehogs, 210, *211*
Marmota monax, 130
Matched filter, receivers, 62, *64, 65*
Mating system, woodcock, 145, *146*
Medical and Biological Telemetry Devices, 1978 (MPT 1312), 22
Medical and Biological Telemetry Devices, Class I, 1968 (W6802), 22
Meles meles, 34, 282, 291—299, 301—323
Melopsittacus undulatus, 121
Microelectronics, 47—59
Microtus, 215
Microtus agrestis, 197—205
Microtus pennsylvanicus, 181, 204
Microwiring, 55, *56—58*
Migratory fish, 95—105
Mink, 231—257
Morphology, red and Arctic foxes, 260, *260*
Mortality sensing, transmitters, state of the art, *10*
Movement patterns
Arctic fox, 266
diving and flying birds, 107—128, *112—114, 122, 123*
free-living birds, 129—137
fresh water fish, 95—105
goshawks, 156
greater horseshoe bats, 168, *169, 171*
marine fish, 75—93
mink, 244, *246, 247*, 252
red fox, 266
voles, 197—205
MPT 1309, equipment standards, regulatory control, 22
MPT 1312, 1978, equipment standards, 22, 31

Multi-channel tracking system, receivers, 67
Mustela erminea, 249
Mustela nivalis, 249
Mustela vison, 231–257
Myocastor coypus, 34, 109
Mytilus edulis, 279

N

Night viewing, badgers, 292, 294
Night time radio location, hedgehogs, 214
Nocturnal movements, lions, 335
North America, 1–18

O

Odobenus rosmarus, 108
Odocoileus hemionus, 181
Odocoileus virginianus, 182
Ófeigsfjördur, Iceland, 275, *276*
Ondatra zibethica, 109
Operational licences, regulatory control, 23
Optimum tracking duration, mink, 236, *237*
Oryctolagus cuniculus, 34, 233, 274
Otter trawl
    acoustic telemetry, 76–80, *77*, *81*, *82*
    efficiency, 80, *81*

P

Panthera leo, 325–340, 341
Panthera pardus, 182, 341–352
Panthera tigris, 350
Passerines, 274, 279
Performance, radio tags, fish tracking, 97, *98*
Performance checks, tag system, 31–45
Peromyscus leucopus, 181
Phacocherus aethiopicus, 327, 348
Phalacrocorax carbo, 116
Phase comparison, single-channel tracking system, receivers, 67
Phasianus colchicus, 123

Phoca vitulina, 109, 279
Phyllostomus hastatus, 163
Physiological variables, diving and flying birds, 107–128
Picea abies, 154
Picea sp., 194
Pigeons, *234*
PIM transmitters, 107–128, 130
Pinus sylvestris, 154
Plaice, 78–88, *81*, *83*, *84*, *87*
Pleistocene, 259
Pliocene, 259
Plums, 274
Pluvialis dominica fluva, 119
Polar bears, 7, 11
Population size, hedgehogs, 215
Position calculation, voles, 199
Post-release activity, radio-tagged mink, 235
Posture monitoring package, goshawk, *132*
Posture sensing package, free-living birds, *132*, 134
Power output, transmitters, state of the art, 7
Power sources
    state of the art, 5
    transmitters, state of the art, 5
Preening, goshawks, 155, 156
Pressure measurement, transmitters, state of the art, *9*
Prey species, lions, *328*, *329*
Pride association, lions, 333, *334*
Pride size, lions, 332, *333*
Processing
    hybrid/microcircuit technology, 53
    silicon technology, 49
Ptarmigan, 283
Pulse duration, transmitter tag, design and performance, 41
Pygoscelis papua, 117

Q

Quercus sp., 194, 241

R

Rabbits, *234*, *240*, 241, 248, 252, 253
Radiated power output, transmitter

Radiated power output *cont.*
    tag design and performance, 39,
        40
Radio-frequency bands, regulatory con-
        trol, 20
Radio location
    hedgehogs, 213
    grey squirrels, 190
    woodcock, 142
Radio packages
    effects on animal studies, 181
    effects on mink, 235
    effects on woodcock, 143
Radio Regulatory Department, Home
        Office, 19–30
Radio tags
    fish, 95–105
    woodcock, 139–152
Radio tracking
    badgers, 291–299, 301–323
    hedgehogs, 207–230
    leopards, 341–352
    lions, 325–340
    Arctic foxes, 259–289
    voles, 197–205
Range
    leopards, 343, *344, 346*
    lions, 325–340, 343, *344*, 345,
        346
Range diameters, grey squirrels, *192*
Range sizes, badgers, *294*
Ranging patterns, lions, 333, *335*
Rangifer caribou, 7
Rangifer tarandus, 265
Rats, 31
Recapture rate, grey squirrels, *184*
Receivers
    automatic scanning, 139–152
    grey squirrels, 175
    signal-to-noise enhancement, 61–
        73
    state of the art, 12, 17
    tag system, 34, *36*, 41
    voles, 198, *198*
    woodcock, 142
Red foxes, 259–289
Redunca redunca, 327
Reedbuck, *347*
Regulatory control, telemetric devices,
        19–30
Relative mobility ratio, home range,
    mink, 245

Reliability, transmitters, state of the
        art, 4
Remote data recording, state of the
        art, 13
Repetition rate, transmitter tag, design
        and performance, 41
Respiratory cycle, flying birds, *123*
Respiratory frequency, diving and fly-
        ing birds, 107–128, *122*
Resting, goshawks, 155, 156
Rhinolophus ferrumequinum, 161–
        173
Ringed greater horseshoe bats, *166*
River Exe, 233, *234*
River Fowey experiment, radio track-
        ing, fish, 96, *97*
River Teign, *233, 240, 246, 247*
Rodents, 181, 198, 274, 349
Roding, woodcock, 140, 142
Roosting, greater horseshoe bats, 161–
        173, *168, 171*
Route map superimposition, home
        range area, hedgehogs, 224
Rubus ulmifolius, 274
Rupicapra rupicapra, 265
Rwenzori National Park, Uganda,
        325–340, *326, 327*

                        S

Salix spp., 233, 241
Salmo salar, 34, 97
Salmo trutta, 97
Salmon, 96, 100, 101, 103, 105
Satellites, transmitters, state of the
        art, 11
Scats, radio-tagged mink, 236
Scent marking, hedgehogs, 219
Sciurus carolinensis, 34, 175–196
Scalopacidae, 140, 279
Scolopax rusticola, 130, 139–152
Sea trout, 100, 103, *104*
Sea turtles, 11
Seals, 7, 265
Seasonal changes
    home range, hedgehogs, 225
    home range, mink, 254
Seaweed, 265, 279
Sector scanning sonar system, acoustic
        telemetry, 89

Selective tidal stream, transport, acoustic telemetry, 83

Semi automated radio tagging, 197—205

Sensors, transmitters, state of the art, 7

Sett distribution, badgers, 304, *304, 305*

Shared radio frequencies, regulatory control, 20

Sharks, 98

Sheep, 265

Shellfish, 75

Signal-to-noise ratio, enhancement, receivers, radio location, 61—73

Silicon technology, 49, *50, 52*

Single-channel tracking system, receivers, 66, *66*

Slapton Ley, 233, *238, 246*

Social groups, badgers, 301—323

Social organization
  Arctic fox, 277
  badgers, 291—299
  red and Arctic foxes, 266
  red fox, 272, *273*

Solar power sources, transmitters, state of the art, 5

Sole, 78

Somateria mollissima, 279

Spectral output, transmitter tag, design and performance, 39, *41*

Speed
  hedgehogs, 222
  plaice, *84*

Sporobolus, 327

Squirrels, 175—196

Starlings, *347*

Steinbock, *347*

Sturnus vulgaris, 115, 120, 121

Surveillance techniques, grey squirrels, 190

Swans, 124

Syncerus caffer, 327

System realization, matched filter, receivers, 63

T

Tags
  fish, 99
  free-living birds, 129—137

Tags *cont.*
  grey squirrels, 179, *180*
  pressure measurement, state of the art, 7
  temperature measurement, state of the art, 7

Tail mounted transmitters, posture-monitoring, goshawks, 131, *132*

Telemetering and Telecontrol Licence, 24

Territories
  Arctic fox, *281*, 281
  hedgehogs, 217, *218, 220*
  red fox, 275

Test licences, 23

Testing
  silicon technology, 51
  transmitters, state of the art, 5

Thalarctos maritimus, 265

Themeda, 327, 328

Thermistor package, flight monitoring, free-living birds, *131—133*, 133

Thick film, hybrid/microcircuit technology, *54*, 55

Thin film, hybrid/microcircuit technology, 55

Thuja sp., 194

Tracking
  fish, 101
  freshwater fish, 95—125
  open sea, acoustic telemetry, 80
  plaice, 80, *83, 84*

Transmitter and Receivers for Use in the Bands Allocated to Low Power Telemetry in the PMR Service (MPT 1309), 22

Transmitter attachment
  hedgehogs, 212
  woodcock, 141

Transmitter tags, design and performance, 32, *32, 36*

Transmitters
  free-living birds, 129—137, *131*
  grey squirrels, 175, *178*
  implantable, diving and flying birds, 107—128
  red and Arctic foxes, *269, 270*
  state of the art, 2, *2, 8*, 17
  two channel, diving and flying birds, 111, 121
  variable pulse rates, 129—137
  voles, *200*, 201

Transmitters *cont.*
  woodcock, 141
Transportable radio tracking system, 197—205
Trap range diameter, grey squirrels, *186*
Trapping, grey squirrels, 194
Trees, 135
Triangulation
  radio tracking voles, 200, 203
  radio tracking woodcock, 143
Trout, movement, *104*
Tufted ducks, 107—128
Tuning, transmitters, state of the art, 3
Turbot, 78
Two channel transmitter, barnacle geese, 111, 121

U

Uganda, 325—340
Ungulates, 349
United Kingdom, equipment performance specification, tag system, 38
Uria aalge, 279
Uria lomvia, 279
Use intensity, home range, mink, *243*, 244, 251

V

Vegetables, *311*
Voles, 197—205, 274
Voltage characteristics, transmitter batteries, posture-sensing packages, free-living birds, *134*

Vulpes alopecoides, 259
Vulpes vulpes, 34, 249, 259—289, 302

W

W6802, equipment standards, regulatory control, 22
Waders, 279
Warthog, 328, *347*
Water voles, 198
Waterbuck, 328
Weight changes
  radio tagged grey squirrels, *183*, *184*
  radio tagged mink, 235
Whales, 11
Whip aerial, tag system, 37
Whiting, 78
Wildebeest, *347*
Wing strapped transmitters, flight monitoring, woodcock, 130, *132*
Wireless Telegraphy Act 1949, 24
Woodcock, 129—137, *132*, *133*, 139—152

Y

Yagi-array, aerial, tag system, 38
Yew berries, 312
Yield, silicon technology, 51

Z

Zebra, *347*